Work at Sea

Work at Sea

*The Evolution of
Shipboard Technology*

IVER P. COOPER

McFarland & Company, Inc., Publishers
Jefferson, North Carolina

Portions of this work were previously published in the *Grantville Gazette*. For particulars, see the explanatory note at the end of the bibliography.

Use of released U.S. Navy imagery does not constitute product or organizational endorsement of any kind by the U.S. Navy.

Library of Congress Cataloging-in-Publication Data

Names: Cooper, Iver P., author.
Title: Work at sea : the evolution of shipboard technology / Iver P. Cooper.
Description: Jefferson, North Carolina : McFarland & Company, Inc., Publishers, 2026 | Includes bibliographical references and index.
Identifiers: LCCN 2025037119 | ISBN 9781476696539 (print) ∞
ISBN 9781476654850 (ebook)
Subjects: LCSH: Shipbuilding—History. | Ships—Equipment and supplies. | Seafaring life. | Marine engineering. | Ships—Technological innovations. | Navigation—History. | BISAC: HISTORY / Maritime History & Piracy | TECHNOLOGY & ENGINEERING / Marine & Naval
Classification: LCC VM15 .C688 2025
LC record available at https://lccn.loc.gov/2025037119

ISBN (print) 978-1-4766-9653-9
ISBN (ebook) 978-1-4766-5485-0

© 2026 Iver P. Cooper. All rights reserved

No part of this book may be reproduced or transmitted in any form or by any means, electronic or mechanical, including photocopying or recording, or by any information storage and retrieval system, without permission in writing from the publisher.

Front cover image: *The Atlantic Telegraph*, Sir William Howard Russell (Irish, Tallaght, Dublin 1820–1907 London); illustrator: Robert Charles Dudley (British, 1826–1909); publisher: Day & Son, Ltd., London; printer: Bradbury, Evans, & Co. (British, London), circa 1866, 43 × 30.6 × 3 cm (gift of Cyrus W. Field, 1892, The Met, Open Access).

Printed in the United States of America

McFarland & Company, Inc., Publishers
Box 611, Jefferson, North Carolina 28640
www.mcfarlandpub.com

For Lee, Louise, and Jason

Table of Contents

Preface	1
1. Finding Your Way (Navigation, Part 1)	3
Navigation by Terrestrial Signs	3
Terrestrial Latitude and Longitude	7
Globes and Maps	8
Rhumb Lines and Great Circles	9
Dead Reckoning	10
Navigational Use of the Compass	12
Altitude and Azimuth	24
Measuring Altitude	24
Measuring Azimuth	33
Measuring Time	33
2. Guided by the Sky (Navigation, Part 2)	36
Early Navigator Training and Certification	36
Celestial Navigation	38
Celestial Latitude and Longitude	39
Sources of Error in Celestial Navigation	41
Star Data	42
Astronomical and Nautical Almanacs (Ephemerides)	43
Solar Positions	44
Planetary Positions	45
Lunar Positions	47
Navigational Use of the Pole Star	47
Determining Latitude	48
Determining Longitude	49
Modern Celestial Cross-Fix	53
Log and Trig Tables	54
Sight Reduction Tables (SRTs)	55
LORAN and GPS	55
3. From Oars and Amphorae to Container Ships (Cargo Transport)	58
Propulsion and Shipboard Labor	58
Early Shipping Containers	63

Deck and Hold Stowage	64
Ship Design: Hatches and Cargo Ports	65
Modern Cargo	66
Intermodal Transport: ROROs and LASHs	67
Intermodal Transport: Container Ships	67

4. "Haul Taut!" (Lifting Machines, Part 1) — 70

Machines	70
Pulleys and the Block-and-Tackle	71
Ropes and the Like	80
Use of the Block-and-Tackle	84

5. "Rig the Capstan!" (Lifting Machines, Part 2) — 87

The Load: Anchors and Anchor Cables	88
Winch	88
The Windlass	89
Reel Winches	99
The Capstan	101
Derricks	109
Heavy Lift Derricks	114
Derricks Versus Cranes	114
Deck Cranes	115

6. "Helm a-Starboard!" (Steering Gear) — 118

The Steering Oar	118
The Rudder	118
The Tiller	120
The Whipstaff	121
The Steering Wheel	128
Gears	134
Chain and Belt Drives	135
Axiometer (Helm Indicator)	135
Gear Train-Type Steering Gear	136
Screw-Type Steering Gear	138
Powered Versus Manual Steering	140
The Flettner Rudder	140
Automatic Piloting	141
Steering Propulsors	144
Flettner Rotors	147
Cycloidal Propellers	148

7. Going Fishing (Fishing Vessels and Fishing Gear) — 151

Fishing Gear	151
Fishing Vessels	163

Table of Contents ix

8. Flags and Blinkers (Visual Communication at Sea) — 172
    - Communication and Coding — 172
    - Communication Protocols — 175
    - Limitations of Visual Communication — 175
    - Sail Signals — 176
    - Flaghoist Signals — 177
    - Intraship Flag Signals — 186
    - Myer Wigwag — 187
    - Mechanical and Manual Semaphore — 188
    - Heliograph — 189
    - Visual Communication—Night — 191
    - Pyrotechnics — 192
    - Signal Fires and Lamps — 194
    - Other Light Signaling Methods — 196
    - Range and Speed of Signaling — 198

9. "Ahoy There!" (Nonvisual Communication at Sea) — 200
    - Physics of Audio Communication — 200
    - Direct Communication of Speech — 201
    - Speaking (Hailing) Trumpet — 202
    - Ear Trumpet — 205
    - Voice Tubes (Speaking Tubes; Voice Pipes) — 206
    - Sound-Powered Telephones — 208
    - Electric Megaphone — 211
    - Nonverbal Audio Communication — 211
    - Communication by Radio — 213

10. Underwater Activities — 215
    - The Dangers of Diving — 215
    - Diving Tables and Hyperbaric Therapy — 225
    - Free ("Breath-Hold"; "Apnea") Diving — 218
    - Diving Accessories — 219
    - Locating a Wreck — 222
    - Lifting a Wreck — 223
    - Salvaging Cannon — 225
    - Preservation and Deterioration — 226
    - Salvage Law and Contracts — 227

11. Diving Bells, Diving Suits, and Submersibles — 230
    - Diving Bell — 230
    - The Lethbridge Diving Engine and the Armored Diving Suit — 235
    - Diving Helmets and Flexible Diving Suits — 236
    - Tethered Manned Submersible — 238
    - Early Autonomous Submersibles (Submarines) — 238
    - Wheeled Submarines — 240

Table of Contents

12. Fresh-Air Supply for Underwater Exploration — 242
    Breathing 101 — 242
    The Fresh-Air Delivery Problem — 243
    Pumping Air — 246
    Self-Contained Underwater Breathing Apparatus (SCUBA) — 253
    Compressed Air Tanks in Submersibles — 254
    Oxygen Generation in Situ — 255
    Mixed Gas Diving — 256

Conclusion — 257

Appendix 1. The Grip of Traction-Winches — 259

Appendix 2. The Material Properties of Ropes in Nautical Use — 262

Appendix 3. The Physics of the Ship's Rudder — 264

Bibliography — 265

Index — 295

Preface

For millennia, ships have been used to fish, to transport passengers and cargo, and to bring divers to particular locations to harvest sea life or to salvage wrecks. Ships are thus places of work. Anchors must be raised or dropped; rigging adjusted; and cargo brought on board, stowed, and ultimately off-loaded. The ship must be navigated and steered to its destination despite the vagaries of wind and wave. It may be necessary to communicate with other ships or with those onshore. Fishing and diving gear and methods had to be devised in order to exploit underwater resources.

This book examines these aspects of work on and under the sea from scientific and historical perspectives. In particular, it examines innovations in both hardware and methodology, and the underlying science. In general, its focus is on how things work, rather than on their social or environmental consequences.

While it is not necessary for the enjoyment of the present book, readers may also be interested in an earlier book of mine, *Poseidon's Progress: The Quest to Improve Life at Sea* (McFarland, 2024). It addresses the evolution of shipboard devices for ventilation, water purification, food preparation, sanitation, heating, air conditioning, and pumping out water.

Although there is some discussion of machinery on warships in the present work, the specific hardware used for aiming and loading naval guns is covered in my *Arming the Warship: Naval Weapons Technology and Gunnery from the Spanish Armada to the Cold War* (McFarland, 2024).

While this book does discuss the impact of the propulsion system on labor aboard, it does not go into detail on engine or propeller design, except in the case of steering propellers.

1

Finding Your Way

(Navigation, Part 1)

Jack London, sailing in the Pacific, noted that every step of the navigational art must be performed correctly, or else the captain will hear "'Breakers ahead!' some pleasant night, receive a nice sea-bath, and be given the delightful diversion of fighting the way to shore through a horde of man-eating sharks" (London 1911, 242).

We begin our study of the evolution of the navigational art by considering how ships were navigated by observation of terrestrial signs.

Navigation by Terrestrial Signs

Landmarks. The simplest form of navigation is to take note of prominent landscape features and their bearings. Ideally, you take cross-bearings (simultaneous bearings of two different landmarks, lying in different directions), because that fixes your position. Lighthouses and buoys, of course, can be considered artificial landmarks.

Knowledge of landmarks was initially confined to local sailors. However, it became customary for long-distance mariners to draw "profiles" of the coasts they visited in their logbooks (Taylor 1957, 168). These sketches could be passed on to allied captains.

The "know-how" of sailors was distilled into guidebooks known in ancient times as *periplouses*, and, later as portolans, rutters, and waggoners (Ash 2007). They could include charts, profiles, and logbook summaries. The most famous of them all was Lucas Wagenaer's *Mariner's Mirror* (1584). Sailing directions are primarily directed to coastal navigation.

Soundings. In shallow waters, including the North Sea and Baltic Sea, it was common to navigate by soundings. This involved dropping a sounding line, a knotted rope with a lead weight, to the bottom. The number of knots passing over the side gave the depth. The weight could be "armed" with tallow to pick up sediment from the sea floor. Sailing instructions would tell mariners what to expect. For example, they might say that you have reached the shallow region between Cape Clear and the Isles of Scilly when "at 72 fathoms [you] find fair gray sand" (Aczel 2001, 12–3, 134–5).

The modern lead line is distinctively marked so that the leadsman can recognized the marks by feel even in the dark: two strips of leather at two fathoms, three at three, a white cotton rag at five, red woolen bunting at seven, and so on up to a cord with three knots for 30 fathoms. Lines for offshore work may be 100 fathoms long (Mixter 1967, 9).

For significant depths it was somewhat awkward to measure how much line had been let out. "In 1802 the British clockmaker Edward Massey invented a mechanical device that was attached to the sounding line; as the device sank, a rotor turned a dial which locked in place when the line hit bottom.... The line could then be reeled in and the depth read from the dial" (Webb 2023, 1.4). This was essentially similar to the "patent log" that was used to measure distance traveled and thus determine ship speed (described later), except it looked at vertical rather than horizontal distance. "By 1811 the Navy Board had purchased 1750 of Massey's sounding machines" (Morrison-Low 2017).

For sounding in deep water, the traditional lead line, made of hemp, was inadequate. In 1872, Lord Kelvin (William Thomson) replaced it with "pianoforte" wire, which was stronger. The wire was wound up on a drum and the falling weight caused it to unreel. A dynamometer was attached to the wire and if the strain seemed excessive a brake could be applied. His machine was installed on USS *Tuscarora* and it sounded to over 4,000 fathoms in the Kuril-Kamchatka Trench (Theberge 2014).

Unlike the old hemp rope sounding line, the wire line could be used when the ship was in motion (Thompson 2005, 719). However, the line was deflected from the vertical by the ship's motion. This could be corrected for with "depth tables" if the ship's speed was known and equipment standardized (Leggett 2016, 151).

While the wire could be rewound by hand, by 1880 it was being hauled in by steam (Sigsbee 1880, 18), and in 1907 Lord Kelvin received a patent on the use of an electric motor for rewinding (Read 2021).

Lord Kelvin experimented with several different mechanisms for recording the depth. His 1876 patent described a chemical readout attached to the sinker; water pressure would cause water to rise up a narrow-bored, air-filled glass tube open at one end, and wash away a colored chemical coating the inside, or react with chemicals coating the inside to cause a color change (Thomson 1876). For example, if the coating is red silver chromate, the saltwater reacts with it to form silver chloride and sodium chromate, and these are partially washed away (Hodgins 1911). Pressure tank measurements have shown that the sounding tube indicates the water pressure, and thus the depth, with "considerable accuracy" (Graham and Stewart 1958).

Unfortunately, these tubes were single-use devices and expensive by contemporary standards (Read 2021). For that reason there was also a mechanical readout attached to the drum. This counted revolutions and it was up to the ship's officers to convert that into fathoms (United States Hydrographic Office [USHO] 1874, 6). The chemical tube was used for calibration and confirmation.

In 1885, Lord Kelvin patented a mechanical readout attached to the sinker; water pressure forces "a graduated piston rod" into a cylinder "against the resisting force of the helical spring ... and moves a light marker ... along the rod to record the depth" (Thomson 1885).

An alternative to the chemical tube was patented by Tanner and Blish in 1899. This was a glass tube that was ground on the inside; ground glass was clear when wet

1. Finding Your Way

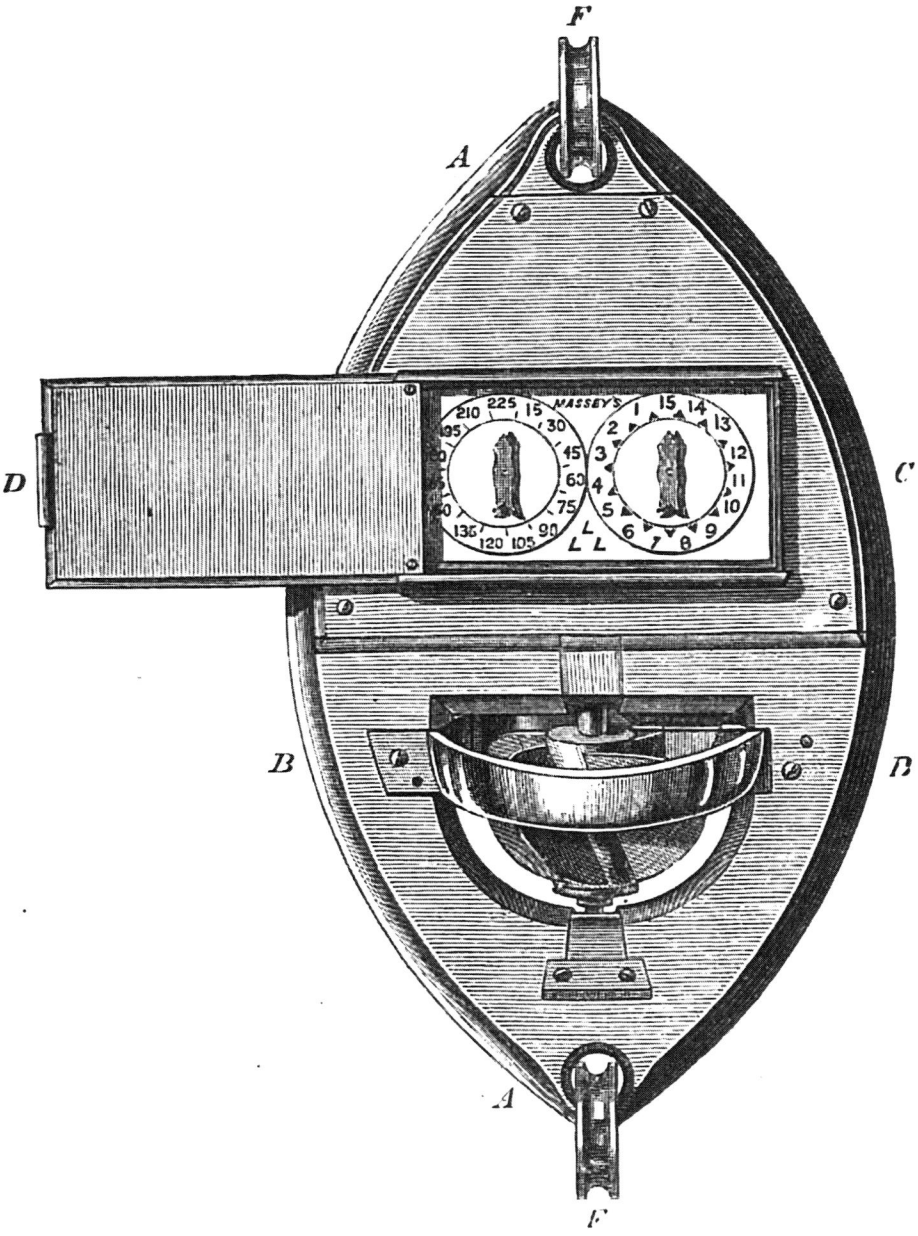

Massey's sounding machine (Thomson 1873, 225). This is more accurately described as the sounding cylinder, as it is not the machine that lowers the line. The sounding line is attached to the lower thimble F, and a weight is suspended from the lower thimble F. While not visible in this cutaway view, there are four oblique vanes on the outside of the device, "in such a position that as the machine descends the axis revolves by the pressure of the water against the vanes. The revolving asis communicates its motion to the indices on the dial-plate c," and the "index on the right-hand dial passes through a division for every fathom of vertical descent." The left-hand dial moves by one division for every 15 fathoms (224-226). For a picture of the machine used to lower this sounder, see Fig. 52 (p. 16) of Nares (1876).

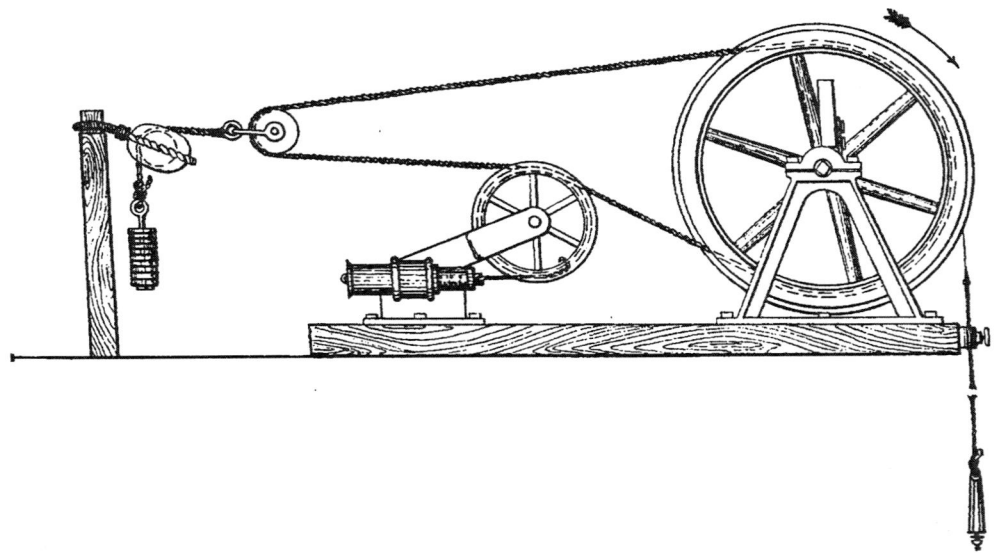

"Showing the general form and working of Sir William Thomson's sounding machine as used on board the 'Blake' during her first season in the Gulf of Mexico (rigged for paying out)" (Sigsbee 1880, Plate 6). The drum was one fathom in circumference, and "on the axle is a worm which engages a counter or register, to mark the revolutions of the reel." The wire is wound around the drum, and the drum has a friction score at one end. To the rear of the drum is a "dynamometer-pulley," and the resistance it provided was regulated by weights on its tail block. Thomson attributed the failure of previous attempts to use pianoforte wire to the failure to apply an adequate resistance. An endless belt ("friction rope") passes around the dynamometer pulley and the friction score of the reel, and it is engaged by a driving wheel (53-54).

and cloudy when dry, and thus showed a demarcation; once dried out, it could be reused (Knight and Baldridge 1921, 150; Tanner and Blish 1899).

A twentieth-century alternative to the sounding line is the fathometer, a SONAR-based echo sounder (Toghill 2003, 37). This relies on the principle that the speed of sound in water is about 4,800 feet per second, and the time interval from emitting the ping to hearing the return (proportional to twice the depth, if the ship is still) may be measured (Mixter 1967, 9).

Other signs. Those who live on the sea tend to notice subtle cues as to where they are. These include the color of the sea, the typical currents and winds, bird and fish movements, and clouds that hover over islands and perhaps even reflect the color of the land below (Taylor 1957, 59–60; Calahan 1952, 82).

The Polynesians made the study of these signs an integral part of their sophisticated system of navigation. Swells could be used both to steer in the open sea and to find land as a result of how it distorted swells. In the Santa Cruz Islands, there was the Hoahualoa (from the southeast), the Hoahuadelatai (from the east-northeast), and the Hoahuadelahu (from the northwest). These "probably originated from the southeast trades, the northeast trades …, and the northwest monsoon, respectively" (Lewis 1994, 127–8).

Near land, not only are swells refracted around islands and reflected from them, and diffracted as they–pass between two islands, but also interference patterns can be formed in the waters within island chains (224–52). In the Marshalls, "stick charts" were used to teach swell patterns (245–9).

In the Carolines, there was much reliance on observation of birds to locate islands (216). While the presence of land-roosting seabirds always indicates that land is near (how near, depending on the species) (212–14), it is only "in the early morning when the seabirds fly out to their fishing grounds and toward evening when they return home again … that their flight paths indicate the direction of land" (205–6).

Cloud signs were used in the Gilberts, but it was not a simple matter of locating a stationary cloud. Color and shape were also important in distinguishing a "land cloud" from a towering cumulus forming over open ocean (216–21).

Currents were also mapped. The Gulf Stream was discovered by Juan Ponce de León (1512) by its effect on his ship's speed en route to the Caribbean, and it was subsequently used by the Spanish treasure fleets for part of the return trip from Havana to Cadiz (Mazur 2022, 105). It was later recognized that the Gulf Stream was warmer than the surrounding waters, and in 1799, Jonathan Williams proposed that one could follow it simply by "thermometric navigation," without the mariners having "to heave the lead or observe the heavenly bodies" (Casey 2016).

Terrestrial Latitude and Longitude

While we may not have thought so in high school, one of the great intellectual inventions of mankind is the coordinate system. For example, if a city is laid out as a grid, we can send someone to the intersection of, say, North Tenth Street and East Third Avenue.

Latitude and longitude are the dimensions of a gridded spherical surface coordinate system first devised by the ancient Greeks (Brown 1977a, 49, 51). The earth is not a perfect sphere, but for our present purposes, it is close enough. Imagine the earth as a hollow, see-through globe, with you hovering somehow at the center. If your body were aligned with the earth's axis, you could identify any point on the earth's surface by two angles, one measuring "up-and-down" relative to "level" (latitude) and the other "left-and-right" relative to "front" (longitude).

For each of these angles, we need a reference, a "zero." For latitude, it is the earth's equator, the intersection of the earth's surface with an imaginary plane perpendicular to the axis. Any point on the equator is zero degrees latitude. Angles are traditionally measured in degrees; by ancient convention, a circle is divided into 360 degrees (each degree, symbol °, is divided into 60 arc minutes, symbol ', and each minute into sixty arc seconds, symbol "). Above your head would be the north pole, defined as 90 degrees north latitude. Below your feet would be the south pole, at 90 degrees south latitude, sometimes represented as -90 degrees.

Except at the poles, the points on the earth's surface that have a particular value of the latitude form a circle on the earth's surface; the circles are parallel to each other (that is, they maintain a constant distance), and hence are also known as "parallels" (e.g., the 49th parallel, part of the border between Canada and the western United States).

The "lines" (really, half-circles) of constant longitude are called meridians. For longitude, we have to pick an arbitrary zero. Hipparchus proposed using a meridian that passed through the city of Rhodes (Brown 2010). Currently, the zero longitude (prime) meridian is one established by an 1884 international treaty and passes through the Royal Observatory at Greenwich, England. Longitude is measured as being so many degrees (up to 180) east or west of the prime meridian.

On a globe, the "lines" (circles) of latitude will always cross the "lines" (circles) of longitude at right angles. (A map may distort this relationship.)

If two points are on the same meridian (constant longitude), but one degree of latitude apart, that's a distance of about 69 miles. It would be the same distance, regardless of where you were, if the earth was a perfect sphere. Thus, an error of one degree latitude corresponds to 69 miles; an error of one arc-minute ('), 1.15 miles; an error of one arc-second ("), 100 feet.

If two points are on the same parallel (constant latitude), but one degree of longitude apart, the distance between them would be a maximum of 69 miles (at the equator). The further away they are from the equator, the shorter the distance would be.

The true length of a degree of latitude was measured with high precision (error about 0.55 percent) and published in 1637 by Richard Norwood (*Encyclopaedia Britannica* 1911 [EB1911]/Navigation 19:289). The relative length of a degree of longitude, given the latitude, could be determined from the table of "meridional parts" in Edward Wright's *Certaine Errors in Navigation* (1599) (288).

Thanks to land observations, even in the early sixteenth century the latitudes of many ports were known. Those given in the *Regiment of the Astrolabe* (1509) are accurate to 30', sometimes even 10' (Taylor 1957, 166).

Globes and Maps

Globes, like the earth, are spherical. Maps are flat. As you can verify by trying to flatten out the skin of an orange while keeping it as a single piece, some creativity is required to flatten out a spherical surface.

The technical term for the mathematical manipulation by which points on a spherical surface are converted to points on a flat surface is "projection." Any map projection is going to distort certain properties of the earth's surface and, hopefully, preserve others. Projections can preserve direction from a central point (azimuthal projection), distance from a central point (equidistant), local shape (conformal), area (equiareal), and so on. You need to use the right map projection for a particular purpose.

The Mercator projection, empirically developed by Gerardus Mercator and first presented in 1569 (Monmonier 2010, 4), was given proper mathematical form by Wright (1599). It is still used for navigation (Brown 2010, 134–9). It is best suited for showing a small section of the globe at one time, since its scale varies with latitude (Greenhood 1964, 128, 131). Its advantage is explained in the next section.

Rhumb Lines and Great Circles

A rhumb line (loxodrome) is a path on the spherical earth that corresponds to following a constant true compass bearing (azimuth), or, to put it another way, to crossing every meridian at the same angle. If a map uses a Mercator projection, rhumb lines appear as straight lines. Parallel sailing is a special case of rhumb line sailing in which one sails along a parallel (line of latitude), thereby crossing every meridian at right angles.

Except for the special case of the loxodrome along the line of latitude, the full path of the loxodrome, if heading away from the equator, spirals toward a Pole. This was first pointed out by the mathematician Pedro Nunes in 1537 (Randles 1989, 122–23).

A great circle (orthodrome) is a circle on a sphere that has the same diameter as the sphere, and thus divides the sphere into two hemispheres. The equator (zero latitude) is a great circle, and the meridians are portions of great circles (with constant longitude). However, these are special cases, and great circles can connect points that differ in both latitude and longitude.

If a map uses a gnomonic map projection, great circles are shown as straight lines. On a Mercator projection, they are curves.

The shortest distance between any two points on the surface of a sphere is a portion of the great circle that connects the two points. Unfortunately, traveling on a great circle path requires continual correction of one's compass heading. Great circle sailing can also carry one to a higher latitude than is desirable (too much ice and fog). Also, if the ship is wind-driven, the great circle route may be disfavored because of unfavorable winds or currents (Davies 2014, 86).

A great circle route may be approximated by a series of short rhumb lines connecting waypoints that lie on or close to the great circle.

Nunes suggested following an approximation of a great circle route by changing course every hour (Randles 1989, 125) and, in another 1537 text, proposed a method of calculating the necessary correction to be made at each new degree of latitude (126). This was not adopted by his contemporaries. Randles (1989, 125) pointed out that the accuracy of the instruments available for measuring latitude and heading "would have rendered the whole operation impracticable."

Recognizing both the advantages and disadvantages of great circle navigation, John Davis called it a "rare practise" and chose to "overpass it" (1595c). However, in the mid-nineteenth century, there were attempts (e.g., Seaton 1850; Towson 1852) to develop methods of calculating the great circle route that would be acceptable to mariners; the bases of the various methods are reviewed by Littleshales (1899).

One successful use of Great Circle sailing was in 1852, on the *Marco Polo*'s run from Liverpool to Melbourne. Ships normally made this run by way of Cape Verde and Cape Town, and the average passage time was 110 days. John Thomas Towson calculated that the Great Circle route was 1,000 miles shorter. Captain Forbes took Towson's Great Circle tables and Matthew Maury's charts of prevailing winds and sea currents on board, and made a 76-day passage to Melbourne (Thorp 2017, 67).

However, "on a short voyage of 500 miles or less the savings of great circle sailing are so slight that those routes are not customarily used" (Nelson 1912).

Nowadays, great circle navigation is made trivial by computer-driven automatic steering and the Global Positioning System (GPS).

Composite sailing is a combination of great circle sailing (or approximation thereof) to and from "the highest latitude deemed safe" (Proctor 1888, 14) and parallel sailing in between.

Dead Reckoning

In dead reckoning, the navigator plots the last known location on a chart and extrapolates the present location based on the ship's subsequent heading(s), speed(s), and time elapsed.

The Spanish called dead reckoning *navegacion de fantasia* (Gurney 2004, 19), and Edward Wright (1599) referred to the estimated position as "the point of imagination" (Williams 1992). Dead reckoning estimates of longitude were sometimes more than 400 miles astray (Wakefield 2005, 165).

The *traverse board* was a device used to keep track of the courses steered. Every half hour, a peg would be placed in one of 32 holes, each representing one point of the compass. There were eight such concentric circles of holes, thus recording an entire four-hour watch (Phillip-Birt 1971, 191).

Of course, steering a particular course didn't mean that the ship necessarily moved in the expected direction. The helmsman could be lax, the ship's steering arrangement could be inaccurate, and the ship could be forced off course by powerful winds and currents.

The prudent navigator attempted to estimate "leeway" (the extent to which the ship was forced off course) by looking at the angle between the wake and the heading (Williams 1992, 22).

Moreover, even if the ship was placed on the desired compass bearing, that bearing might not be the desired true bearing, by reason of errors in correcting for magnetic variation and deviation (discussed later) or of determining true north from the sky.

Logging speed. For measuring speed, the sailor used a log. The *common log* was a piece of wood tied to a knotted line. The log was thrown out behind the ship, and the line was allowed to run out. One sailor counted the knots as they passed over the rail, while another watched a sand glass. The count continued until the sand glass emptied. The first written description of this method was in William Bourne's *A Regiment for the Sea* (1574) (Williams 1992, 39 n. 3), and the log was in general use, at least by the English and Dutch, in the 1620s (Swanick 2005, 100).

The sailing term "knots" refers to the fact that sailors estimated their speed, in nautical miles per hour, as the number of knots run out per "glass." A knot is one nautical mile (6,076 feet, about one arc-minute of latitude) per hour. Earlier schemes overestimated speed (perhaps deliberately), but in the late eighteenth century, sailors used a knot spacing of 47.25 feet and a 28-second glass (Gurney 2004, 25; Phillip-Birt 1971, 196).

There are some other obvious problems with this method. The log might be caught in the ship's wake, and the line might not pay out properly. There might be little delays in calling out the end of the time interval. The knot counter might miscount, or have trouble estimating an intermediate value. The speed of the ship might change, after the fact, as a result of shifts in wind and current.

An alternative form of the common log was the "Dutchman's log": Throw a chip off the bow and time how long it takes to pass by the stern (Mixter 1967, 10).

To get the distance run, the navigator multiplied the speed (presumed constant) by the time elapsed. Measuring shipboard time in the early seventeenth century was a rather chancy proposition, typically involving sandglasses.

While the water clock (clepsydra) was known in the ancient world, sandglasses first appear in the historical record in the fourteenth century. The possible fills included sand, "powdered marble, tin/lead oxides, and pulverized burnt eggshells." These materials had to be "carefully washed, dried and sieved." The sandglass itself was tightly sealed so that water would not seep in and "cause caking of the filling and consequent irregular running" (Mills 1996).

The "period" of a sandglass depends on the particle fill and their total volume, the shape of the sandglass, and the diameter of the aperture separating the two compartments. Early sandglasses were designed to empty in half an hour. It was not "until the late 16th century" that "½ minute versions" were "combined with a 'log' to give a ship's speed through the water" (Mills 1996).

It eventually became at least theoretically possible to use a pocket watch or stop watch rather than a sandglass. "The pocket watch of 1600 had only one hand, namely, the hour hand. The concentric minute hand came into general use about 1700 or a little later" (Milham 1923, 238). The second hand was not customarily added until 1780 (Milham 1923, 548), and the ability to count seconds was needed if the watch were to be used in conjunction with a log line to determine a ship's speed.

Stop watches, designed to measure short time intervals accurately, were also developed but were quite expensive (and perhaps "overkill" for shipboard use, given all the other uncertainties, such as the stretching of the logline and the effect of currents). The "Physician's Pulse Watch" was made by Samuel Watson in 1695, for the physician Sir John Floyer, and was accurate to one-fifth of a second (Seiko 2024). However, it was not commercialized (Gibbs 1971).

In 1810, Moore suggested use of a stop-watch to check the accuracy of the "half-minute glass" (Moore and Dessiou 1810, 133).

The common log was ultimately replaced by the *patent log* (more aptly called a screw or propeller log). This was a towed rotator, with spiral fins (Toghill 2003, 36). The passing water caused it to spin, and the rotations were mechanically communicated to a mechanical counting device by means of an endless screw (Massey 1820, 3). The patent log had to be calibrated by testing it on a run of a known length. Preferably you carried out two runs in opposite directions, so as to reduce the effect of any local current.

The patent log came in two basic forms. In the earlier form, the rotator and the "register" were in close proximity, so the device had to be hauled inboard to read the output. Later, the two parts were separated by a long flexible shaft, and only the rotator was thrown overboard, the register remaining fixed inboard, for example, to the taffrail.

A steamship engineer could construct a power curve relating ship speed to engine speed (revolutions per minute, RPM) by carrying out similar runs at each of several engine speeds. Then the engine tachometer could be used as a log (Mixter 1967, 10–11).

The pitometer log measures the difference between the total and static water pressure. The total pressure (which includes the effect of the ship's forward motion) is measured forward and the static pressure sideward. That difference, the dynamic pressure, equals one-half the product of water density and the square of the water velocity (the relative

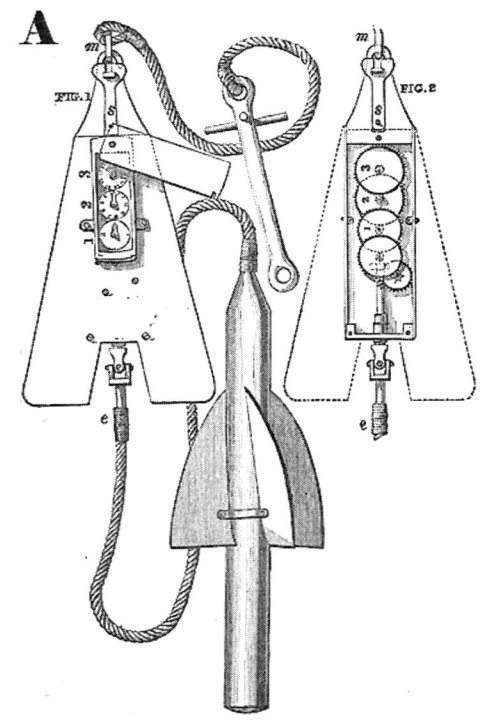

velocity of the water to the ship). An early form of the pitometer log was Napier's Pressure Log (Harbord and Goodwin 1897, 258), patented in 1871.

Plotting. When dead reckoning is figured as if the earth is flat, that is called "plane sailing." For a dead reckoning plot to be accurate over long distances, you need to use a Mercator projection chart, or to correct your eastings and westings for the changing length of a degree of longitude. The corrections were carried out with the previously mentioned table of meridional parts. But in the late seventeenth century, Sir John Narborough said, "I could wish all seamen would give over sailing by the false plane charts and sail by the Mercator's chart … but it is a hard matter to convince any of the old navigators" (Williams 1992, 43–6).

Navigational Use of the Compass

The compass has two purposes: determining which course is being steered, and providing a reference point for the measurement of azimuth (horizontal direction) in celestial navigation. An error of 3° in setting the course of the ship results in a positional error of one mile for every 20 miles run.

The Magnetic Compass

The standard magnetic compass has a magnetized needle that only swings horizontally. However, there are also "dip" compasses that can pivot vertically, too.

1. Finding Your Way

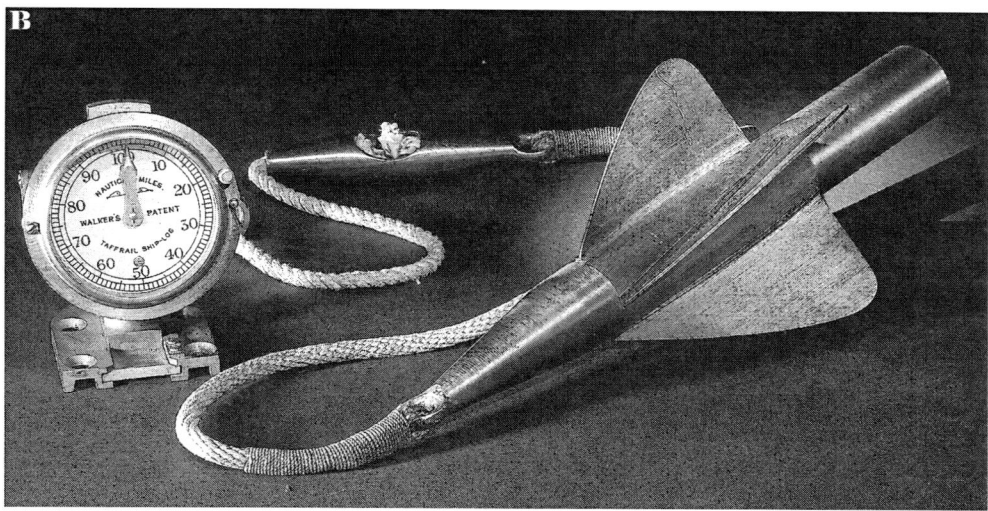

Opposite: A. Massey's patent log, as depicted in Fig. 55 of Nares, *Seamanship* (1876). Nares does not identify the labeled parts. However, Fig. 2 appears to attempt to show the internal mechanism of the log of Fig. 1. *Above:* B. Walker's log and rotor, patented in 1881 in the United States (U.S. Patent 238,187) and October 30, 1878, in Britain (4,369). This is a photograph of his U.S. patent model (National Museum of American History, Smithsonian Institution. Catalog 308558, accession 89797).

The marine compass typically has a rotatable compass card, marked with the compass directions. At least one magnetized needle is attached to the underside of the card (unlike the Boy Scout compass, in which the needle turns and the card is stationary). The earth's magnetic field causes the needle, and with it the compass card, to turn on its axis until the needle is properly aligned with the local magnetic field.

Needles were first magnetized by stroking them with a natural magnet (lodestone). In the mid-eighteenth century, Gowan Knight used a lodestone to magnetize steel bars, which he then combined to form a stronger magnet. With that, he repeated the process, ultimately creating a magnetizing machine with two "magazines," each consisting of 240 bar magnets and weighing over 500 pounds (Gurney 2004, 105). In the nineteenth century, it was recognized that one could magnetize steel rods by inserting them into a current-carrying coil.

Increasing the number of needles made the compass more sensitive, and it thus performed better when the sea is quiet, but then it oscillated too much when the waters were rough (Walker 1863, 72).

Curiously, compasses weren't routinely tested until the nineteenth century. After the 1707 Scilly disaster, the Royal Navy inspected its compass inventory and found that only three out of 145 were working properly (Wakefield 2005, 45).

One type of compass is "dry": The card pivots on a vertical pin, inside an empty bowl. One or more magnetized needles are fastened to the card. The bowl, which was covered with glass, itself "hung in gimbals (double pivoted rings) to keep the compass card as horizontal as possible" (Gurney 2004, 25).

The epitome of the dry compass is, perhaps, the "Admiralty Standard Compass (Pattern 1)," introduced in 1840 (Gurney 2004, 208–10). The dry card was 7½ inches in diameter, and was suspended on an "iridium pivot," with a "sapphire cap," inside a "black-lacquered copper bowl." (Iridium, an extremely hard, albeit brittle, and corrosion-resistant metal, was discovered in 1803, and used on a fountain pen point in 1834 [*Wikipedia*].) Perhaps most significantly, it had "four needles" rather than one, arranged symmetrically on either side of the pivot. The arrangement used was one calculated by Edward Sabine as eliminating disturbance by the ship's roll (Gurney 2004, 208–9). In contrast, a single heavy needle would cause the card to turn to align the needle with the ship's roll (123–4).

Around 1876, William Thomson increased the diameter of the dry card to 10 inches and braced the perimeter with an aluminum rim, in turn attached to the boss by 32 silk threads. The card carried eight needles instead of four, and nonetheless weighed only 180 grains, versus 1,500 for Pattern 1. The reduction in weight was expected to reduce pivoting friction (Gurney 2004, 241), and thus to improve responsiveness. By an extensive lobbying campaign, Thomson persuaded the Admiralty to adopt his compass in 1889, and it continued to be installed in new ships until 1906, despite its "unsteadiness in a seaway, under gunfire, and from the vibration of ships driven at speed" (Gurney 2004, 256–7). This was, in a way, a case of history repeating itself, because in 1752 all "Royal Navy ships bound abroad" were equipped with Gowin Knight's compass even though then–Captain George Rodney had pronounced it "impossible to steer by" (Gurney 2004, 123).

The earliest form of wet compass was simply a magnetized needle in an open, water-filled bowl, developed in China by the eleventh century (Gurney 2004, 36), and in Europe by the end of the twelfth century (Gurney 2004, 34). However, it had the practical problem that the water would slosh back and forth as the ship pitched and rolled (Gurney 2004, 39). Also, in colder climes it could freeze.

Nonetheless, it was apparently used by East Asian mariners. In the late sixteenth century, Japanese pirates adopted the European dry card compass, and the Chinese soon followed suit (Gurney 2004, 37–38). However, William Barlowe later reported that in the East Indies, sailors used a six-inch needle floating on water inside a bowl (Gurney 2004, 265).

In 1813, the watchmaker and silversmith Francis Crow patented an innovative wet compass in which the needle was sealed inside a floating copper "lens." The lens floated on an alcohol solution, which had a lower freezing point than that of water. It was constrained from below by a weight and a ring confining the movement of the latter. On top, it had an inverted hollow cone, engaging a downward pointing pivot on the underside of the glass cover. The top of the lens was painted to show compass directions (Gurney 2004, 267). There was also provision for the expansion and contraction of the liquid as a result of temperature change (Gurney 2004, 266).

The wet compass was "very steady in a sea-way, and but little affected by heavy rolling, shock, or vibration." But it was less sensitive to the earth's magnetic field than the dry compass (Hall 1904, 74).

Perhaps for that reason, the Admiralty did not adopt a wet compass (Dent's) until 1845, and even then it was considered a rough-weather backup for the dry card compass (Gurney 2004, 268). In a trial, Dent's was one of the two steadiest, and it outperformed its main competitor (Preston's liquid compass) by being "more sensitive in indicating any change in the direction of the ship's head" (Johnson 1852, 79).

Around 1876, the United States Navy replaced the dry compass with a wet one designed by Edward Ritchie and patented in 1862. This was rather similar to Crow's, but "substituted another pivot in place of Crow's weight" (Gurney 2004, 267, 269).

In the form exhibited at the 1867 Paris Universal Exposition, the needles were inside "an equal armed cross formed of hollow cylinders of thin metal," attached to a circular ring bearing the "divisions of the compass." Since the cylinders were buoyant, "the weight is so adjusted that the pressure of the system upon the pivot is reduced to a few milligrams." Also, the "moving system is of so nearly the same specific gravity as the liquid in which it is immersed as to make both participate equally in movements caused by the rolling and tossing of the vessel" (Blake 1870, 3:604).

It was not until the advent of high-speed torpedo boats and torpedo boat destroyers that the wet compass eclipsed the dry one in the Royal Navy, although even then it was a slow process (Gurney 2004, 263–4; Williams 1992, 136–7).

In 1905, Louis Chetwynd introduced a wet compass that, like Crow's but unlike Dent's, had a "reduced diameter." This avoided the disturbance of the compass "card" by "liquid swirl," the liquid dragged by adhesion to the bowl when the ship (and thus also the bowl) makes a sharp change in course (Gurney 2004, 270–1).

The compass fluid damps the movement of the compass "card," lubricates the card pivot, and renders the card buoyant (which reduces its friction with the pivot). In the twentieth century, hydrocarbon oils became an alternative to ethanol–water solutions (Harris 2010, 32).

Putting a dome top on a compass magnifies the compass card, reduces swirl error, and obviates the need for gimbals, since the card can rotate into the dome (Gurney 2004, 271). The compass card may also be made concave, making it easier to read at a distance, and the bowl and dome may together form a sphere, increasing stability (Maloney 2006, 437).

The magnetic compass is subject to a number of inherent errors (the earth's variation and ship's deviation), so mariners speak of three different kinds of directions: compass, magnetic (compass direction adjusted for deviation), and true (magnetic direction adjusted for variation).

Variation and deviation are explained below.

Magnetic Variation

The magnetic compass works, ultimately, because (1) the earth has a liquid iron outer core, (2) the molten iron is in constant motion, and (3) at least some of that motion is attributable to the rotation of the earth. The result is that a magnetic field is generated that, very loosely speaking, has one pole (place where a "dip" compass would point straight down) near the earth's true North Pole, and the other near the true South Pole.

However, a magnetic compass needle does not actually point toward the nearest

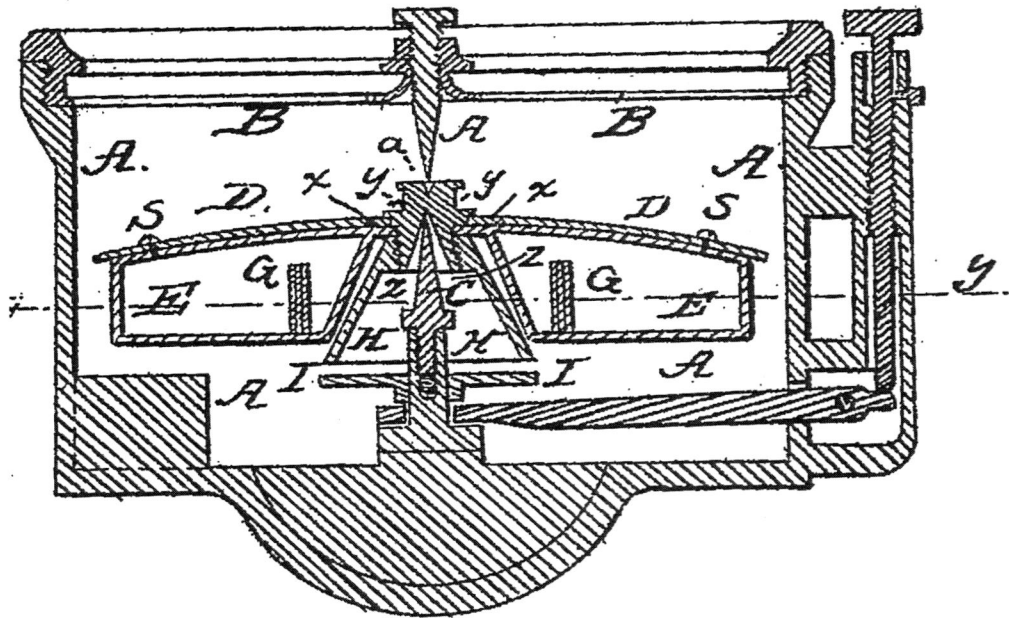

Patent drawing Fig. 2. for Ritchie's Mariner's Compass (1862). The compass card D turns on pivot C within the bowl A. The bowl has a glass top B. On the underside of the card is an airtight vessel E containing the magnets G. The vessel is perforated on the bottom so the pivot passes through it to the cap a. On top of the vessel is the "hollow frustrum of a cone, H," which serves to secure the cap and card and the air vessel together." The buoyancy of E is such as to "cause it to nearly fit the Card D, ... the pressure remaining on the pivot, C, being graduated so as to be of as small an amount as will suffice to cause the card bearing of cap to remain in contact with the pivot, and turn thereon with a trifling amount of friction." There are further illustrations and description of the Ritchie compass in Taunt (1883, 175–80) and Harrington (1882, 71-78).

magnetic pole. The magnetic field of the earth is complex,* and the needle will point "in the direction of the horizontal component of the magnetic field where the compass is located, and not to any single point" (NCEI 2024). And what we really care about is how that direction differs from the direction of the geographic (true) pole.

The difference in direction between magnetic north (south) (the magnetic compass direction in the absence of deviation; see later discussion) and true north (south), expressed as so many degrees to the east or west of true north (south), is called variation (or declination), and differs depending on where on the earth the compass is situated. Mercator tried to explain variation by postulating first one (1546) and then two (1569) north magnetic poles (Natural Resources Canada 2008).

Variation is unaffected by heading, and compensation with countermagnets is not possible. It also varies with location (and time). It was thus essential in the days of navigation by magnetic compass, especially when sailing great distances, to keep

---

* Some of its complexity may be efficiently modeled by spherical harmonics; see Pickering (2004).

track of the magnetic variation so that the correct course could be steered. The Cape of Good Hope is called Cape Aguilhas ("Needles") by the Portuguese because of the way the compass misbehaves in its vicinity (Walker 1863, 1).

You had to be careful where you bought your compass. For example, in northern Europe, compasses frequently had hidden offsets (needle at angle away from north on card) of 6–11°, to compensate for magnetic variation. On the other hand, Italian-made compasses lacked these offsets (Gurney 2004, 63). An unsuspecting soul who bought a northern compass and then tried to use it in the Mediterranean could get an unpleasant surprise.

Magnetic variation was first indicated on a European chart in about 1504 (Williams 1992, 26). Nonetheless, one of the reasons for the loss of the English fleet off the Scillies in 1707 was that their navigators didn't make allowance for the magnetic variation in the region (7.5°W at the time) (Gurney 2004, 95–6).

Determining a compass's variation requires taking the compass bearing of an object for which the true bearing is known:

- Celestial object—The most commonly used celestial objects are Polaris and the rising or setting sun. While Polaris is always very close to true north, the sun moves about, so you need to compute or look up its azimuth for a particular day and time.
- Landmark—If you have an accurate chart, and your ship's position is known, take the bearing of a landmark shown on the chart.
- Place line—If your position is not known, sail so that two landmarks shown on the chart line up. Preferably, the landmarks are far apart.

An example of calculating magnetic variation was given by Hariot in 1595. The azimuth of sunrise was measured with the meridian compass, and the simultaneous solar declination was estimated from successive noon values in the *Book of the Sun's Regiment*, and that was used as an entry, together with the ship's latitude, into Hariot's "Table of Amplitudes," arriving at the true azimuth of the sun. The variation was the difference between the true and observed azimuths (Taylor 1957, 221).

The Chief Pilot of the Portuguese India Fleet, De Castro, made numerous measurements of variation around 1540 and asserted that it could be measured with an accuracy of 0.5° on smooth water and 2° when the ship was rolling (Taylor 1957, 183). The first map of magnetic declinations was made by Edmund Halley in 1699.

Gilbert's *De Magnete* (1600) taught that the variation was fixed. However, within a couple of decades, scientists collected evidence that Gilbert was mistaken. For example, Borough found that the declination at London in 1580 was 11°4'E, while in 1622, Gunter said that it was only 6°13'E. The discrepancy was at first ascribed to experimental errors. Sometime in 1633, Henry Gillebrand began to suspect, based on new observations, that the declination had continued to trend westward, and he became sure of this in midsummer 1634 (and published his findings in 1635). This is explained at length in EB1911/Magnetism, Terrestrial, which offers numerous tables showing the change in declination in different parts of the world.

This "secular change" is just as geographically diverse as magnetic variation itself. Even outside the polar regions, it can be as fast as a 20° shift in one year. One silver lining is that for a specific location the rate of change is fairly close to constant (Bloxham 1992). Hence, local maps (such as the USGS quadrangle maps) can be published that state both the current variation and the annual rate of change, and they are then useable for a few decades for local compass correction.

Daily and seasonal fluctuations also occur. At Cheltenham, West Virginia, the westernmost declination is at 2 p.m., and the easternmost at 8 or 9 a.m. If time of year is considered, the range is from 6°E on a summer morning, to 4.8°W on an equinoctial afternoon (Sipe 1974, 77).

It is possible to construct maps of magnetic declination for historical periods using a combination of archaeomagnetic data (itself possibly derived from mariners' observations) and a standard geomagnetic model (Jackson et al. 2000; Pickering 2004). The National Centers for Environmental Information have a Historical Magnetic Declination webpage that displays "historical isogonic [constant magnetic declination] lines and magnetic poles calculated for the years 1590–2025."

Magnetic Deviation

The errors in magnetic compass bearings that are attributable to the ship and its contents are called deviations. They can vary depending on where the compass is located and the direction of the ship's heading.

The earth's magnetic field induces transient magnetism in soft iron, and the resulting deviation is greatest when the ship is on an easterly or westerly course. Even in a wooden ship, there are iron items. João de Castro's 1538 observation of variation was "troubled by the proximity of artillery pieces, anchors and other iron" (Gurney 2004, 139).

These "soft iron" deviations change as the ship moves north or south (changes magnetic latitude). The force induced in "horizontal iron" (such as a beam) is greatest at the equator and least at the poles. The reverse is true for vertical iron, and its direction reverses when the ship crosses the magnetic equator. Vertical soft iron in early nineteenth-century sailing ships included "hanging knees, nails, and bolts in the deck, the capstan spindle, anchor flukes, stanchions, chain plates, belaying pins, rudder stock" (180).

In wooden ships, the deviation is greatest when the ship is on an easterly or westerly course (Walker 1863, 67); this is the result of asymmetrical vertical soft iron, forward or aft of the compass (NGIA 2004, 13). Bear in mind that the compass is by the helmsman, at the rear of the ship.

Downie, master in HMS *Glory*, 1790, wrote: "I am convinced that the quantity and vicinity of iron, in most ships, has an effect in attracting the needle ... the needle will not always point in the same direction, when placed in different parts of a ship... [T]wo ships, steering the same course by their respective compasses, will not go exactly parallel to each other yet when their compasses are on board the same ship, they will agree exactly" (Walker 1863, 11).

A small amount of iron close to the compass can be as disturbing as a large mass further away. A belt buckle, moved closer than 12 inches, can cause a deviation. So can a ballpoint pen at 5 inches, or a wristwatch with a metal band a foot away, a knife at 2 feet, or a metal handle axe at 4 feet (Sipe 1974, 84–5). With some qualifications, the magnetic field strength is inversely proportional to the cube of the distance (Sipe 1974, 83).

As one might expect, deviation became a greater concern in the nineteenth century when iron hulls were introduced. The deviations experienced on an iron ship can exceed 50°! (Gurney 189, 200, 217). When the steel is hammered, bent, riveted, or welded, the earth's magnetic field imprints it, converting it into a "subpermanent" magnet that records the direction the ship was "headed" when built (Mixter 1967, 55–6). This deviation is maximized when the subpermanent dipole is at right angles to the compass needle.

Deviation is measured by "swinging the ship": placing the ship on each standard heading and comparing the compass bearing to the true one. The known variation (unaffected by heading) is taken into account, and the residual error is the deviation.

There are two basic approaches to dealing with deviation. The nineteenth-century British Navy method was to never assume that the compass was correct; rather, routinely swing the ship. (This method was first proposed to the Admiralty by Matthew Flinders in 1812 [Gurney 2004, 171].)

The Merchant Marine approach was to judiciously place countermagnets so as to counterbalance the deviation (Gurney 2004, 255–6; Toghill 2003, 24–5; Williams 1992, 131–6). Some correction, at least, is desirable on iron ships, since large deviations can cause the compass needle to become sluggish or erratic.

The countermagnet method developed rather slowly. In 1805, Matthew Flinders proposed placing "a vertical iron bar ... close to the binnacle" (Gurney 2004, 169). The "Flinders bar" corrects for the "induced magnetism of the ship's vertical soft iron" (Gurney 2004, 279). It did not come into actual use until decades later (Gurney 2004, 182).

In 1831, William Fairbourn placed a reference compass on shore and moved a piece of iron around the *Lord Dundas*'s compass until the two agreed (Gurney 2004, 189), In 1837, the Astronomer Royal George Airy corrected the deviation of the four compasses on the *Rainbow*, and the binnacle and telltale compasses on the *Ironside* (Gurney 2004, 200–2), and in 1839, he addressed the compasses of the gunboat *Nemesis*. In the latter case, he used three bar magnets to correct for the "semicircular deviation" caused by magnetism of the hard iron of the ship during construction, and iron chain links to correct for the "quadrantal deviation" resulting from "the induced magnetism of horizontal deck beams" (Gurney 2004, 200–6). In 1861, the Liverpool Compass Committee recommended the use of the Flinders bar (Gurney 2004, 213).

Until the middle of the century, it had been customary to fix corrective magnets in the deck. But in 1854, John Gray patented a binnacle that contained "easily adjustable magnets and iron spheres" (Gurney 2004, 233, 231). A similar binnacle was popularized by William Thomson two decades later (Gurney 2004, 241–2).

Arguably, there was a third approach, proposed by William Scoresby in his *Arctic Regions* (1820): placing a compass aloft, remote from the iron of the hull (Gurney 2004, 221–2).

Magnetic Dip

The first compasses had a needle that could only pivot horizontally. In 1581, Robert Norman discovered that if the needle were permitted to move vertically, it would dip (Walker 1863, 9–10). This magnetic "inclination" of the needle varies across the world. The needle will point straight down at the magnetic poles, and is flat at the magnetic equator (a wavy line ranging perhaps 10° north and south of the true equator). The unreliability of magnetic compasses in the polar regions is a consequence of dip; the magnetic force on the needle is then primarily vertical, and the needle may be more responsive to ship movement than to the tiny horizontal magnetic force.

In 1602, Gilbert and colleagues suggested that dip could be used to determine latitude when the sky was overcast (Taylor 1957, 247). This was a forlorn hope; points of equal dip are not uncommonly 20–30° latitude apart. Dipping needles are nonetheless useful in correcting the deviation observed when a ship heels over (rolls) (BOIMA 1921, 102).

Gimballed ("Cardan") Suspensions

A gimballed suspension is used to keep an object from tilting despite vehicular rotational motion. They have been used on shipboard for magnetic compasses, gyrocompasses, artificial horizons, chronometers, and oil lamps.

The suspension features two or three concentric rings. The object is attached to the innermost ring. Each ring is pivotably connected to the ring outside it, and the axis of rotation of each ring is perpendicular to that of the other ring(s). With three rings, the object is insulated from yaw, pitch, and roll.

For magnetic compasses, two rings, with axes oriented to keep the face level despite pitch and roll, are the norm. The compass bowl may yaw, thus showing the change in the ship's heading as the compass card continues to point the same way.

Harrington wrote, "The well-known gimbal action is not to be overlooked as a fundamental condition of steadiness … Without this or its equivalent no other provisions of compass-steadiness would be of any avail, and even the existence of such an instrument as the marine compass impracticable" (1882, 71).

Philon of Byzantium (circa 230 BC) described "censers and inkpots with articulated suspension," and in 1245, Villard de Honnecourt "sketched a small circular oven which was suspended on circular rings" (Seherr-Thoss et al. 2006, 1). A doodle by Leonardo da Vinci in the Codex Madrid I (1504, folio 13v) shows a three-ring suspension (Rubio 2024, Fig. 9, MDG-5.10) holding a "pail." In 1550, Geronimo Cardano reported that the sedan chair of Emperor Charles V was mounted in triple gimbals, to give him a smoother ride (Forrester 2013, 834–5).

Gunther (1923, 1:301) claimed that the first depiction of a magnetic compass in a gimballed suspension was from 1604, but gimballed compasses were in use earlier. The earliest compass (circa 1570) in the National Maritime Museum (NMS, Greenwich) "is mounted in a brass gimbal ring" (NMS 1570). Sixteenth-century compass "balances" have been found by archaeologists (one from a Manila galleon, one from a Basque whaler, three from the *Mary Rose*, and one from a Dutch expeditionary vessel). The two gimbals of the Manila galleon compass were found still fastened together (Rivera and Junco Sánchez 2023).

Fluxgate Compass

Unlike the classic magnetic compass, the fluxgate compass uses two coils of wire, wrapped around a core of magnetizable material (e.g., ferrite), to detect the direction of the horizontal component of the earth's magnetic field. An alternating current is passed through the primary coil, and this induces a current in the secondary coil (Avis 2012). The local magnetic field "causes shifts of phase depending on its relative alignment with each coil. By measuring both phase shifts, the direction of the magnetic field in two dimensions can be computed" (Siegwert et al. 2011, 117).

It was derived from the fluxgate magnetometer (developed 1928), which was initially used for "detecting submarines and for airborne mapping of Earth's magnetic field" (Avis 2012). The modern form uses "solid state electronics" (and thus has a "low sensitivity to vibration, shock and temperature changes") and provides a "digital output." This output may be "automatically corrected for deviation" and, with GPS lookup into a chart of magnetic variation, for variation, too, thus providing a true direction (Avis 2012).

Gyrocompass

A gyroscope is a rotor designed so that it can spin rapidly, and a free gyroscope is one mounted so that its axis can point in any direction. Bohnenberger built and described such a machine in 1817 (Wagner 2005). The axis will continue to point in the spin-up direction unless it is disturbed by an external force acting on a line perpendicular to the spin axis and off center (a torque). If that happens, then as a result of the combination of the applied force and the inertia of its spinning rotor, the axis tilts at right angles to the line of the applied force; this is called precession. Thus, if a rotor was turning on a vertical axis, and a force was applied to the rim to cause it to tilt down toward north, it would actually tilt toward east or west (depending on the direction of rotation).

The gyroscope has "gyroscopic inertia"—that is, it resists a change in the orientation of its spin axis. The rate at which it precesses is directly proportional to the applied torque and inversely proportional to the spin rate and the rotor's moment of inertia (a function of both its mass and how the mass is distributed—preferably, near the rim).

A gyrocompass is a controlled gyroscope, one equipped with a mechanism causing it to precess until its spinning axis lies in both the horizontal plane and the perpendicular north–south (meridian) plane (the meridian is the "line" on the earth's surface connecting its present location and the two geographic poles). This mechanism exploits the force of gravity, which always acts in the meridian plane (Burger and Corbet 2016 [1963], 19).

The pole-seeking mechanism in essence makes the gyroscope frame act as a

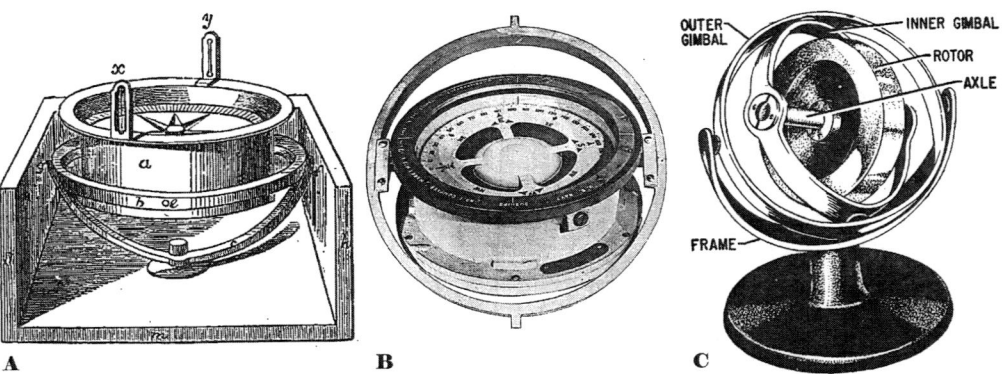

A. Late-nineteenth-century bearing compass (Angell 1879, 44). The bowl a turns on pivot e (the opposite pivot is hidden by perspective). The ring b turns on pivots g, f. The half-ring c possibly turns on the unlabeled pivot. The sight vanes x, y are for taking the bearings. As drawn, the bowl and rings do not have a full range of motion. B. U.S. Navy 7.5-inch magnetic compass (Bureau of Naval Personnel 1966, 20). You are looking down; the inner black ring to which the compass bowl is attached is tilted away from you, as if the ship is pitched. The silver ring's pivots, at the top and bottom of the photograph, rest in the binnacle. C. Schematic of gyrocompass suspension (Bureau of Naval Personnel 1965, 61). The rotor is the actual gyroscope; once spun up, its axle will attempt to point in its original direction. There are two gimbals, compensating for two axes of rotation. If the frame could rotate on its pedestal, this would become a three-axis suspension. An animation on *Wikipedia* shows how the gimbals of a three-ring suspension can move without disturbing the central object: https://upload.wikimedia.org/wikipedia/commons/d/d5/Gyroscope_operation.gif. The reader may also appreciate the "early modern compass" depiction on Wikimedia Commons. While the date stated is "1570," no source for the image is given.

pendulum, for example, by attaching a weight to the top or bottom, in line with the spin axis (Burger and Corbet 2016 [1963], 19–20). Assume the gyroscope is spun up with the spin axis horizontal. If the spin axis is tilted relative to the local horizontal plane, gravity acting on the weight will create a torque to reduce the tilt. Because of the spin, the rotor will precess in a direction perpendicular to the applied torque. Since the tilt was in a vertical plane, the precession is in the horizontal plane.

Now, why would the spin axis tilt in the first place? The reason is the rotation of the earth. The earth rotates from west to east. As the earth turns under the gyrocompass, the frame rotates with it. The rotor axis doesn't initially change which way it is pointing, so (unless it is already aligned north–south) it appears to tilt relative to the horizontal plane (it is actually the frame that tilted while the rotor remained still). Since the gravitational force is perpendicular to that plane, it creates a westward torque by its action on the weight, and if the rotor is spun in the correct direction, this causes a northward precession of the north "pole" of the rotor axis.

The gyro needs to precess continuously to the west "as fast as the earth is carrying the gyro off to the east." Hence the gyro will have a slight tilt to the rotor axis to create this precession (Bureau of Naval Personnel [BNP] 1946, 17B4).

The pole-seeking mechanism by itself will not permit the gyrocompass spin axis to settle on the north–south direction; rather, it will hunt back and forth across the meridian. Thus, a practical gyrocompass must also include a damping mechanism to reduce the oscillations (Burger and Corbet 2016 [1963]).

Gyrocompasses do not need to use a solid weight to create the pendulum behavior, and indeed that construction is impractical at sea. "Inter-cardinal rolling [northeast to southwest, or northwest to southeast] causes a compass to swing in its gimbals, with the result that the pendulum would be subject to acceleration forces which would cause a continuous torque about the vertical axis of the compass" (Sperry 1944, 10a).

Instead, the gyroscope frame may hold bottles in line with the horizontal spin axis of the rotor, but at opposite ends. These bottles are connected at the bottom so liquid can flow between them. When the rotor axis is horizontal, the liquid levels are the same and the system is balanced. When the spin axis is tilted, some liquid is transferred and the system is unbalanced, creating a torque about a horizontal axis, which in turn causes precession. While oil was used in the Brown gyrocompass, the preferred liquid is mercury because of its higher density (Burger and Corbet 2016 [1963], 20–21). Because mercury moves within the system, it is sometimes called a "mercury ballistic."

Since the weight of the mercury is "distributed equally above and below the gyro-axle, ... no torque about the vertical axis of the compass is introduced by the swinging of the compass in its gimbals" (Sperry 1944, 11a).

This system can experience damping because it takes time for the liquid to flow between the two bottles. The flow rate is affected by the viscosity of the liquid and the diameter of the connecting pipe.

If the vessel is moving in a direction with a northerly or southerly component, the movement will cause the spin axis to settle in a direction that deviates from the meridian. However, this deviation can be precisely calculated if the ship's velocity (speed and course) and latitude are known.

The first practical gyrocompass was invented by Hermann Anschütz (1906), and, in a revised form, it passed sea trials on the *Deutschland* (1908) (Anschütz & Co. 1910, 9). The rotor was suspended from a stalk topped with a float and a compass card; the float was in a steel bowl containing mercury, and the bowl was carried on gimbals (Anschütz & Co. 1910, 25–26). The rotor casing included an air inlet, and the rotor acted like a centrifugal blower. Below the rotor casing was a pendulum, and its movement controlled a plate dividing the air outlet. If the rotor axis were not horizontal, the pendulum would swing to one side, the two streams of air would become unequal, and this would create a damping torque (Anschütz & Co. 1910, 32–33).

That pendulum was not the pole-seeking mechanism. Rather, "the center of gravity of the floating system is below its metacenter," and thus the whole float–stalk–rotor casing subsystem acted as a pendulum (Chalmers 1920, 140).

The spin rate was 20,000 rpm (Anschütz & Co. 1910, 27) and the "directive force" achieved was "some fifteen times as great as in the case of a good Magnetic Liquid Compass" (Anschütz & Co. 1910, 39). The gyrocompass's rotor position could be read and converted to an electrical signal, with the latter then being transmitted to receivers elsewhere in the ship (Anschütz & Co. 1910, 39–40).

Corrections were made manually for the ship's latitude (Anschütz & Co. 1910,

35), north–south speed (Anschütz & Co. 1910, 37), and changes in north–south speed (Anschütz & Co. 1910, 38).

The U.S. Navy adopted the Sperry gyrocompass. This used a "mercury ballistic" as the control element (Sperry 1944, 3). There were various damping expedients, depending on the model (Burger and Corbet 2016 [1963], 30). The most important innovation here was an automatic correction apparatus. It did need to know the ship's speed and latitude (at least within "3 knots and 3° respectively" [Sperry 1944, 11]), but then it corrected "for all courses and change of course of the ship" (Sperry 1912, 31, 42–43).

The gyrocompass has the advantage of pointing to true north; it is not subject to magnetic variation or deviation. A properly adjusted gyrocompass is usually accurate to one degree or better. However, it requires constant electrical power (to keep the gyro spinning), and its accuracy decreases as the ship moves above 75 degrees latitude (Maloney 1985, 43). Of course, the latter problem is also experienced (for different reasons) with magnetic compasses, which tend to go haywire in polar regions.

Altitude and Azimuth

For celestial navigation (discussed in the next chapter), the mariner needs to be able to describe the positions of celestial objects. From the observer's standpoint, the easiest system is to measure the angle between the object and the horizon (altitude) and between the horizontal direction to the object and true north (azimuth). While both altitude and azimuth are needed to completely define the position of an object in the sky, many celestial navigation problems are solved just by use of multiple altitude readings. Azimuth is used mostly for correction of compasses.

Measuring Altitude

To measure altitude, you need a reference, either "down" (established by a plumb line) or "level" (the true horizon). Unfortunately, at sea, plumb lines sway, and horizons are obscured by mist, waves, and spindrift (Calahan 1952, 154).

Renaissance navigators had several devices to use for measuring altitude. The first was the mariner's astrolabe (a scaled-down and possibly simplified version of the astronomer's planispherical astrolabe*). This was disc-shaped, and had a circular scale and a radially mounted, rotatable arm (alidade) holding a pointer and a sight at each end. The astrolabe was suspended from a cord so it would be vertical. You eyed the star through the two sights and then read off the altitude from where the pointer crossed the scale. For a sun sight, you let the sun shine through one hole and illuminate the other.

One person steadied it, a second took the sighting, and a third read off the

---

* This should not be confused with the "spherical astrolabe," the "armillary sphere."

altitude. The larger the astrolabe, the more precise were its measurements. A mariner's astrolabe might be six or seven inches in diameter; for the six-incher, "each degree at the circumference" would be "about a millimeter" (Knox-Johnston 2013). Graham (1999) thought this could be (mentally) divided in half, and Knox-Johnston, more optimistically, into quarters. Vasco da Gama had a 24-inch astrolabe in 1497 (Stimson 1985, 576), but such large astrolabes were cumbersome to use. Sometimes the navigator made a landing just so the astrolabe could be used more easily (Miller 2024).

The two problems that beset the navigator attempting to use the astrolabe at sea were the pitching and rolling of the ship, and the force of the wind on the astrolabe. It was recognized that while "broad astrolabes" were "truer," they would be buffeted more by the wind. Hence, according to Blundeville (1594), the "Spaniards doe commonly make their Astrolabes ... narrow and weightee, which for the most part are not above five inches broad and yet doe weigh at least four pound and to that end the lower part is made a great deale thicker than the upper part." In contrast, the English sea astrolabes were "six or seven inches broad," but "very massive and heavie" (Stimson 1985, 578). While it was nominally disc-shaped, "cutouts" reduced the windage of the instrument (Vanin 2022).

Knox-Johnston conducted a practical test of the accuracy of the astrolabe by using it and dead reckoning on a boat attempting to duplicate Columbus's 1492 voyage. The noon positions determined by astrolabe were radioed home and were compared to the true positions reported by an onboard Argos satellite transmitter. The "index error" of the astrolabe was determined before the voyage began. The average error in latitude was 13.63 nautical miles.

There are several possible sources of astrolabe error (De Hilster 2014; Brooks 1999). Using a replica of the Valentia astrolabe (182 mm diameter), de Hilster's (2014, 46) "land-based observations have a standard deviation of about 8 arc minutes, while at sea the standard deviation was approximately half a degree."

Another device was the simple quadrant, not to be confused with the later Davis quadrant. It was first used at sea in 1461. It was a quarter-circle arc, with sights along one straight edge. A plumb line hung from the vertex. One sailor sighted through the two holes while another read off where the plumb line crossed the scale on the rim of the arc (Williams 1992, 35).

According to John Davis (1595), the astrolabe and quadrant were difficult to use on shipboard, except when the sea was calm (Phillips-Birt 1971, 139). You can imagine the astrolabe, or the plumb bob of the quadrant, careening wildly as the ship was buffeted by waves.

By the late sixteenth century, these devices had been largely superseded by the cross-staff (forestaff), consisting of a staff and a sliding transom (the cross-piece or vane). You put your eye at one end of the staff, and slid the transom toward or away from you along the staff, until the top of the transom was aligned with the star and the other end with the horizon. (Since you looked somewhat like an archer, this was called shooting the stars.) The staff was graduated so you could read off the location of the cross-piece. While the cross-staff could be used by one person, it was awkward to keep both the horizon and the celestial object aligned simultaneously.

The cross-staff was equipped with one to four vanes, of different lengths. In a common implementation, the staff had a square cross section, and each of its four sides bore a different scale. These varied, but one arrangement was 3–10°, 10–30°, 20–60°, and 30–90° (Seller 1694, 163). The four scales corresponded to the four vanes, from shortest to longest. Only one vane was used at a time. The largest would be used when the star was high in the sky, and the smallest when it was close to the horizon (which would be the case for the North Star when a ship was near the equator).

The length of the staff and cross-pieces dictated its region of usefulness; Swanick (2005, 76) says that it could only be used to sight objects between twenty and sixty degrees altitude, which would be degrees of latitude in the case of the Polestar. The cross-staff was best used when the altitude was substantially less than 60°, since for higher elevations, the corresponding graduations on the staff were small (Phillips-Birt 1971, 128, 145).

Digges's *Prognostications* also warned navigators that there is a parallax problem with the cross-staff: It would yield the correct altitude only if the eye were at the center of the staff (Taylor 1957, 206). Since the staff wasn't transparent, that was impractical, so the user had to make a downward correction to the nominal altitude. Hariot actually calculated the necessary individual correction for Raleigh and certain other English explorers, but on average it was about 1.5° (Taylor 1957, 220).

A modern test of a cross-staff replica, carried out at Lumberton, Mississippi (latitude 31.6°), resulted in readings of 32, 31.5, and 33° latitude (calculated as altitude of Polaris) from one scale and 31.5, 32, and 31.5° for the other. This was done, remember, on land. It looks like the angular accuracy of a cross-staff was 0.5–1.5° (Cookman 2001).

John Davis invented the simple back staff. This was a "back sight" instrument. That is, you observed not the sun itself, but the shadow it cast (which saved your eyesight, and also meant that you could simultaneously sight the shadow and the horizon, but also meant that you couldn't sight the stars) (Calahan 1952, 157). The original back staff was similar to the cross staff, albeit used in reverse. His *The Seamans Secrets* (1594) described two forms.

The first replaced the straight cross-piece of the cross-staff with a 45-degree sighting arc ("transversary") that was above, and with a base that slid along, the (horizontal) staff. This carried a brass plate with a slit at one end. You aimed the staff so you saw the horizon through the slit. (While not shown in the figure, it appears from the text that there may have been another brass plate with a slit at the center of the staff, so you lined up the horizon with both slits.) Then you slid the sighting arc until its shadow fell on the slit. The sun's altitude could not exceed 45° (Davis 1626, pdf 117; Davis 1595a, 1595b).

The second had a straight shadow-casting "half-cross" above and nearly perpendicular to the staff and a sighting arc below the staff and closer to the horizon end. The staff was a yard long, and the half-cross, "not 14 inches." Davis asserted that "this staffe does contain the whole 90 degrees," although that is not apparent from the figure (Davis 1626, pdf 120; Davis 1595b).

The Davis Quadrant was named after John Davis and sometimes falsely attributed to him. It was called a quadrant because it could measure angles up to

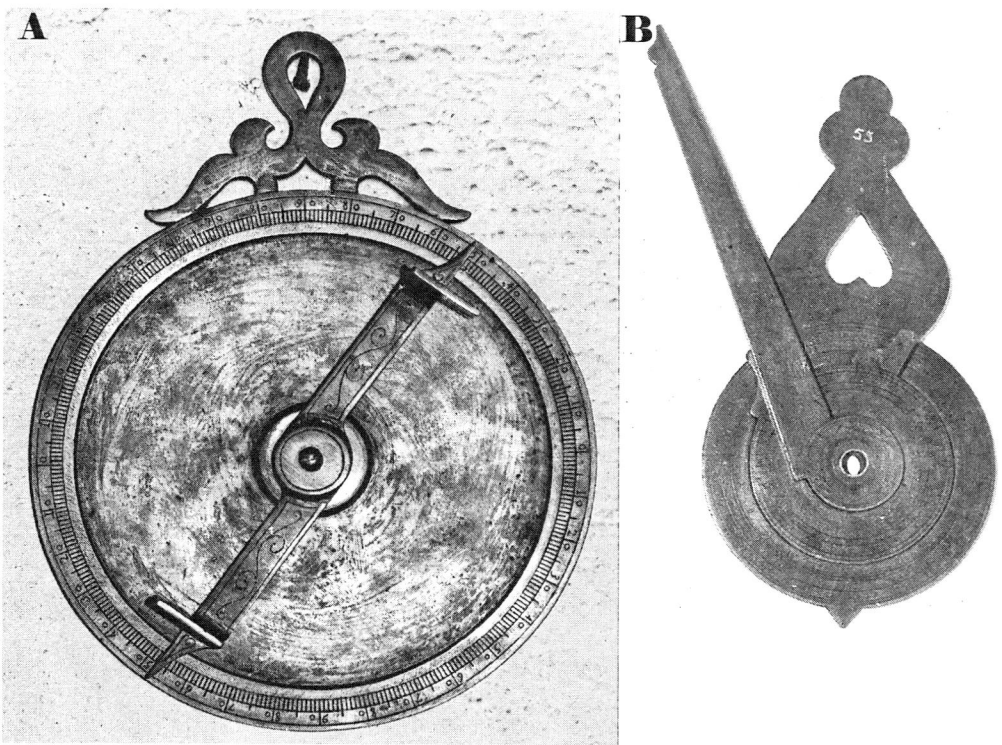

A. Sixteenth-century Spanish mariner's astrolabe, photographed at Museo Maritimo, Barcelona, Spain, 1966. Note the scale and the pointer (alidade) with two sight vanes (Naval History and Heritage Command, Catalog NH 115364). B. Seventeenth-century nocturnal. This one was dual purpose; it could "determine the nocturnal hours (front side) and the corrections in altitudes of the pole (back) to obtain the latitude." According to the Naval History and Heritage Command, it was designed so the navigator would align the pointer with Kochab in Ursa Minor and, unusually, Dubhei in Ursa Major. Courtesy of Museo Maritimo, Barcelona, Spain, 1966 (Naval History and Heritage Command, Catalog NH 115367).

90°. It had two concentric fixed arcs, one atop the other, of 30° and 60° width, respectively. Both arcs had scales. It also had three vanes with slits, two of which slid along the respective arcs, and the third fixed at the "center" defined by the arcs. You looked at the horizon through the lower (sight) vane and pivot (horizon) slits, and you adjusted the upper (shade) vane so the sunlight passed through the upper vane slit onto the pivot (horizon) slit (Davis 1595b; Brown 2010, 186; Cline 1999).

The term "back-staff" was used generically to described both the older forms and the Davis quadrant; the latter's key distinguishing feature was the double arc. A two-arc back-staff was depicted, albeit schematically, in George Waymouth's *The Jewell of Artes* (circa 1604). In 1643, George Fournier provided a more detailed illustration and said it was in use in England. The instrument didn't receive the name "Davis Quadrant" until around 1680 (De Hilster 2011).

The scale of the Davis quadrant might have 0.5° divisions, and quadrants could be read to perhaps 0.25° (Taylor 1957, 215). However, in 1631 (Miller 2024), Pierre Vernier described the Vernier scale, by which a scale could be read to the nearest arc minute. The even more accurate micrometer was invented at the end of that decade (Gascoigne, 1639) but was ignored until much later.

The double reflection octant and sextant. The first double reflection instrument was the Hadley octant (1731). It had an operating principle markedly different from that of the prior instruments. There were two mirrors. The first, the fixed horizon mirror, was only partially silvered. You sighted the horizon through the unsilvered portion. There was also a rotatable index mirror, which was attached to a pointer. You rotated the index mirror until you saw the celestial object's second reflection in the silvered half of the horizon mirror, then checked the pointer against the scale. The Hadley octant had a magnified scale, giving it an accuracy of 1–2 arc-minutes in Admiralty tests (Taylor 1957, 257; Calahan 1952, 158, 164).

The term "octant" arose because the frame was a one-eighth (45°) slice of a circle. Because of the double reflection, it could still measure an angle of 90°. In 1757, Campbell suggested that if you wanted to measure the angle between two celestial objects, such as the moon and the sun, it was desirable to have an instrument with a greater angle of action (Williams 1992, 98). That led to the creation of the sextant, for which the frame was one-sixth of a circle (60°), and which thus could measure an angle of 120°.

Subsequent improvements to the sextant included:

- increasing the arc (so angles greater than 120° could be measured)
- silvered glass mirrors (the original ones were of speculum metal, and tarnished)
- larger mirrors (larger field of view)
- the tangent screw (to adjust the index arm)
- sun-shades
- Vernier scales (and associated magnifiers)
- micrometer (for even finer adjustments)
- low-expansion frame material
- mountable monocular (for light amplification)
- spirit levels
- mountable artificial horizon

Altitude Measurements on Land

A captain could decide to make a landing to obtain a more accurate location fix. On land, the horizon may be hidden by mountains. Hence it is necessary to provide an artificial horizon. This took the form of a pool of mercury. Mercury, being a liquid, would naturally flow to form a horizontal surface. The observer would simultaneously sight on both the celestial object and its reflection; its altitude would be half the angle between them.

Use of mercury was not without its problems. Mercury is highly toxic and therefore had to be handled with care. Also, the artificial horizon could be disturbed by wind, smoke (from a fire 300 feet away), or ground vibrations (e.g., a horse galloping 500 yards away). Topographers would dig a trench around the pan holding

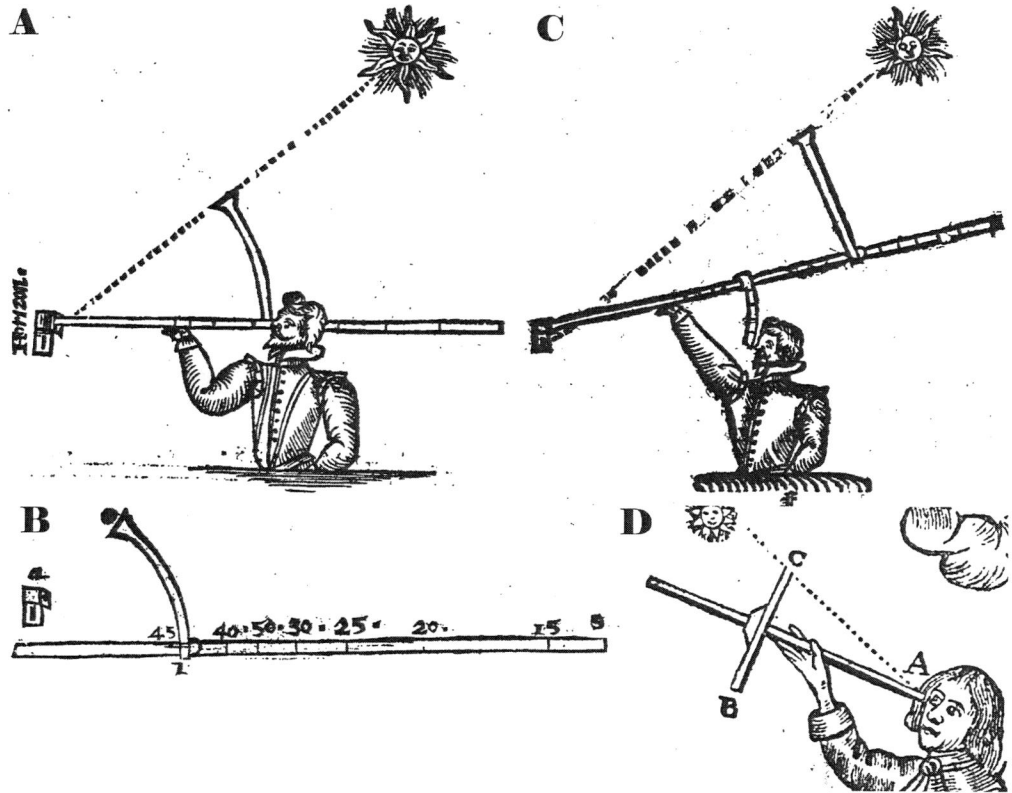

A. The Davis 45° back-staff in use (Davis 1626, pdf 117). B. The scale of the Davis 45° back-staff (pdf 116). C. The Davis 90° back-staff in use (pdf 120). D. For comparison, the use of the cross-staff (Seller 1694, 182). Note that the celestial object is sighted on the line AC and the horizon on the unmarked line AB. (The scale of the cross-staff is shown by Davis [1626, pdf 114].)

the mercury, to isolate it, and the observer would stand outside the trench (Schafer 2001).

Altitude Observational Errors

The apparent position of a celestial object may differ from its true (geometric) position for a variety of reasons, some dependent on the observer and the instruments used, and others on atmospheric and astronomical phenomena.

Sextant construction errors. To have accurate measurements, you need accurate scales. Scales were initially made by hand and eye. Later, "dividing engines" were devised for accurately dividing a scale into its units. The Ramsden dividing engine (1775), which was developed with financial support from the British Board of Longitude (National Museum of American History [NMAH] 1775), allowed the sextant to be halved in size, without loss of accuracy (Gurney 2004, 112).

Sextant calibration errors. The sextant must be recalibrated on at least a daily

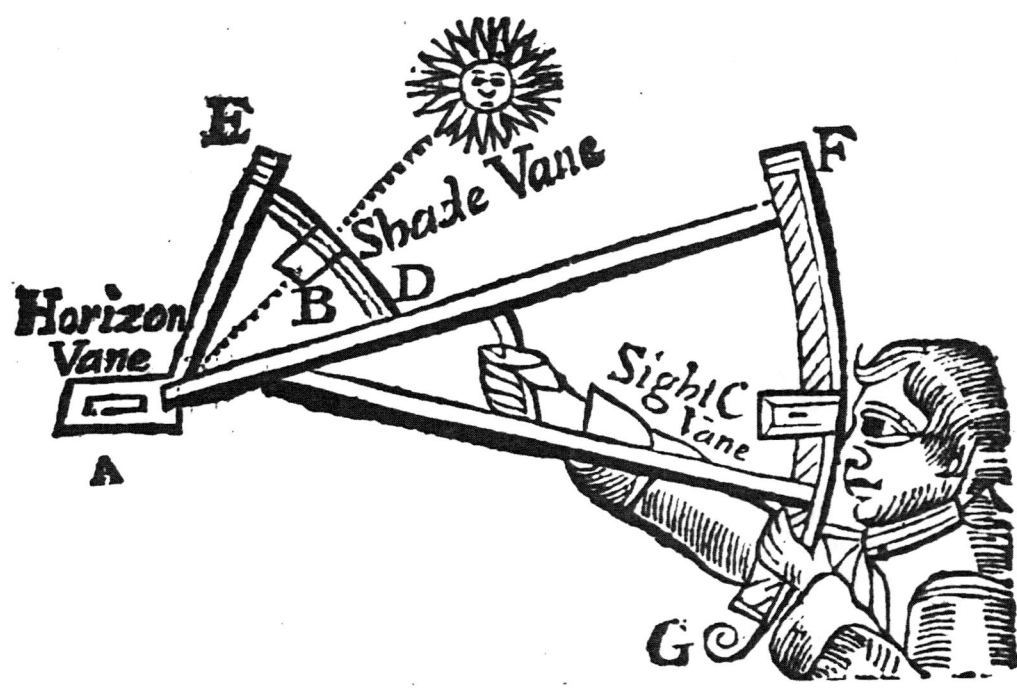

The Davis quadrant (Wakeley et al. 1755, 159). Wakeley cribbed this illustration from Seller's *Practical Navigation* (1694). Cf. Cline (1999).

basis, to check for and remove index error (the two mirror faces aren't parallel), side error (the horizon mirror not perpendicular to the plane of the sextant), and perpendicularity error (the index mirror not perpendicular to the plane of the sextant). Recalibration is necessary because the sextant is affected by changes in temperature and, of course, accidental knocks.

The sextant must, of course, be designed to allow these errors to be corrected. According to Toghill (2003, 30–1), the only error that can be tolerated (and taken into account in calculations) is index error, and then only if the error is less than 5 arc-minutes.

The 1784 edition of *Falconer's Mariner Dictionary* doesn't even mention the sextant. It is briefly mentioned in the 1800 edition (194). However, Burney's 1815 edition discussed the sextant and its adjustments (449).

Sextant reading errors. The observer must take the reading when the image of the object is just "touching" the image of the horizon (the moon and stars can be difficult to "land" on the horizon properly) and the sextant is absolutely vertical (Toghill 2003, 33, 91).

If the horizon is ill-defined, the altitude in turn is fuzzy. Usually, the stars and planets are observed during "civil twilight," when they aren't lost in solar glare but there is still a horizon. The best time is perhaps 20 minutes before sunrise, or after sunset (Schlereth 1975, 100–1).

Dip. Because the eye is elevated, and the earth is curved, the natural horizon (where the sea and sky meet) is lower than the celestial horizon. All sextant altitudes must be corrected for dip, or they will be overestimated by several arc-minutes. The dip (in radians) is the square root of $2h/R$, where h is the height of the eye and R is the radius of the earth (Young 2020). Expressed in arc-minutes, with eye height in feet, it is 1.06 times the square root of the eye height (Freiesleben 1950, 270).

The first dip correction table was constructed, for eye levels of 5–40 feet, by Thomas Hariot in 1584. His calculations are consistently about 1 arc-minute too high (Taylor 1957, 219–20). Edward Wright's dip table was published in 1599, and he calculated that if the eye level were 20 feet, the dip was 5 arc-minutes, 5 arc-seconds (Wright 1657, Chapter XV). The correct value is 4.74 arc-minutes.

Refraction. Light doesn't move in a straight line, but rather in the path that takes the least time. Since light moves more quickly through warm air than cold air, and the air nearest the earth's surface is warmest, the light you see, unless it is from directly overhead, has taken a curved path, favoring the warm air, in order to reach your eye. As a result, it comes from an apparent direction that is lower than the true direction of the object which you are observing.

This effect is greatest when the object is low in the sky; indeed, you can see the sun even after it has set below the celestial horizon. In the *Nautical Almanac*, for a star on the horizon, the refraction correction is -34.5 arc-minutes; at 10° apparent altitude, -5.3; at 45°, -1.

The *Nautical Almanac* assumes an air temperature of 50°F and pressure of 29.83 inches mercury. The refractivity of air changes if either temperature or pressure changes, but this needs to be taken into account only for low-altitude observations. In the subsidiary table, the maximum correction for unusual temperature or pressure, for apparent altitude on horizon, is -6.9 arc-minutes. Above 8° apparent altitude, it's less than 1 arc-minute (Maloney 1985, 316).

The astronomer Tycho Brahe (1546–1601) published the first table of atmospheric refraction, determined by observation. He reported stellar refraction to be 30 arc-minutes at the horizon, 10 at 5°, 3 at 15°, and nonexistent from 20° up. For the sun, he was able to detect refraction only up to 45° (Heilbron 1999, 128).

The first theoretical model of atmospheric refraction was advanced by Cassini in 1666. It assumed that the atmosphere had a constant density up to a particular height, and then came to an abrupt halt. Cassini's model predicted that there was one arc-minute refraction of a star at an altitude of 45°, contrary to Brahe's teachings. Cassini was right about that, but wrong about the structure of the atmosphere; the higher you go, the more rarefied is the air.

The "international standard atmosphere" model assumes that the air is in hydrostatic equilibrium and obeys the ideal gas law, and that in the troposphere, there is a constant decrease in temperature with "geopotential" altitude (altitude "corrected" for the decline in gravity with distance from the earth's center of mass). Algorithms exist for calculating the refraction as a function of apparent "altitude" under standard atmosphere conditions at sea level, and for correcting this for nonstandard temperature and pressure (Siranah 2018).

Aberration. This phenomenon was identified by Bradley in 1729. If the observer

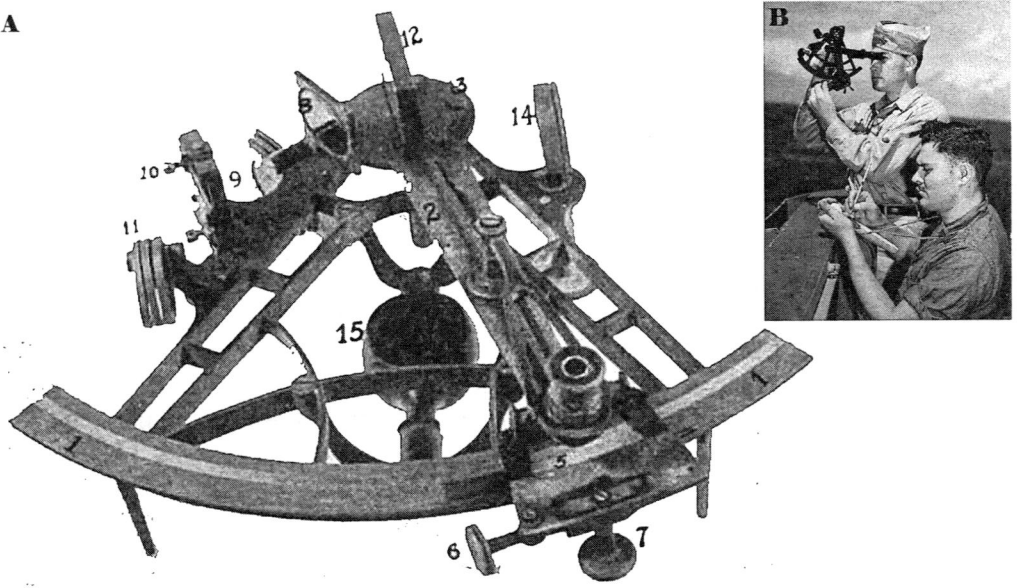

A. Parts of an early-twentieth-century sextant: arc (1), sliding index bar (2), reading glass (3), Vernier shade (4) [not shown here], Vernier (5), index tangent screw (fine advance) (6), clamping screw (7), shade glasses (8), half-silvered horizon glass (9), adjusting screws (10), back shade glasses (11), index glass (12), adjusting screws for mirror (13), telescope collar (14), handle (15) (Sheppard 1922, 211-12). B. Sun sighting by sextant on the USS *Alaska* (CB-1), circa March 6, 1945, during the Iwo Jima Operation. Sighting taken by Chief Quartermaster John P. Overholt, and notes taken by Quartermaster Third Class Clark R. Bartholomew (Official U.S. Navy Photograph, the National Archives. Catalog number 80-G-K-3736).

is moving away from the true line of sight to the star, the latter will appear to be displaced in the direction of the orbital motion. The principal source of aberration is the motion of the earth around the sun. The maximum displacement is 20.5 arc-seconds (Pasachoff and Filippenko 2019, 499). (The rotation of the earth can also cause aberration, but only, at most, 0.33 arc-seconds [Williams 1992, 95].)

Parallax. We use geocentric equatorial coordinates to describe the positions of celestial objects because they simplify calculations. However, a person on the earth's surface would see the sky from a slightly different angle than that of an imaginary observer at the center of a transparent earth. The angular separation of their lines of sight is called parallax. Lunar parallax was first measured by Hipparchus and was well known to Renaissance astronomers.

The maximum lunar parallax occurs when the moon is on the horizon, and it disappears when the moon is at zenith. The parallax also varies with the distance of the moon from the earth, so that in the horizon case it is 54–61 arc-minutes. Hence, if you are using the moon for celestial navigation, you have to take parallax into account. The sun is much further away, so its maximum parallax is 0.15 arc-minutes (Mixter 1967, 234).

Semidiameter. The stars and planets can be treated as point sources, but the sun and moon have discernible disks. To use the astronomical tables for the moon or sun, you need to know the altitude of the center of the body. Davis, in giving instructions for the use of the cross-staff, said to set "the upper edge of your Transversary, half the body of the Sun" (1626, pdf 41).

However, finding the center is easier said than done, and the normal practice is to measure the altitude of the lower or upper limb. They are both about 15.7 arc-minutes from the center, and vary as the distance to the moon or sun changes (by 2 arc-minutes for the moon and 0.6 for the sun) (Mixter 1967, 234). Hariot (1595) told Raleigh to use a correction factor of 16 arc-minutes (Taylor 1957, 221).

Augmentation. The moon is closer to the observer (by slightly more than the radius of the earth) when it is at zenith than when it is at the horizon, and hence looks larger, altering the semidiameter correction (Mixter, 236).

Measuring Azimuth

To measure azimuth (bearing), you use an azimuth compass. This instrument was first described in a 1514 Portuguese manual (Wakefield 2005, 40), and it combines a standard compass with an azimuth circle.

The azimuth circle, in its simplest form, is a ring with opposed sights, such as a peephole on one vane and a vertical wire on the other. The ring is turned until, looking through the peephole, the wire is directly in front of the object, and then you read off the orientation of the ring relative to the compass arrow.

That version only allows the navigator to take the bearing of an object close to the horizon, such as a landmark. However, there are more sophisticated forms in which a dark glass reflector is attached to the far vane and is pivotable, so that at an object at any altitude can be "brought down" to the horizon (Maloney 1985, 101–2). A well-designed azimuth circle will have leveling screws and "bubbles" so it can be made perfectly horizontal. Also, the near vane can be equipped with a telescopic sight.

The use of a simple "peep" system to observe the sun would be hard on the eyes, so the modern bearing circle comes with a second pair of "sights": a slit and a mirror on one end, and a prism on the other. The sunlight passes through the slit, and the prism creates a band of light on the compass card.

The Admiralty Standard Compass was also an azimuth compass, with an azimuth ring graduated in half-degrees, "two vernier scales …, a prism with dark shades, and a folding sight vane" (Gurney 2004, 208).

Celestial navigation usually makes more use of altitude than azimuth. That is probably because of issues of accuracy. Mixter (1967, 45) says that azimuths can be measured only to 0.5° in quiet water, 1° with the slightest roll, and 2° or more at sea.

Measuring Time

On premodern ships, short time intervals were measured with a sandglass. A 28-second glass was used for logging speed and a half-hour one for governing the ship's daily schedule (a bell was sounded every half hour).

The *nocturnal* was disk-shaped, with a fixed outer scale, a rotatable disk with a pointer and an inner scale, and a moveable arm. It was used to determine the orientation of circumpolar stars relative to the Pole Star (alpha Ursae Minoris), and thus (given the time of year) to find the local sidereal time. Usually, the "Guard Stars" (Kochab and Pherkad, the far side of the "bowl" of the "Little Dipper," more formally known as beta and gamma Ursae Minoris), were observed (Waters 1988, 305). You turned the rotatable disk to point to the time of year on the outer scale. You looked at Polaris through the central hole and lined up the arm (which might have sight holes or notches) with the Guard Stars. The arm then crossed the inner scale on the movable disk, indicating the time (DeVoy 2023). While it could only be used at night, and then only if the stars in question were visible (i.e., not in the Southern Hemisphere), it had an accuracy of perhaps 15 minutes (Swanick 2005, 108; EB1911/Navigation). "The first printed description" was in Peter Apian's *Cosmographica* (1524) (DeVoy 2023), and these stars were still in use in the eighteenth century, as evidenced by surviving examples (Oestmann 2001) and by Nathaniel Colson's *The Mariners New Kalendar* (1752, 67).

There was also the *planisphere*, an example of which is depicted in Gunter's *The Description and Use of the Sector* (1623). The basic principle was that you set the date, matched the planisphere to what was observed in the sky, and read off the time. In the simplest form, the "sky" was represented rather abstractly by radial lines corresponding to various bright stars. The *volvelle* was rotated so the line of the star then on the meridian (due south if in the Northern Hemisphere) matched the date, then the observer looked up the time. There was also a pictorial type, with simplified constellations (Turner 1980, 67).

The nocturnal and the planisphere could be combined into a single device. A "planispheric nocturnal" was taken from the wreck of the *LaBelle* (1686). It includes a planisphere with 27 constellations inscribed, some located in the southern celestial hemisphere (Swanick 2005, 155–67).

Chronometers are used to determine the time at a point of known longitude (where the time was set), and the difference between local time and chronometer time is indicative of the local longitude. In Jules Verne's *Mysterious Island*, Harding reports when it is local noon (based on the length of a stick's shadow) and Gideon Spilett reads off the time on his watch (set to standard time in Washington). (Conveniently, the date of the observation was April 16, when standard and true time were identical.) I discuss the use of chronometers in more detail in the next chapter.

Prudent seventeenth-century sailors were mindful of the "four L's": Lead, Log, Latitude, and Lookout. The Lead was the sounding line, which not only warned whether the ship was in danger of running aground but also gave a clue as to its location if it was roaming familiar coastal waters. The common Log gave the ship's speed, and hence was essential for "dead reckoning" the movement since the last celestial observation. The Latitude, computed from observations with astrolabe, cross-staff, and so on, helped fix the position. Finally, the Lookout was needed to spot hazards that either were not shown on the maps or were unsuspected because of faulty navigation.

The "Mariner's Creed" warned sailors that if they neglected any of the four L's, they would "some day surely perish" at sea.

Thanks to innovation, the sounding machine replaced the hand-thrown lead; the towed "patent log," the chip thrown overboard; and the sextant, the cross-staff and its ilk. For that matter, chronometers replaced nocturnals, and gyrocompasses eclipsed their magnetic forebears. But even with all of these improvements in navigational instruments, there was still a need for a sharp-eyed sailor in the crow's nest of the sailing ship, or on the navigation bridge of a more modern vessel.

2

Guided by the Sky
(Navigation, Part 2)

In *Mr. Midshipman Hornblower*, the tyrannical senior midshipman, Mr. Simpson, given a navigation problem by the sailing master, computes the ship's position as being in Central Africa. The Captain acidly praises him for discovering the source of the Nile. Hornblower, the most junior midshipman, is the only one with the correct answer; "everybody else had added the correction for refraction instead of subtracting it, or had worked out the multiplication wrongly, or had (like Simpson) botched the whole problem."

The errors made by Hornblower's peers differ only in degree from the real-life errors that were made by countless navigators, sometimes with the result that the ship ran aground, or sank.

The sixteenth century marked the beginning of the transition in Europe between navigation based solely on "personal knowledge about the coasts, winds and currents," and that which required also the ability "to calculate a position" based on astronomical observations (Perez-Mallaina 1998, 232). The transition was not an easy one; a satire claimed that three pilots on board a ship from the Indies calculated the ship's position and came up with three different answers, one even concluding that they were "sailing on land" (Perez-Mallaina 1998, 231).

The navigators of the Spanish *flota* were regarded with scorn, and on occasion, the denouement to the stranding of a ship's crew was the assassination or execution of the navigator (Marx 1977, 71).

Early Navigator Training and Certification

Portugal. It was long believed that the first school for navigators was established by Prince Henry in 1418, at Sagres, Portugal, and he brought foreign experts there to teach and perhaps taught there himself. Twentieth-century Portuguese scholars have questioned the story. To begin with, Prince Henry wasn't given land at Sagres until 1443 and didn't order construction of a villa there until 1460. The only navigation expert that contemporary sources report Prince Henry brought to Portugal was the Majorcan cartographer Jacome, and they don't indicate that he went to Sagres. In short, the school was mythical (Randles 1993; Russell 2000, 6–7).

That said, the Portuguese were pioneers in the development of astronomical navigation, in part because of the need to sail well out to sea for the most efficient use of prevailing winds on voyages to sub–Saharan Africa and India. The oldest known book with "nautical rules" was the *Guia Náutico de Munique* (1509), which despite its title was published in Lisbon. Early Portuguese texts described both the "Regimento do Norte" (the use of the altitude of Polaris to obtain latitude in the Northern Hemisphere) and the more complex "Regimento do Sol" (the use of the altitude of the sun to obtain latitude). Tables were prepared so that sailors only had to perform additions and subtractions to obtain the altitude (Canas 2017).

Spain. The House of Trade in sixteenth-century Seville employed cosmographers, who were college graduates, schooled in mathematics and astronomy. They authored books on navigation and (together with certified pilots) examined prospective pilots on their knowledge of navigational instruments and calculations, but did not go to sea themselves (Perez-Mallaina 1998, 232–3).

The utility of those books was limited to those pilots who were literate. Only about three-quarters of sixteenth-century Spanish pilots even knew how to sign their name. (Admittedly, that was better than the general run of Spanish sailors, where only one-fifth had achieved that minimal level of literacy [Perez-Mallaina 1998, 231].)

Around 1550, those wishing to be certified as pilots had to "provide evidence that they were at least twenty-five years old, of good character, not foreigners, and had at least six years experience sailing to the Indies." Three witnesses, all certified pilots, who had sailed with the candidate had to testify that they thought him to be competent. The candidates were tested on both local knowledge (of major ports, watering places, etc.) and long-distance navigation (ability to read a map and find latitude based on the sighting of the sun and the Pole Star) (Sandman 1999).

The education of a prospective pilot probably began with informal lessons, while at sea, from the ship's pilot (Perez-Mallaina 1998, 78–9). Nonetheless, before 1552, "to prepare for the exam, almost all of the would-be pilots hired one of the local cosmographers ... to tutor them in the necessary knowledge." It appears that there were also some group classes, spread over "several days preceding the exam," that were "well-attended." (Possibly the sixteenth-century equivalent of an SAT prep course.) But it was not unheard of for the would-be pilot to bribe members of the tribunal to pass them despite some flaw in their background or navigational knowledge, with the bribe being disguised as a tuition payment (Sandman 1999; Perez-Mallaina 1998, 57).

After an investigation into a bribery scandal, "the individual cosmographers were forbidden to give classes." Instead, a new teacher was hired, but initially the class curriculum was more theoretical and also more detailed, lasting a full year. This of course was a financial hardship for prospective pilots, as the sailor wishing to attend would have to have money saved both to pay for the class and to cover the cost of living. The class was therefore shortened to three months. In the decades that followed, there was continuing argument between the cosmographers and the pilots as to the balance between theoretical and practical content.

England. In the late sixteenth and early seventeenth centuries, the leading

English practical textbooks were William Bourne's *Regiment for the Sea* (1573, 1631), which was based on Martin Cortes's *Arte de Navigar* (1551); Thomas Blundeville's *Exercises* (1594, 1597, 1606, 1613); John Davis's *Seamans Secrets* (1594); Edward Wright's *Certain Errors in Navigation* (1599); Richard Polter's optimistically titled *The Pathway to Perfect Sailing* (1605, 1613); and Thomas Addison's *Arithmetical Navigation* (1625). For those more mathematically inclined, there was Robert Tanner's *Brief Treatise of the Use of the Globe Celestial and Terrestrial* (1620). If you were more interested in navigational instruments, you could consult Anthony Ashley's *Mariner's Mirrour* (1588), William Barlow's *The Navigator's Supply* (1597), and various books by Edmund Gunter (1632, 1624, 1630 and 1636) (Swanick 2005, 57–67).

As in the case of the Spanish and Portuguese works, the utility of these books depended on the literacy of the reader. In 1603–36, 63 percent of ship's officers and 37 percent of common seamen signed their own depositions. This measure of literacy increased to 94 percent of officers and 76 percent of ordinary sailors in 1650–76 (Patarino 2012, 178).

To actually carry out the calculations required for oceanic navigation, a knowledge of arithmetic was required, and to understand how the "rules" were deduced, the higher mathematics—algebra, geometry, and trigonometry—were called for. Susan Rose asserts (2004, 180) that "mathematics was not taught in schools in Tudor England and those who needed to keep reckonings of various kinds probably picked up the most basic kind of arithmetic 'on the job.'"

However, that isn't quite correct. While grammar schools primarily taught the classics, mathematics did have a limited role in their curriculum. In the timetable of a typical Elizabethan grammar school for 1598, "arithmetic is listed at the end of the Saturday afternoon period in classes I and II. The higher classes (older pupils) appeared not to have any arithmetic at all" (Allen 2021, 22). Moreover, arithmetic might be taught by "special tutors," or "as extra curricular tuition, particularly in the grammar or public schools" (Allen 2021, 24). Robert Recorde's arithmetic book, written in English rather than Latin, was published around 1542.

McCourt (2017) suggests that in the seventeenth-century seaport towns, "driven by their need to fill increasing demand for seaman who could navigate, many of the [local grammar] schools began to introduce mathematics beyond the basic arithmetic that had been taught to date."

In 1673, a mathematical school was founded at Christ's Hospital, London, and its original purpose was to "provide young people to be apprenticed to the captains in the Navy and hence tended to be biased toward arithmetic and navigation in the widest sense" (Allen 2021, 31).

Space does not permit the examination of the evolution of navigational training in other maritime powers, notably France and the Netherlands.

Celestial Navigation

Celestial navigation is navigation by reference to the apparent position of one or more celestial objects.

The Polynesians practiced a form of celestial navigation that determined not the location of the vessel, but its proper course. This depended on the observation that a particular star "always" rises and sets at the same location on the horizon, even though the timing of the rising and setting varies from day to day.

Rising stars were used if the course were easterly, setting ones if westerly (Lewis 1994, 82). Strictly speaking, a star had to be somewhat above the horizon to be visible, perhaps 15 degrees (Lewis 1994, 97). How far a star could rise and still be useful for steering depended on whether it rose vertically or obliquely, which depends on the star's declination (discussed later) and the observer's latitude (Lewis 1994, 83). In the course of a night's voyage, one might follow a succession of stars ("star path"), preferably ones rising at the same point (Lewis 1994, 83). Some navigators used the stars to point out the "geographically direct route," others the "course actually steered" (allowing for currents) (Lewis 1994, 99). The star used could be one at an angle to the course, even astern (Lewis 1994, 94). The stars used are "often of quite small magnitude" (Lewis 1994, 102); Delta Aquilae, used in the Carolines (Lewis 1994, 104), has an apparent visual magnitude of 3.4 (*Wikipedia*; lower numbers correspond to brighter stars).

The European form of celestial navigation was devised to determine the ship's location.

Such navigation is possible because the apparent position of a celestial object is dependent on the location of the observer. It is complicated because the apparent position is also dependent on the rotation of the earth about its axis and on the revolution of the earth about the sun, and thus on the passage of time. (For the moon and the planets, it is also affected by their own motion about the earth and sun, respectively.)

Celestial Latitude and Longitude

The sky looks a bit as though it were painted on the inside of a giant sphere (this is mimicked by a modern planetarium). In fact, Omar Khayyam referred to "that inverted bowl they call the sky." The stars (including the sun) appear to move across the "celestial sphere" as a result of the rotation of the earth and of the revolution of the earth around the sun. The apparent motion of the moon and planets is a composite of their true motion and the apparent motion generated by the earth. If we want to use the heavenly bodies as guides to navigation, then we need to be able to describe their positions in the sky.

Celestial coordinates are typically given in one of several different ways, depending on who is using them. First, we have to decide the "origin" (where we imagine we are standing when we measure the positions). The choices are heliocentric (measured from the center of the sun), geocentric (from the center of the earth), or topocentric (from a point on the earth's surface). The difference between geocentric and topocentric measurements is likely to be noticeable only for the moon.

Second, we need to choose a coordinate system: the plane from which to measure the angular distance up-or-down, and the direction from which to measure the angular distance left-or-right. There are three principal coordinate systems: equatorial, ecliptic, and "Alt-Az."

WAVE Specialists (T) 2nd Class Ruth Ingerslew, Patricia Baldwin, and Sally King, in July 23, 1945, photograph, studying a model of "the Earth's relation to the celestial sphere" (Official U.S. Navy Photograph. Naval History and Heritage Command, Catalog NH 97519).

Equatorial Coordinates

At a planetarium, you could project a grid onto the screen, representing the "lines" of celestial latitude and longitude. In the equatorial system, the plane of reference is the celestial equator: the projection, into the sky, of the terrestrial equator. Likewise, the North Celestial Pole (NCP) is directly above the earth's North Pole, and the South Celestial Pole below the earth's South Pole. Celestial latitude, measured from the equator, is called "declination," and is measured in degrees, arc-minutes ('), and arc-seconds (").

Just as we needed a terrestrial prime meridian from which to measure terrestrial longitude, we need to arbitrarily fix a location for the celestial prime meridian in order to determine right ascension. In 1950, it was defined as passing through the "first point" in the constellation Aries.

There's no east or west celestial longitude; it is measured either eastward from the first point as 0–24 hours ("right ascension," RA), or westward as 0–360° ("sidereal hour angle"). RA is stated either like declination or, because of the relationship of longitude to local time, in hours, minutes (m), and seconds (s) (one hour RA is 15°; one minute RA is 15'; one second RA is 15").

For any celestial object, there will be a point on the earth's surface such that the object would be directly overhead. That's called the "sub-point" (or "geographical point," GP). As the earth rotates and so on, the GP will move.

Ecliptic Coordinates

Geocentric equatorial coordinates work well for the sun and the stars, at least in the short term (years as opposed to centuries), but for the planets, it helps to carry out computations in ecliptic coordinates. The earth's orbital plane is called the ecliptic, and a line drawn through the center of the earth and perpendicular to the ecliptic defines the North and South Ecliptic Poles. Depending on what you are trying to compute, you can use geocentric or heliocentric coordinates.

Because of precession (the wobbling of the earth's axis relative to the plane of the earth's orbit), the equatorial coordinates of even the sun and stars change slowly with time. (One full precession cycle takes about 26,000 years.) Celestial North has to be defined on the basis of the orientation of the earth's axis, relative to its orbit, as of a particular time ("epoch"). Precession causes the NCP to revolve around the North Ecliptic Pole. Thanks to precession, the celestial prime meridian passes through the constellation Pisces.

The earth's orbit itself is perturbed by the rest of the solar system, resulting in changes in the orientation of the major axis and the orbital plane relative to the rest of what I will loosely call "distant outer space." These changes are just too small and too slow to worry about here.

Horizontal (Alt-Az) Coordinates

A navigator's observation of a celestial object isn't likely to be recorded, initially, in equatorial coordinates, but rather in terms of the object's altitude and azimuth. The altitude is the vertical angle between it and the "celestial horizon," which in turn is a distant imaginary circle, centered on the observer and level with the observer's eye, and in a plane perpendicular to the zenith line (from the observer to the point directly overhead, opposite the direction of gravity). The azimuth is its horizontal direction, an angle measured from the direction that points to the North Geographic Pole. The imaginary semicircle running across the sky from north to south is the observer's meridian.

Alt-Az coordinates are relative to an observer on the earth's surface, and thus are topocentric. After correcting for observational errors, they can be converted into other coordinates.

Sources of Error in Celestial Navigation

There are several kinds of error that can occur. The first are observational errors, wherein the "read" position, in Alt-Az coordinates, doesn't correspond to the actual position of the object at that time. Or there can be an error in determining the time at which the observation was made.

Second, there can be an error in the prediction of the celestial coordinates. If the navigator is using a published star atlas or catalogue, then this could be an error on the part of whoever computed the published coordinates, or on the part of the navigator, in taking the value from the table, and perhaps in updating it as needed.

Finally, there can be a sight reduction error, that is, an error in the use of the observation and the reference data to compute the latitude and longitude of the ship.

It does no good to worry about computing star positions to the correct thousandth of an arc-second if your observational instrument is only accurate to the nearest degree. Hence, in improving the art of navigation, scientists and seafarers needed to tackle sources of error in their order of importance. Nunez's *Defense of the Sea Chart* (1537) said that there was no point in correcting for the meridian of observation, in using solar declination tables, unless the longitude difference was at least six hours, because of the grosser errors resulting from the imprecision with which the astrolabe measured altitude (Taylor 1957, 181).

I am going to ignore sources of error that are always smaller than one arc-minute. Usually those mean an error of about a mile on the ground, but if you are using the "lunar distance" method to measure longitude, a one arc-minute error in lunar distance corresponds to a 0.5° error (up to 35 miles) in longitude.

Star Data

If the ecliptic or equatorial coordinates of a star are known, and the star is bright enough and observable on a particular route, it can be used for navigation. One early useful compilation was the star catalogue of Tycho Brahe. His "cat D" (1598) provides ecliptic coordinates (nearest 0.5') for 1,004 stars. Tycho was well aware of precession (see next section), and since the catalog was the fruit of years of observation, all star positions were corrected to what they should be for epoch "1601.03."

Tycho's accuracy was excellent. Rawlins (1993) compared Tycho's positions to those predicted by combining the Yale Catalog (1982) with "Newcomb's traditional precession constants" (see "Precession" section). For his 100 "select stars" (the bright stars likely to serve as navigational beacons), the error in either equatorial coordinate was *never* as much as 6 arc-minutes. The mean error was 1.62 in right ascension and 1.48 in declination. The greatest weakness of Tycho's data is that his observatory was in Denmark, and hence his coverage of Southern Hemisphere stars is poor.

Flamsteed's star catalog, published posthumously in 1725, was accurate to 10 arc-seconds (Wakefield 2005, 51).

Precession

Star catalogs become outdated with time, primarily as a result of axial precession, the slow change in the orientation of earth's axis. The positions of the North and South Celestial Poles (the apparent fixed points in the star fields, around which the other stars appear to move as a result of the earth's rotation) change. Thus, Polaris hasn't always been the "Pole Star."

The apparent position of poles moves in a circle against the star field. The cycle is about 26,000 years long, and the precession is thus at a rate of about 50 arc-seconds per year.

Star data can be roughly corrected for precession by assuming that precession occurs at a constant rate (it actually varies slightly), and then carrying out the appropriate spherical trigonometry calculation. Tycho and Kepler both corrected older data; the value used by Tycho was 51 arc-seconds/year (Rawlins 1993, 17).

Proper Motion

For some navigational stars (Rigil Kent, Arcturus, Polaris, Zuben-ubi), proper motion (the real motion of the star relative to the solar system) can create an error noticeable to astronomers. The magnitude of proper motion, which is an angular change in the apparent position of the star, depends on both the star's speed and its direction of motion relative to the line of sight (Reis 2004). It is so slow that it is of no real significance to navigators, unless they are using copies of old star charts. Jacques Cassini showed that from 1584 to 1738, the change in ecliptic latitude of Arcturus was about 5 arc-minutes (Verbuni et al. 2019)—an average of 1.46 arc-seconds/year.

Astronomical and Nautical Almanacs (Ephemerides)

An ephemeris is an almanac that tabulates the positions of an astronomical object as of different times. The difference between the nautical and astronomical almanac is one of emphasis. A modern nautical almanac will list predictions only for the sun, moon, and the "navigational" planets and a stellar reference point, the constellation Ares. It will also identify the locations of the navigational stars (57 nowadays) relative to Ares. An astronomical almanac will cover the other planets and moons, and will provide coordinates for many additional stars. In either case, the solar, lunar, and planetary predictions are usually good only for a few years, unless you have a computerized version.

Some Internet sources would have you believe that the first nautical almanac was published in 1767. That was merely the first one with "lunar tables" for calculating longitude. The 1545 almanac of Martin Cortes was a long-term (1545–1580) almanac in which the solar declination was calculated by combining values for the month/day and the year, to obtain the zodiacal position of the sun, and that was then used to find the actual declination. In contrast, the almanac of William Bourne (1576) featured a simple look-up, but was useable for a much shorter period. A more recent almanac was Davis's *Seaman's Secrets* (1594). It provided a table of the sun's declination for noon each day for the years 1593–1597 (EB1911/Navigation). Digges's *Prognostications* (1553) had a table of the sun's altitude for every hour of the day at latitude 51.5°N (Taylor 1957, 187).

Some of the early modern almanacs also had star data; *Mariner's Mirror* (1588) offered the declination and right ascension coordinates for 100 "fixed stars" (Taylor 1957, 209).

A modern nautical almanac provides the declination and the Greenwich Hour Angle (GHA), to the nearest 0.1', for several celestial objects useful in navigation. The GHA is essentially the angle between the celestial meridian of the object, and the celestial meridian over Greenwich in England. Values are given for every hour (Greenwich Mean Time, GMT) of every day for the sun, moon, and planets, and for every day for the first point of Aries. There is a correction value given for each day for the sun and planets, to allow for interpolation between whole hours. The moon's movement is so irregular that a separate correction value is given for each hour.

To get the GHA for a particular star, you add the star's sidereal hour angle (SHA) to the GHA of Aries. (The GHA of Aries changes 1°/day because, thanks to the earth's orbital movement, the earth doesn't have to quite complete a full rotation to face the same star a second night. The changes in the SHA of a star is the result of precession and thus is very slow.)

The major concern with regard to the early modern manuals is accuracy. For example, the errors in Bourne's solar declination tables ("Regiment of the Sea," 1574) were about 10 arc-minutes. The problem was that Bourne, not knowing Kepler's laws of planetary motion, had miscalculated the apparent solar movements (Siranah 2021).

While that was a "model error," computational errors were common. According to Bowditch, Tables 1 and 2 of Moore's *Practical Navigator* (1800) had 3,500 errors. Astronomer-Royal Maskelyne's "Requisite Tables" were equally faulty, with 1,024 mistakes in Table 21 (Calahan 1952, 215).

Errors can also occur in using tables. Unlike a computer program, a table can't give celestial positions for every location at any instant of the day. If the observation isn't for the location and time of day assumed by the table, then for greatest accuracy you must interpolate between table values. Sometimes the seamen didn't bother to do that.

Solar declination tables were calculated as of the local time at a particular location. If the ship was at a different longitude, then its local time was different, and the navigator should make a longitude correction before using the declination, as taught by Hariot. Wright (1599) said that by ignoring longitude, the mariner might be "deceived sometimes 10, or 12, [arc] minutes (or more in a long voyage) in taking the Sun's declination" (Wright 1599, Chap. XXVIII). We know that Drake did not correct his declinations for longitude (Siranah 2021).

It is worth noting that premodern seamen made calculations using the abacus (Swanick 2005, 42) and Gunter's line, sector, and scale (basically devices for graphical solution of trigonometry and log problems).

Solar Positions

The apparent motion of the sun is a direct consequence of the real orbital motion of the earth. There is a systematic error in many seventeenth-century predictions of the sun (and hence of the planets) because of Tycho's erroneous value (0.018) for the eccentricity of the earth's orbit (Gingerich and Welther 1983, xix). Cassini (1667) recalculated it as 0.017. Duffett-Smith says that it was 0.01675104 in 1900 and provides formulas for calculating it for other times (1988, 86).

There were also errors in the maximum solar declination, which is determined by the earth's axial tilt (which is slowly variable). Medieval texts gave the value as 23°33', and this value was used as late as 1581, but in the 1590s the correct value was 23°29.5'. It was variously determined by Reinhold (23°28.5'), Tycho (23°31.5'), Wright (23°30'), and Hariot (23°31').

For 1593–1596, Hariot's solar declination tables had a maximum error of 4' (Roche 1981).

Planetary Positions

The apparent motion of a planet results from the combination of the real motions of that planet and the earth. The only planets used for navigation are Venus, Mars, Jupiter, and Saturn. Their advantage was that they are bright; their disadvantage was that it was more difficult to predict their positions than those of the "fixed" stars.

In the early modern period, planetary appearances might have been predicted using the solar system models of Ptolemy, Copernicus, Tycho, or Kepler. Kepler, for example, predicted planetary positions through 1637 in the Rudolphine Tables (1627). Lorenz Eichstadt produced sequels in 1634, 1637, and 1639. Kepler's Rudolphine Tables competed with the 1632 ephemeris of the Copernican Philip van Lansberge and the 1622 *Astronomia Danica* of the "Tychonian" Christen Longomontanus.

Andrea Argoli based his 1621 ephemeris on pure Copernican theory (adjusted circular heliocentric orbits). In 1634 he published new tables that followed the "Tychonian" model (all planets except the earth circularly orbit the sun). Argoli's predictions for Mars (1650s) were within 10' arc. His accuracy was less for other planets: Saturn (~40'), Jupiter (30'), Venus (2°), and Mercury (9°). For the "sun," it was 8' (Gingerich and Welther 1983, xi–xx).

What astronomers learned in the century or so that followed was:

(1) Kepler was on the right track; the planets are, to a first approximation, in elliptical orbits with the sun at one focus (his first law), they don't move at a constant velocity (his second law), and the periods are related to the size of the orbits (his third law).

(2) The Keplerian laws aren't really laws; they are a corollary to a special case of the real law governing planetary motion—Newton's law of universal gravitation. (Kepler's laws can be derived if one assumes that there are just two bodies in the universe and one is much more massive than the other.)

(3) The other bodies "perturb" the orbit of interest, changing (usually slowly) its orbital elements (size, ellipticity, orientation).

(4) There is no exact solution to the problem of the gravitational interaction of three or more bodies (the "N-body" problem), but a pragmatic solution is obtainable by approximation methods.

With just the Keplerian laws, Kepler successfully predicted (in 1627) the transit of Mercury in 1631. The theory of gravitation, in its turn, made it possible for

Edmund Halley to recognize that the comet seen in 1682 had previously been observed in 1531 and 1607, and that it would return in 1758.

An elliptical planetary orbit is defined by five orbital elements (and a sixth indicates where the planet is in that orbit at a particular time). Two describe the orbital geometry and the other three the orientation of the orbit in three dimensions relative to Earth's orbit.

The six elements necessary to specify an elliptical orbit can be determined from three observations. However, there are many different routes of getting from the observational data to the orbital elements, and they vary in terms of accuracy and computation time.

When Kepler determined the orbital elements of Mars (*Astronomia Nova*, 1609), he knew that the orbital period of Mars (the time for it to revolve around the sun) was about 687 days. Hence, he could triangulate its true position by considering the apparent position of Mars on dates (from Brahe's data for 1585–1595) that were an integer multiple of the orbital period apart (when Mars was in approximately the same position but the earth was in different positions). He made the simplifying assumptions that earth's orbit was circular and that it was in the same plane as the Martian orbit. Kepler's value for the eccentricity of Mars's orbit was 0.09265 (Murray and Dermott 1999, 20). The value in 1900 was 0.093309 (EB1911/Planet) and in 2000, 0.09341 (Pasachoff and Filippenko 2019, 648).

Historically, the major post–Newtonian contributions to orbit determination were those of Euler (1744), Lambert (1761–71), Lagrange (1778), Laplace (1780), Olbers (1797), and Gauss (1809) (Dubiago 1961, 7–14). The point to note is that some very heavy hitters studied the problem and that it still took a century to get from Newton's *Principia* (1687) to the Gaussian method.

The observations needed to be far enough apart so as to "see" the orbit from significantly different perspectives, but close enough together so that the elements hadn't had time to be significantly perturbed. Once the elements were known, additional observations could be used to try to figure out how that orbit is being perturbed. Then one could calculate a more definitive present orbit, and that in turn allowed the later detection of smaller or less frequent perturbations.

The basic concept of perturbation was one familiar even to seventeenth-century mathematicians; the epicycles engrafted on the Ptolemaic (and Copernican) models of the solar system perturbed the basic "circular" orbits of the planet in such a way as to account for discordant observations.

When pertubation theory was applied, each orbital element, instead of being constant as in the Keplerian model, became a complicated function of time. This is called an "analytical model."

In theory, if you know the masses, positions, and velocities of all significant bodies in the solar system at the same point in time, you can instead use "numerical integration" to determine their positions and velocities in the past and in the future. In essence, you calculate the gravitational forces on each object, and determine how their positions and velocities change over a small time interval. Then you calculate the forces acting over the next time interval, and so on.

This wasn't feasible until high-speed computers were developed.

In practice, you are combining observational data, of varying reliability, from different dates. The initial state for the simulation is determined using an analytical model. The simulation is run, generating an ephemeris for a period for which observations exist. The initial state is tweaked until the predicted ephemeris is a "best fit" to the observations. The simulation can be extended to make predictions concerning the past and future perambulations of the bodies.

The accuracy of the predictions will depend on the accuracy of the starting data and on use of a sufficiently small time step. The smaller the interval, of course, the more computation is necessary.

Among professionals, the trend has been to use numerical integration to generate a "background ephemeris," and then find the analytical expressions that best fit the data. At the end of the twentieth century, the "gold standard" for the planets was the VSOP87 "semi-analytical" model, in which analytical (polynomial and trigonometric) expressions were fitted to the DE200 ephemeris (covering 1600–2169) generated by numerical integration. VSOP87 is believed to be accurate to 0.05 arc-second for the modern period and to 1 arc-second over a period of several thousand years. Unfortunately, it also contains thousands of correction terms for each planet (Bretagnon and Francou 1988, Table 6). Amateur astronomy programs typically use, at best, a simplified version of VSOP87.

Lunar Positions

The moon's orbit about the earth is only approximately elliptical, because of the effects of the sun, the earth's equatorial bulge, the planets, and so on. It is thus incredibly difficult to predict.

The average lunar motion is about 30'/hour, but three anomalies (eccentricity, evection, and variation) were known by the early seventeenth century. Flamsteed thought that the lunar theory of 1683 was capable of predicting lunar position with an accuracy of at most 12', and Kollerstrom believes that Newton's 1702 theory was accurate to 7–8'. The theoretical accuracy would be degraded by computation errors. For example, in 1695–1701, a French almanac had lunar longitude errors in 1695–1701 that sometimes exceeded 30' (Kollerstrom 1995; cf. Williams 1992, 79).

At the end of the twentieth-century the "gold standard" for the moon was the "semi-analytical" ELP2000. ELP2000–85 provided lunar longitude accurate to 0.0004° for 1900–2100 and 0.0054° for 1500–2500 (Walker 1997).

Navigational Use of the Pole Star

In the Northern Hemisphere, observation of the Pole Star, Polaris, allows you to determine your latitude, as well as the direction of true north (as distinguished from compass north). Of course, it is important that you know your constellations. "Columbus's celestial navigation was almost invariably unfortunate, a litany of wildly wrong latitudes caused by his mistaking other stars for Polaris" (Phillips-Birt 1971, 178).

If Polaris were in fact located at the North Celestial Pole (NCP), then the altitude of Polaris would be your latitude, and its bearing would be the direction of true north. Petrus Peregrinus's *Epistola de Magnete* (1269) recognized that Polaris moved in a small circle about the North Celestial Pole. Polaris is now less than 1° from the NCP. In 1601, Polaris was at 87°08.7' declination, 5°56.6' RA—about 3° from the NCP (Rawlins 1993, 99-S2). Wright (1599, Chap. XXXI) said that the distance was 2°52'.

The *Regimento do astrolabio e do quadrante* (Lisbon, 1510) evidences that downtimers knew how to use the "rule of the north"—based on the orientation of the "Guard Stars"—to find the latitude from the altitude of Polaris. If they formed a vertical line, then Polaris was at the same altitude as the celestial north, and no correction was necessary. If they were horizontal, then you had to add or subtract several degrees, depending on whether they pointed west or east (Taylor 1957, 46).

Determining Latitude

Pole Star Altitude

I have already alluded to use of the Pole Star to find latitude. That doesn't work in the daytime, or in the Southern Hemisphere.

Meridian Sight

The second method requires observing the altitude of a celestial object when it crosses the observer's meridian. For the sun, this will occur at local noon. (A circumpolar star will cross the meridian twice a day, and the crossings are called upper and lower culmination.) Knowing the declination of the sun for the date in question, simple arithmetic yields the latitude.

Of course, that requires both an almanac with a declination table, and the ability to recognize when local noon has arrived. The sun ascends during the morning, and descends during the afternoon. Local noon is the moment at which the solar disk seems to hang motionless in the sky. "A commission set up by King John II of Portugal in 1484 ... produced a manual of solar declination tables" (Stimson 1985, 575).

Obviously, if you don't measure the altitude at precisely local noon, the computed latitude will be in error. However, at moderate latitudes, around local noon the trajectory of the sun is fairly flat, that is, it "hangs." For 40°N, the maximum change of altitude is 0.1 arc-minutes at 2 minutes before or after local apparent noon, and 0.6 at 4 minutes. Sail up to 80°N, and the altitude changes are 0.4 and 6 arc-minutes, respectively (Mixter 1967, 304).

Double Altitude

Sometimes the weather doesn't permit a meridian sighting of the sun. If you make two successive observations, and you know the time interval between them, and the sun's declination (almanac), you can calculate the latitude without bothering to observe the sun at local noon. For this purpose, the watch doesn't need to keep accurate time over the long term; it just needs to be able to measure a time interval of an hour or two. The measurements should be close to when the object reaches the meridian, and the procedure is more prone to error when the meridian crossing is high in the sky (Bowditch 1826, 128). A complication is that the first altitude must be corrected for the estimated movement of the ship.

Ex-Meridian Altitude

The weather may be so bad that you can only make one sighting, but close to noon. You can still compute the latitude from the altitude, albeit less reliably. The first item you need is a well-regulated watch: one that keeps good time and that was recently set (perhaps the preceding morning), based on celestial observations, to the local time. The second is an estimated latitude (Bowditch) or longitude. And you need almanac information.

Equal Altitude

This is a special case of the double altitude method. You take a timed morning sighting, and then, in the afternoon, you time when the sun drops back to the same altitude. (You keep the sextant set at the original altitude and let the sun swim into view.) The time midway between, suitably adjusted, is considered the time of local noon (Williams 1992, 111). Polter (1605) objected to the use of equal altitudes because the declination changes between readings, even though the change is only 1 arc-minute/hour at most (Taylor 1957, 218).

If the sun is hidden all day, or poorly located for use of the double altitude method, one may instead observe a star (if there is a visible horizon), a planet, or the moon, but the latter changes its celestial position rapidly, and this poses computational complications.

Determining Longitude

All methods of determining longitude require comparing local time with the simultaneous time at a reference meridian (e.g., Greenwich). The difference in time (in hours), multiplied by 15°, yields the difference in longitude.

No celestial object hovers over a single point of the earth's surface (although Polaris comes close). During the course of a day, as a result of the earth's rotation, the geographical position (GP) of a star traces a circle on the earth's surface. That circle is at a fixed latitude (determined by the declination of the star), but the longitude of the GP can only be determined if you know the local time. You need to know the longitude of the GP if you want to calculate the longitude of the ship.

Local Time

The simplest way, in theory, to know the local time of an observation is to carry it out when the sun has "hung" in the sky (reached meridian altitude), which is, approximately, local noon. A watch time can be corrected, after the fact, to local time by using the equal altitude method (see the earlier section "Determining Latitude") to determine the watch time at which local noon occurred (Preston 2000, 172). At sea, it was more common to shoot the sun when it was bearing east or west and use its altitude, together with computed latitude and estimated longitude, and the sun's declination, to calculate the time of observation (Bowditch 1826, 155). Star positions can also be used to estimate a local time.

If local noon is determined by a sighting, the time since noon can be tracked by means of an hourglass or, better yet, a simple timepiece. (Even a timepiece that was not suitable for keeping accurate time over the length of a voyage might be reasonably accurate over the hours between a noon-sight and a twilight-sight.) At night, local time could be determined to perhaps the nearest quarter hour using a "nocturnal" (see Chapter 1).

Reference Time

The reference time may be determined either by observing some celestial event (which happens essentially simultaneously for both the reference observatory and the ship's location) or by inspection of a chronometer set previously to the reference and that has kept "consistent" time (it loses or gains time in a predictable manner) since then.

The celestial events that have been used for longitude determinations include jovian moon eclipses, lunar eclipses, lunar occultations, and particular angular separations of the moon from the sun or stars.

Jovian moon method. In theory, a reference time could be determined by noting when the moons of Jupiter passed into or out of its shadow, and comparing it to the times stated in an ephemeris computed for a location of known longitude. When a predicted immersion or emersion was observed, a clock was set to the ephemeris time. The next day, the observer noted the clock time at which the sun peaked (local noon). One then calculated the longitude, hoping that in the course of a day the clock wouldn't lose or gain too much time from the true reference time.

The ephemeris for Paris was calculated by Cassini in 1668, and by 1696 Cassini had published a map of the world that used longitudes determined by this method. Unfortunately, the method was impractical on shipboard. The necessary telescope (15–20 feet long) had a narrow field of view, so it would be difficult to keep the moons under observation while the ship pitched and rolled. If you were using a pendulum clock, then there was the further problem that the clock wouldn't work properly, even over the relatively short time interval between the two necessary observations. The experienced astronomer Halley tried, but concluded that the Jovian eclipses were "absolutely unfit at sea" (Mentzer 2002; Wakefield 2005, 86–7).

Just as well. Cassini's tables were in substantial error because he failed to consider the effect of the finite speed of light (discovered by Roemer in 1676) on the time of observation of Jovian eclipses (Wakefield 2005, 164).

Lunar eclipse. A lunar eclipse occurs when the moon passes through the earth's shadow. It is observable from anywhere on the night hemisphere, and begins and ends at the same time for all observers. If you have an almanac giving the time the eclipse begins or ends for a reference site of known longitude, you can compare that reference time to the local time. Unfortunately, lunar eclipses occur only a few times a year, are difficult to time, and in practice yield an accuracy of only perhaps 0.5–1.5° (Espenak and Meeus 2011; Oliver 2000).

Lunar occultation. The moon takes about 29.5 days (its synodic period) to travel 360° in celestial longitude, so its change in celestial longitude is about half a degree per hour. In contrast, the stars have essentially fixed celestial longitudes. Hence, the movement of the moon, relative to the stars, could be used to judge the reference time.

Initially, it was proposed that astronomers predict the times that the moon would "occult" (pass in front of) various stars. Unfortunately, while we think of the moon as large, even when full its angular size is just half a degree—that of "a penny held two meters from your eye" (Kimball 2015, 22). On a ship at sea, you are normally going to be able to identify only the brighter stars, and the odds are not great that, on a particular night, one of these will be occulted by the moon.

Lunar distance ("lunars"). Hence, astronomers instead predicted the "lunar distance," the angular distance between the moon and a celestial reference point (a star or the sun), for different hours of the day, day after day. (Although the sun moves against the sky, its celestial longitude can be predicted with accuracy.)

The lunar distance method was proposed by Regiomontanus (1475) and Werner (1514), but it wasn't seriously pursued until the 1700s. There are several methods of calculating the "lunar distance," with different trade-offs between accuracy and speed; Bowditch's 1802 method was the first one considered practical by mariners. That method was dropped from *Bowditch's American Practical Navigator* in 1880, and the replacement method was dropped from the appendix in 1914.

Although the rapid movement of the moon makes it a potentially useable celestial clock, it was a somewhat frustrating one in practice. All errors in observation, prediction, and computation are multiplied by the ratio of the earth's rate of rotation (15°/hour) to the rate of change in lunar longitude (~0.5°/hour): 29.5.

Lunars were difficult from an observational standpoint. Normally, a sextant is held vertically and used to measure altitude above the horizon. For lunars, it had to be held obliquely, depending on where the moon and the reference object were located.

The observed angle would be affected, like any other sighting, by dip and refraction. However, to correct the lunar distance, you needed to know the altitudes of the moon and its "partner." That meant, ideally, taking three simultaneous sextant measurements: the two altitudes, and the lunar distance. That was usually impractical, so what was done instead was to measure (1) the altitudes of both objects, (2) then the lunar distance, and (3) the altitudes again. The "before" and "after" altitudes for each object were averaged together to estimate the altitudes at the time of the lunar distance measurement.

Then you had to apply the tables. Their accuracy depended on the astronomers' understanding of lunar movements. Predicting lunar position is complicated because the moon's orbit is strongly perturbed by the sun, so it can't be calculated purely by Keplerian methods. In the 1783 Nautical Almanac, the average error was 14" in ecliptic latitude and 30" in longitude. In 1817, the average latitude and longitude errors were 5' (Williams 1992, 96).

Then the ship's longitude had to be computed correctly. In practice, "longitude by chronometer" (see next paragraph) was perhaps 10 times more accurate than "longitude by lunar distance," because the lunar observations and calculations were so complex and prone to error (Sobel 1995, 162). Preston (2000, 180) says that in the early nineteenth century, it was not unusual for lunars to yield a 30' error in the longitude. Lewis and Clark used the lunar distance method, and their errors in longitude were as great as 185' for moon–star and 76' for moon–sun measurements (Preston 2000, 185).

Chronometers. In 1530, Gemma Frisius pointed out that if one had a good clock, one could set it according to the time at a location of known longitude. Multiplying by 15 the time difference in hours between the reference clock time when the sun peaked (local noon) and noon then gave the longitude difference in degrees from the known longitude.

An alternative to reading the chronometer time at local noon is to take two readings, one before noon and one after, when the sun is at the same altitude. The time of local noon is then the average of the two equal altitude times (Schlereth 1975, 96; Preston 172).

Practical use of either method had to await the development of a ship-friendly clock.

The early modern clocks had a cumulative error of 10–15 minutes/day. "After a few weeks at sea, clock error could correspond to a longitude error as wide as the ocean" (William 1992, 78; Mixter 1967, 263; Wakefield 2005, 136).

The first reliable marine chronometer was designed and built by Harrison in the late eighteenth century. His H-4 (1760) lost only 5 seconds after 81 days at sea. After another two months, its temperature-adjusted total error was still under 2 minutes (Sobel 1995, 120–1).

The Harrison chronometer's principal weakness was the time and expense necessary to build it. The copy of H-4 made by Kendall (1770) cost 500 pounds and took two years to construct. Kendall told the Board of Longitude, "I am of the opinion that it would be many years (if ever) before a watch of the same kind ... could be afforded for 200 pounds." In contrast, a sextant and a set of lunar distance tables would cost a mere 20 pounds (Sobel 1995, 153).

Later clockmakers nonetheless brought costs down by having the less critical parts made by lesser craftsmen. By the 1780s, you could buy an Arnold box chronometer for 80 pounds, or an Earnshaw for 65. There was also the Arnold pocket chronometer, which only gained or lost 3 seconds a day (Sobel 156–63).

It is not necessary that the clock keep perfect time. What is necessary is that its error, if any, be known and predictable. A modern chronometer is set, in port, to approximate GMT, and certified as to its initial departure from GMT, the average rate at which it gains or loses time, and the date of the time check (Mixter 1967, 264).

Most of my sources emphasized how much more difficult it was to determine longitude at sea than on land. However, maritime travel did have the advantage of being fast; an Atlantic crossing took something like two months by sail, less by steam. Hence, the chronometer's time error—especially the unpredictable error—wouldn't have the chance to accumulate to unbearable levels before you reached a port of known longitude and could recalibrate it.

"Lunars" were only rarely used after 1850, in view of the convenience and accuracy of calculations based on improved chronometers. One use was to verify that the chronometers were still in working order. The *Nautical Almanac* stopped publishing lunar distances in 1907.

Nowadays, observatories issue time signals, and they are broadcast over the radio. These may be used to correct chronometer time (Mixter 1967, 267).

Longitude calculation from noon sight and time sight. To calculate longitude, one needed to know both the chronometer time and the local mean time. The local mean time was the sum of the local apparent time and the "equation of time," the "time difference between Mean Time and Apparent Time on the Prime Meridian" (this would be listed in a nautical almanac). The local apparent time was determined from the sun's local hour angle, which was calculated from the sun's altitude, the ship's latitude, and the sun's declination (also taken from the almanac).

"The scheme of determining Latitude by the Noon Sight and Longitude by a 'Time Sight' was simple and well established on all seagoing vessels in the beginning of the 19th century. However, it had one basic flaw: an error in the obtained Latitude would also result in an erroneous Longitude. When the Altitude was taken around Noon, … the error on Longitude can be considerably larger that the error on Latitude" (Siranah 2008).

Modern Celestial Cross-Fix

Earlier celestial navigation methods required making an observation at a special time (e.g., local noon), which simplified the calculations. Latitude and longitude were determined separately. But you can determine both ship's latitude and longitude simultaneously if you know the reference time (per chronometer) and take two (at most three) sextant readings ("time sights") on a celestial object for which the position is tabulated in a nautical almanac.

From the almanac and the reference time, you know the geocentric position (GP) of the celestial object, that is, the point on Earth's surface lying directly under that object. The sextant reading defines a circle about the GP, whose radius equals 90° altitude. If a second mariner simultaneously takes a sextant reading of a second object, you get a second circle, and the ship must be at one of the two intersection points. Usually your dead reckoning from the last known position will tell you which of the two is right, but if it doesn't you can take a reading of a third object to be sure.

If the sextant readings aren't simultaneous, the older reading is "advanced" (moved, based on ship's course and speed) so that they are effectively simultaneous. This is called a "running fix."

It is inconvenient to plot these circles, which are usually very large, on a globe or a chart. Fortunately, a small enough arc of a large circle can be approximated by a straight line, and the position is then where the two "lines of position" (LOPs, Sumner lines) cross. These can be plotted on a large-scale map (EB1911/Navigation).

Even a single LOP can be useful if there is uncertainty as to the ship's latitude. One can calculate two positions, based on assumed latitudes bracketing the apparent one, and draw an LOP through them. "If this line runs in the direction of the port, or point of land, towards which the ship is sailing, by steering along the Sumner line the vessel will reach her destination" (Tizard 1920). That was, in fact, what Sumner did when he first calculated an LOP (Siranah 2008).

Graphic methods work best when the two LOPs meet at a substantial angle, and the objects can be picked to ensure this. The sun and moon will generate LOPs crossing each other at an angle of 45° or better perhaps 10 days a month (Schlereth 1975, 77). Or you can pick stars that are in different quadrants of the sky.

As described in EB1911, Saint-Hilaire suggested (1875) a major improvement in the 1847 Sumner method. This involved using an assumed position to compute an expected altitude and azimuth for the GP, then plotting the LOP perpendicular to the computed azimuth line, moving the LOP toward or away from the assumed position to account for the difference between the expected altitude and the observed (after correction for observational errors) altitude.

The expected altitude and azimuth can be obtained by spherical trigonometry, by solving the navigational triangle (two sightings yields two linked triangles). This is formed by the assumed position (AP), the GP of a sighted celestial object, and the nearest pole (P). For each triangle, we have three sides, whose lengths are:

GP-AP: 90 minus expected altitude
GP-P: 90 minus object declination (looked up in almanac)
AP-P: 90 minus assumed latitude

The angle with vertex at P is the expected difference in longitude (meridian angle) between AP (assumed longitude) and GP (looked up). The angle with vertex at AP is the expected azimuth (or 360°-azimuth) of the GP, if the ship is at the assumed position. The sides AP-P and GP-P and the meridian angle are used to calculate the expected altitude and azimuth.

At local noon, the meridian angle becomes zero and the triangle degenerates into a straight line, vastly simplifying computations.

Log and Trig Tables

Sight reduction (the conversion of observations to positions) is heavily dependent on knowledge of spherical geometry and trigonometry ("trig"). Many different formulas were known by the early seventeenth century. Typically, these required multiplication of trigonometric functions. In the sixteenth century, the trigonometric functions of angles had been calculated to 15 decimal places. Logarithms were important because instead of multiplying trigonometric functions you could just

2. Guided by the Sky

add their logarithms. Napier published logarithmic tables in 1614, and by 1624 they had been computed to 14 decimal places (Williams 1992, 47–54).

Sight Reduction Tables (SRTs)

So that sailors didn't have to do trigonometric calculations, sight reduction tables were prepared. They compiled precomputed navigation triangles, typically covering each possible whole degree value of the meridian angle, latitude, and declination. Unlike almanacs, these are always valid; the math doesn't change.

To use the tables, you assumed a ship position that was at a whole degree latitude and longitude within half a degree of the dead reckoning position, and calculated the meridian angle from the longitude and the almanac listing of the "hour angle" of the object. The declination was in the almanac. Together with the assumed latitude, you used the SRTs to find the expected altitude (nearest 0.1') and azimuth (nearest 0.1°) of the object. You then computed the altitude difference and drew the LOP accordingly.

LORAN and GPS

LORAN (Long Range Navigation) was developed to meet the needs of World War II aviators, but was used by mariners, too. The receiver determined its location by measuring the time-difference-of-arrival (in millionths of a second) from two (or

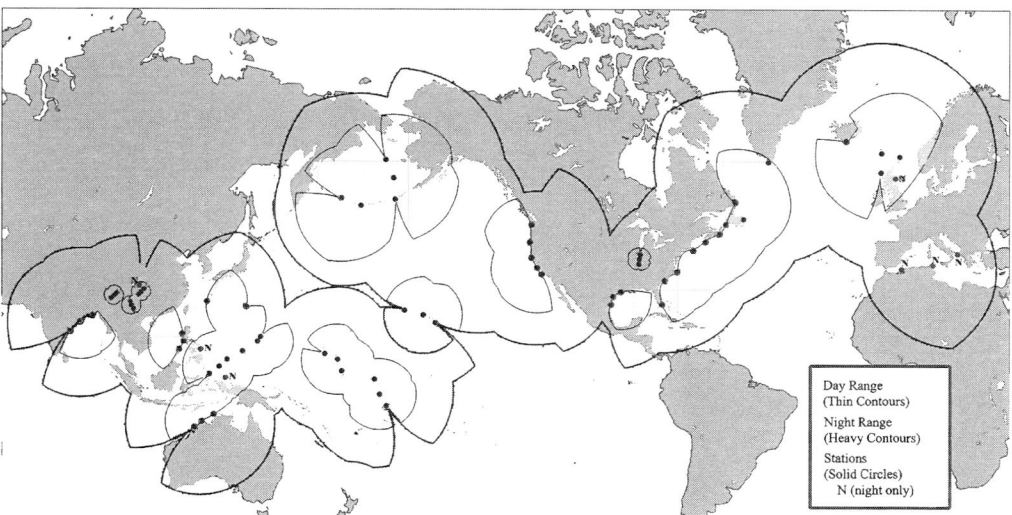

"LORAN Day and Night Coverage, 1945" (Hornish 2024). The coverage at the time was "over 30 percent of the globe." LORAN C reached 70 percent. "The ionosphere and terrain limited daytime coverage, so LORAN was far more effective at night" (Color map created by Ashley Hornish, National Air and Space Museum, Smithsonian Institution. Redrawn and converted to monochrome by the author).

if need be, three) different pairs of synchronized but widely separated ground-based transmitters. The transmitters sent identifying signals, so their locations were known. The lines of position produced by celestial altitude ("distance") measurements were circular; those produced by time interval measurements are hyperbolas (NGIA 2019, Chapter 24).

The LORAN C system was introduced after World War II. Its absolute accuracy was about 0.1–0.25 nautical miles. However, if a vessel visited a buoy and noted the associated LORAN time differences, it could return to that buoy with an accuracy of 60–300 feet. The United States and Canada discontinued LORAN C in 2010 in view of the navy's adoption of GPS for radionavigation (NGIA 2019, Chapter 24).

The GPS (Global Positioning System) uses satellites that broadcast their location and the time of broadcast as measured by an atomic clock. The GPS receiver determines the time of arrival of the radio signal, and calculates the distance to the satellite, taking into account that radio signals travel at the speed of light. With four such distances determined simultaneously, the GPS receiver determines its position in three-dimensional space (the fourth measurement "avoids the need for an atomic clock" in the receiver) (FAA 2022).

3

From Oars and Amphorae to Container Ships

(Cargo Transport)

The first documented cargo transport was around 2600 BC, when the Pharaoh Snefru had 40 Phoenician ships bring cedars from Lebanon to Egypt (Casson 1991, 6). But that was certainly not the first cargo to be carried by sea.

Propulsion and Shipboard Labor

Oared propulsion had the most obvious and direct impact on shipboard labor. The use of oars to row a boat dates back at least 7,000 years (Gu 2022, 131). Even after the development of the sail, ancient ships (and some later ships) carried oars for use when the wind was inadequate or coming from an unfavorable direction.

Peak performance for rowers was determined in 1937 for a U.S. Navy racing cutter rowed over a three-mile course by a championship crew. At their average speed of seven knots, the hydrodynamic drag force was 112 pounds, and since power equals force times velocity, the implied propulsive power was 2.04 hp. "Since there were twelve oarsmen, this means that each was producing about one-sixth effective horsepower." However, this speed "could be maintained for no more than 20 minutes" (Guilmartin 2003, 210–11).

High speed was needed by war galleys, but a merchant vessel could be rowed at a lower, more sustainable speed.

The earliest representation of a sail is from ancient Egypt, around 3200 BCE (Casson 1991, Fig. 1). Sails propel a ship either by drag (the wind fills the sail from behind and pushes it forward) or by lift (the wind flows over and behind the curved surface of a sail, and the difference in air pressure is felt as a sideward lift force). The force on the sail is transmitted successively to its yard, its mast, and the hull.

No human labor is involved in this transmission of forces. However, the wind isn't constant in either force or direction. Labor was needed to adjust the sails to the changes in the wind. If the wind slackened, more sail should be set to maintain speed. If it increased in force, sail had to reduced lest a sail be carried away, or worse, the ship be dismasted. If it changed direction, the sails had to be braced (turned) appropriately.

One could not sail directly into the wind, which led to the practice of tacking, in which one maintained an average course upwind by sailing in a zigzag pattern. However, this required periodic changes in tack from one side to the other,* and that required a major change in the positioning of the sails. It was customary to do this, when possible, at the time of changing of the watch, so both the port and starboard watches were available without interrupting the rest of the watch below (Harland 1985, 93).

There are two basic types of sails, the square sail, so called because it is hung on a yard (a horizontal spar mounted on a mast) that is "square" (perpendicular) to the keel, and the fore-and-aft sail, which may be hung from either a yard (a spar typically mounted obliquely on the mast) or a stay (a line running between masts, including the bowsprit), and is parallel to the keel.

The various nineteenth-century rigs (sail plans) are defined by the number of masts and by which type of sail is used on which mast. The choice of rig had a considerable influence on manpower requirements for handling the vessel.

In the nineteenth century, the term "full-rigged ship" referred to a sailing ship with three or more masts, with all yard-hung sails, save the lowest sails on the rear (usually mizzen) mast, being square sails. In contrast, a "schooner" had at least two masts and all of the sails were fore-and-aft sails.

"It is an important advantage of schooners that even when there is a large sail area, only a small crew is needed to operate the sails, because almost everything can be accessed from the deck" (Schäuffelen 2005, xxi).

"The largest fore-and-aft schooner ever built was the 7-masted *Thomas W. Lawson* at 5218 tons gross, which was operated by 16 men" (Schäuffelen 2005, xxi) and had a sail area of 43,000 square feet (MEL 1902, 564). However, that economy of manpower is not merely attributable to the use of fore-and-aft sails; this schooner was launched in 1902 and was equipped with steam winches for both sail and cargo handling (560–3). "A square rigger of the same size would need at least 35 men; and it is expected that the schooner's sails can be changed in about a third of the time that would be necessary for the same operation on a square rigger" (564).

However, even a square-rigged ship could be designed to reduce manning requirements. In the Baltic Sea, the Dutch *fluit* (*fluyt*, flute) was "the prevalent merchant vessel during the 17th and early 18th century," and "wrecks of fluits have been found almost all around the world" (Treffner 2022, 3). "Compared to other traditional merchant vessels, a fluit required about half the crew to sail, decreasing the operational labour costs considerably … Richard Unger … claims that in Norwegian trade a fluit of 150 tons could be handled by seven men and a boy. That gives a ton-to-man ratio of about 20:1, while at the same time the English could only manage a ratio of about 7:1 and the Germans about 11:1" (Treffner 2022, 15).

---

* The change in tack may be accomplished by swinging the bow into the wind ("tacking") or the stern into the wind ("wearing ship"), and on to the new tack. "Box-hauling" involves initiating tacking, but before the bow faces the wind, the ship is swung the opposite way, i.e., it "wears."

3. From Oars and Amphorae to Container Ships 59

Why was the *fluit* so manageable? According to Unger (1973, 407), the "foremast ... carried a single square sail while the mainmast carried two square sails, the top being small. The mizzen carried a lateen sail and in some cases a square topsail ... On smaller fluyts there would be no main or mizzen topsails. Sail area was kept small and masts short relative to carrying capacity which meant a slower ship but a smaller crew."*

Based on a modern model of the *Zeehaen* (1639) in the Deutsches Historic Museum (2024), that *fluit* (which lacked the mizzen topsail) had a sail area of about 110 m². The original had a cargo capacity of 100 lasts (Hoving and Emke 2000, 53).

A seventeenth-century East Indiaman, in contrast, would have had three sails on its fore and main masts (RMG 1685), although admittedly it would not have been used in the Baltic trade. The replica of the Swedish East Indiaman *Götheborg I* (launched 1738) has a sail area of 1964 m², and a gross tonnage† of 788 (Götheborg 2024). (The original was rated as "340 lasts, which corresponds to a load capacity of approximately 830 metric tons" [Nilsson 2023].)

Unger also asserts that "crew size was also kept down by the extensive use of pulleys and blocks in controlling the yards and sails" (1973, 407). The implication is that the Dutch *fluit* had more of them than, say, an English ship of the same tonnage, but I am not sure how this conclusion was arrived at.

Another approach to reducing crew requirements was taken in East Asia: the junk rig. The junk sail was somewhat like a giant Venetian blind with rectangular fabric panels, connected along their long edges by wooden, slightly inclined slats (battens).

The mast was usually unstayed. Stays, if present at all, ran only from the top, so they didn't interfere with the movement of the slats. (The sail could therefore be turned almost 90 degrees.) The sail was positioned across the mast, with about one-quarter of the sail area forward of the mast, three-quarters behind.

On junk sails, the battens (which are functionally equivalent to Western yards) are traditionally made of bamboo. Western yachtsmen who have experimented with junk sails have tried fiberglass and ABS pipe (Kasten 2023).

The junksail could easily be shortened or even hauled in, much like an upside-down Venetian blind, without sending topmen aloft. It is also self-tacking and self-jibing (the running rigging doesn't need to be adjusted). Thus, it is a very labor-efficient rig. Gougeon and Knoy provocatively wondered "if a clipper hull, had it been fitted with junk-type sails, could have been even faster. It almost certainly could have been manned by about a third of the crew required for the Western-type rig" (1973, 19).

Going aloft to adjust sails, especially at night or in a storm, was dangerous. Not only was there the risk of falling, there was the intense physical exertion involved. "By the mid–1800s, the Admiralty became concerned about the incidence of heart

---

* However, a different picture is painted by Stafford (2022), who says that on fluits, the "masts were high to harness stronger wind for speed but rigged in such a way that the topmasts could be easily lowered in the event of bad weather."

† This is a nonlinear function of ship volume.

trouble in upper yardmen, and instituted a 'breather' to make the journey aloft less strenuous" (Harland 1985, 93). Also, there was a risk to the entire ship if reducing sail took too long and a squall struck.

From the mid-seventeenth to the mid-nineteenth century, the most common method of reducing the topsail was by reefing. Most square sails were equipped with one or more horizontal rows (bands) of reef points (small cords) so that the fabric along the reef band could be lifted and tied to the yard. This was done aloft, by sailors strung out along the yard, their feet on footropes.

There were attempts to change the mechanics of reducing sail to make it faster and easier. Robert Forbes (1841) and Frederick Howes (1853) independently came up with the idea of "splitting" the topsail into upper and lower canvases (Harland 1985, 137). Forbes's rig was used successfully on some clippers, but it had numerous requirements, and it thus "was only possible to fit it to new ships, or to those being totally re-sparred" (Scott 2009, 56).

Howes "used two separate pieces of canvas ('double topsails')," and "the yard of the lower topsail was fixed to the mast, whereas the yard of the upper topsail could be raised and lowered" (Sager 1996, 210). This became the "industry standard." The *Great Republic* (1853), originally fitted with the Forbes rig, switched to Howes's rig in 1855 (Scott 2009, 54), and Forbes used the Howes rig on the *Florence* in 1856 (Scott 2009, 56). "Double topgallants ... appeared as early as 1866" (Scott 2009, 56).

In the 1850s, Henry Cunningham, a paymaster in the Royal Navy, obtained several patents on a method of reducing sail by rolling it up on the yard as the yard was lowered ("roller reefing"). The yard was equipped with a "whelped or cogged grooved boss" or a "cog wheel," which was rotated from below by pulling on a chain.

"Roller reefing was less popular than double topsails because roller reefing gear had more parts that could become jammed or frozen and because it was too easy for the sail to roll unevenly around the yard" (Sager 1996, 210).

Steam power revolutionized sea travel, as it freed the mariners from the tyranny of the wind. However, even after sails were entirely dispensed with (there was a long period of use of sail–steam hybrids), human labor was needed for shoveling coal and maintaining the propulsive machinery.

Mechanical stokers were developed for supplying steam engines with coal more quickly than a human stoker could. Besides this efficiency advantage, human stokers were exposed to great heat and coal dust (hence the term "black gang"), which can hardly have been good for their health.

The first mechanical stoker was that of William Brunton (1819). Coal was fed from a hopper to a revolving circular grate. Jukes (1841) invented a traveling "endless chain" grate, operating much like the moving walkways in modern airports, but depositing ash at the end rather than people (Worker and Peebles 1922, 23–24). The traveling grate could be horizontal or inclined. Jukes also, in 1838, devised an "underfeed" stoker in which the coal was dropped in front of a ram and shoved into the bottom of the furnace (Worker and Peebles 1922, 28), and this approach was refined by Jones (1889) (Worker and Peebles 1922, 29–30). Wood's 1898 stoker used a screw conveyor (Worker and Peebles 1922, 30).

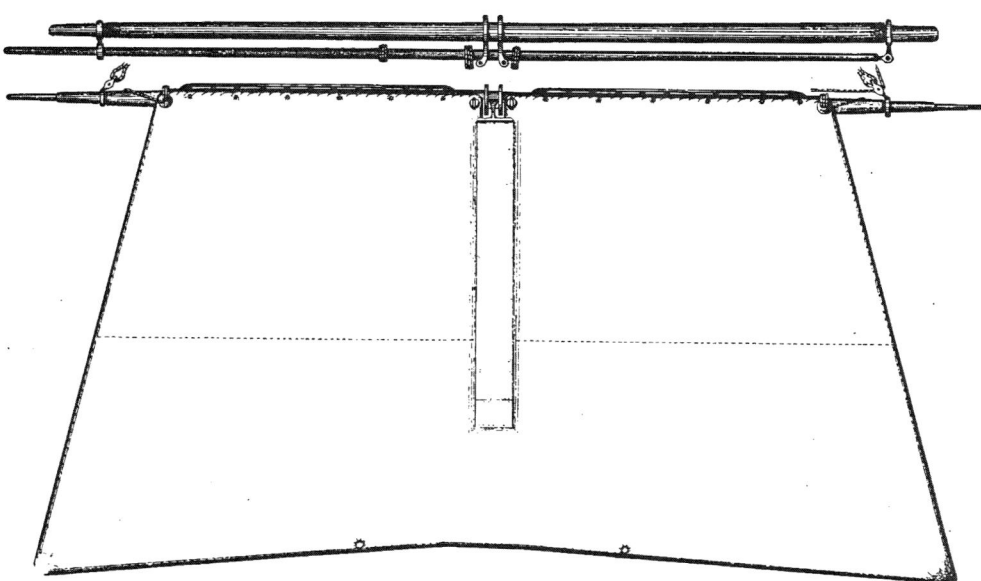

Cunningham roller reefing system; the illustration shows the sail, yard and roller (Cunningham 1853, Plate 2).

The first ship to be equipped with a mechanical stoker was the SS *Enterprise* (1840), which employed the stoker invented by John Chanter (1837). The description is vague, but it appears to have been an inclined traveling grate (Brownlie 1923, 378).

There are two basic types of boilers, the fire-tube boiler (the water is heated by tubes carrying the combustion gases) and the water-tube boiler (the water is carried in tubes through the combustion gas). Unfortunately, even in the 1920s, the fire-tube (Scotch) boiler was dominant in the merchant marine. These were long cylinders that had to be fed at the front end, and ship design was such that the space for the stoker (human or machine) was very limited (Seaboard 1920). Consequently, mechanical stokers were then little used at sea, although "a modern land power plant," using mechanical stokers, generated "steam with one-fifth the labor of the best ships" (Seaboard 1920).

The niche for marine mechanical stokers was essentially limited to ships that had water-tube boilers but were not oil-fired (see later description). While water-tube boilers were used on the *Charlotte Dundas* (1801) and *Clermont* (1807), the contemporary materials and manufacturing technology were not ready for them. It was not until the end of the nineteenth century that these problems were overcome (Lyon and Hinds 1912, 9–11).

In 1899, Milton wrote, "For all the recent vessels of the Royal Navy water-tube boilers have been adopted, but very few have been fitted into steamers of the merchant service" (Milton et al. 4). And he quoted Melville to the effect that "if the battle of Santiago taught nothing else, it certainly made very clear the absolute necessity of water-tube boilers for our modern war vessels" (104).

"Water tube boilers came into extensive use in the mercantile marine during and immediately following the 1914–18 War." Compared to the Scotch boilers, they were lighter relative to heating surface area; they operated at higher pressures and temperatures (thus, more compact and efficient); they were safer; and they could raise steam rapidly (Milton and Leach 2013, 59–61). They also were advantageous for warships; a Scotch boiler, if fed cold water, might take 12 hours to raise steam (Baker 1965, 152).

There was some experimentation with mechanical stokers on steamers with water-tube boilers. One early trial was on the lake steamer *Pennsylvania* in 1899 (*Journal of the Royal United Service Institution* 1899, 1373). In the 1920s, there was detailed analysis of stoker performance on the *Bintochan* (1200 tons deadweight) and *Parigi* (1500) (Muller 1923). Results were favorable, but hand firing continued to be the norm into the 1930s. The *Str. S.T. Crapo*, a coal-fired ship launched in 1927 and still operating on the Great Lakes 60 years later, had an automatic stoker (Thompson 1991, 62, 107). Then came the transition to oil burning.

A steam engine relies on external combustion. That is, the combustion that produces the steam takes place outside the engine proper. Fuel is burned in a furnace, and the combustion gases heat the water in a boiler until it turns to steam. The steam then moves a piston or turbine, and that mechanical motion ultimately drove a paddle-wheel or propeller.

Steam engines could burn liquid fuel as well as coal. In 1868, "a steamer was fitted to burn tar oil" and made "a trial run on the Clyde" (Bertin and Robertson 1906, 122). Oil generally has a higher energy density than coal, and thus increased range (Bertin and Robertson 1906). The principal advantage of oil over coal from a labor standpoint was that it could be pumped, and the pump itself could be driven by steam power. But it was also cleaner burning: There was no coal dust in the air or on every surface, with its consequent health concerns for the crew.

At the time, oil was generally more expensive than coal, but in the Caspian Sea, relatively cheap oil was available from Baku, so it became economical to convert Caspian Sea steamships to oil burning in the 1870s (Bertin and Robertson1906, 125).

In 1912, Britain began to switch the Royal Navy from coal to oil, starting with its *Queen Elizabeth*-class battleships. It did this despite the advantage that its domestic coal fields gave it, because it needed the higher energy density of oil for its ships to "outmaneuver the German fleet" (Dahl 2001, 52). Similar movements were made by the American Navy (54). For the merchant marine, economy was more important than speed, so its transition to oil burning was more gradual.

Motorships rely on internal combustion (IC) engines and usually burn diesel oil. They first appeared in 1903, as an auxiliary propulsion system. The first ship totally dependent on diesel IC was the *Selandia*, launched in 1912 (Marine Insight News Network 2019).

The advantages of engine power extended beyond propulsion, as the engines could also be used to power hoisting machines, as discussed in later chapters. The power could be bled off the main engine, or an auxiliary engine could be used for this purpose. Power could be transmitted to these machines via mechanical, pneumatic, hydraulic, or electric transmissions.

Early Shipping Containers

There are basically three ways of moving an object from one place to another by hand. You can lift it, you can slide it, and you can roll it. If you lift it, the resistive force is the weight of the object. If you slide or roll it, the resistance is called sliding or rolling friction, respectively.

Rolling is easier than sliding, and sliding easier than lifting.

The premiere shipping container of the Classical Age was the amphora. Unfortunately, it had to be lifted. It was eclipsed by the cask. A cask has a cylinder-like shape, differing from the cylinder in that it has curved sides (formed by staves), convex outward. (I refer to casks, not barrels, as a barrel was once a cask of a specific volume. And nowadays "barrel" is often taken to refer to a cylinder.)

Casks came in a variety of sizes. These were standardized by Parliament in 1423, with the largest cask being the tun, a wine barrel with a capacity of 252 U.S. gallons (953.924 liters; 209.834 imperial gallons).* Thus, it was a unit of volume. The effective weight of a full barrel was one long ton (2,240 pounds). The tun was equivalent to two butts (pipes), three puncheons, four hogsheads, six tierces, or eight barrels (Zupko 1985, 423–5). While a ship's cargo capacity was measured in tuns, that does not mean that tun-sized casks were actually placed on shipboard. For a French tun cask, 968 liters, the length was 58.66 inches, the bilge diameter, 43.15, and the head diameter 37.87 (Boudriot 1986, Plate XXXIII).

The "Seaport" shipwreck is of an early nineteenth-century schooner carrying lime in barrels "1.5 × 2.5 feet in size" (Boston 2017). A cylinder of those dimensions would have a volume of 33 U.S. gallons; a traditional barrel (*sensu strictu*) was 31.5 gallons.

What are the cask's advantages when it comes to "doing work" with it—moving it to or from the ship's hold, or stowing it within that hold? And do those advantages exceed those of an ordinary cylinder?

Like a cylinder, a cask is able to roll, and rolling friction is smaller than sliding friction.

However, because of its "bulging bilge construction," the macroscopic surface area over which it comes into contact with the floor is smaller than for a cylinder of the same volume and maximum radius. Twede opines that the friction experienced during rolling is proportional to that contact area, and if that were the case, then that would be one point of superiority over the cylinder.

However, what actually matters is how the mass of the container is distributed (Mungan 2012; Wulandari et al. 2014). The resistance of an object to a change in its speed of rotation is called its moment of inertia and measured about the symmetry axis of an axisymmetric object; this equals the product of its mass, the square of its radius, and the "mass distribution factor" (mdf). The latter is 0.4 for a solid sphere and 0.5 for a solid cylinder; for a cask with circular arc sides, it would be between 0.4 and 0.5.

---

* The U.S. gallon (3.785 L) is almost identical to the British wine gallon from the time of Queen Anne (3.790 L), 1706 Act 5 Anne c27. https://en.wikipedia.org/wiki/Gallon.

Professor Carl Mungan of the U.S. Naval Academy was kind enough to derive for me the pushing force required to roll an axisymmetric object over a surface, as a function of its mass, rolling radius, pushing force angle, surface slope, and mdf. The forces acting on the object, besides the pushing force, are the gravitational force (weight), the rolling friction, which acts to displace forward the point of action of the normal component of the gravitational force (Mungan 2025), and the static friction.

The mdf affects the required pushing force only if the object is accelerating—this is the effect of static friction. The maximum percentage effect of mdf, and thus of the difference in shape between the cask and a cylinder, occurs when the surface and the applied pushing force are both horizontal, and the displacement (and thus rolling friction) is zero. The required force is then proportional to 1 + mdf (Mungan 2024, eq. 7). If the cask had an mdf of 0.45 (midway between sphere and cylinder), the cylinder would require 3.33 percent more force to push at a given acceleration than a cask of equal mass and radius. Rolling friction, and weight if the pushing were uphill, would further dilute this advantage of the cask shape. It therefore seems unlikely that the cask shape was adopted because it was perceived easier to roll.

It is, by the way, easier to roll a cask or cylinder containing a liquid than one of equal mass and size containing a solid: "The solid will rotate with the can, while the liquid does not, so of course the fully solid can is much harder to roll" (Mungan, private communication).

Even if the cask shape didn't make it significantly easier to roll, it might be favored because of its mechanical strength. Arches resist compression. "Like an egg shell, [the cask] is doubly arched, both in length and girth. The bend in the stave's length is the first arch, and the bilge circumference of the stave's width is the second arch" (Sweeney 2010, 97).

Deck and Hold Stowage

Merchant ships sought to take on board as much cargo as possible without endangering seaworthiness.

Barrels are generally stowed on their sides. However, they are packed most efficiently in tiers in a "bilge and cantline" pattern. Within a tier, the barrels are stowed end-to-end and side-to-side, with the rolling axis aligned fore-and-aft. But vertically, they are staggered with the bilges of the barrels in the tier above nestled in the space (cantline) left by the ends of the four barrels in the tier below. Consequently, if the ends were exposed to view, they would form a hexagonal pattern. Alternative storage arrangements were "bilge and bilge, a-burton (across the vessel), and vertical storage," and a mixture of methods might be used to maximize the use of space (Staniforth 1987).

While the packing density of cylinders and barrels is smaller than that of oblong containers (if the space they are all packed into is also oblong), the former do have an advantage when it comes to supporting a vertical load. The hoop shape of a cylinder defines an arch, and an arch resists loads from above better than a horizontal beam (hence, the pervasive use of the arch in architecture). The bulging bilge of

Bilge-and-cantline stowage (Bridger and Watts 1927, 32).

the barrel adds a second arch, perpendicular to the first one, although with a larger radius of curvature.

Ship Design: Hatches and Cargo Ports

Cargo can moved into and out of the hold either vertically, through hatches in the deck, or horizontally, through side ports.

Precisely how it is moved depends on whether the movement is between a ship and a pier, or a ship and a boat—and, if a pier is involved, whether it has cargo handling equipment, such as a "cargo mast," a winch, or a crane, to facilitate the movement.

Side and stern ports. Cargo may be moved through side or stern ports by hand, by hand trucks (like those used by delivery trucks), by dockside cranes or conveyor systems, or with ship's tackle.

On the wreck of a seventeenth-century Dutch *fluit* found in the entrance to the Gulf of Finland, nicknamed the *Swan*, "next to the sternpost on starboard side there is a cargo hatch that could be used for loading long timbers or other tall objects" (Treffner 2022, 33). Beginning in the 1670s, the Dutch increasingly obtained wood, iron, and tar for shipbuilding from the Eastern Baltic (14). The *Swan*'s hatch was 100 cm wide and 70 high. Treffner assumes that "the lower edge of the hatch had to be just above the waterline for the ease of loading of timbers," and hence the hatch would have been "heavily caulked" after it was covered (33).

Some modern ships in coastal, river, and lake trade have side ports. If the tidal range is substantial, trucking the cargo out becomes awkward and a conveyor system is likely to be used (Cunningham 1924, 26).

Deck hatches. By the 1920s, there were vessels with nine holds and an equal number of hatches. The holds and hatches were much larger than those of the premodern period. The *Verbania*, 8,500 deadweight tons, with 429,920 cubic feet cargo capacity, had five hatches, with the largest being 34.5 by 18 feet (Cunningham 1924, 147).

While hatches were necessary to access the hold, they were also possible points of ingress for water. This was addressed by a hatch cover. This could be a simple tarpaulin, but if there was risk of heavy seas, it would be "battened down"—that is, long narrow pieces of wood would be nailed down along the edges to secure it (Burney 1815, 34).

Later, the hatch was raised "above deck level by coamings," and closed with a wooden hatchboard. This was sealed with tarpaulins "secured around the side of the coaming by wedges" (Greenway 2011, 9).

Obviously, once the ship arrived in port, these expedients had to be removed, which delayed unloading the cargo.

In the 1950s, an "articulated steel cover ... mounted on wheels running along either side of the coaming top" was introduced. This cover could "fold up like a concertina for vertical stowage when pulled by a single wire mounted at one end of the hatch." Hatches could thus be opened and closed "by one man in a matter of minutes" (9–11).

Modern Cargo

Cargo nowadays is classified into three categories: bulk, "break-bulk," and container.

Bulk cargo is commodities shipped loose (unpackaged) and poured directly into the hold of a ship designed to carry them. It is usually liquids (oil) or "free flowing" solids (grain, coal, iron ore) (InterlogUSA 2024).

"Break-bulk" cargo was, at one time, everything else. Now it is everything that is shipped in small packages that must be individually handled, rather than in the modern container as carried by container ships (see the next section).

In March 1954, the *Warrior*, a C-2 cargo ship, made a run from Brooklyn to Bremerhaven, Germany, with 5,015 long tons of cargo. This cargo took the form of 194,582 individual pieces, including 74,903 cases, 71,726 cartons, 24,036 bags, 10,671 boxes, and 2,880 bundles. There were only 1,538 drums, 888 cans, and 815 barrels, and there were 53 vehicles on board. In Brooklyn, longshoremen placed many of these items on pallets, for transfer to the ship's hold. In the hold, they "removed each item from its pallet and stowed it" (Levinson 2006, 33–34). Obviously, this article handling was very labor intensive.

The *Warrior* was built in 1943 with 10,565 deadweight tonnage and converted to a containership in 1966, with a capacity of 225 35-foot containers (Colton 2010; Cudahy 2006, 18, 92, 261). Cudahy variously gives its gross register tonnage as 6,065 (18) and 8,673 (261); these are possibly values before and after conversion. Container ships are discussed at the end of this chapter.

Intermodal Transport: ROROs and LASHs

The concept behind intermodal transportation is that goods move faster if they can change mode of transport (roads, rails, water) without being removed from their original containers and repacked into new ones. It is an old concept; in 1844, the French Mail used a primitive gantry crane to transfer a stagecoach body from its bogie to a rail flatcar (Mattes 2007).

One method of implementing intermodal transportation was "piggybacking," for example, driving a truck, trailer, or rail car onto a ship. The first train ferry was operated by the Monkland and Kirkintilloch Railway in 1833, across the Forth & Clyde Canal; the ferry "was equipped with rails and a turn plate" (Maggs 2018). The Edinburgh, Leith and Nehaven Railway ran a "floating railway" across the Firth of Forth from 1850 to 1890. "This was equipped with 'floating slipways' that ensured that the wagons could roll onto the boats whatever the level of the tide" (Happer and Steward 2015). Train ferries were run across the English Channel during World War I (Foley 2021).

While the LSTs (landing ship, tanks) of World War II transported self-propelled vehicles, the emphasis there was on transporting the vehicle, not its cargo. However, converted LSTs were used to carry trailers on the Hudson River in 1947 (Cudahy 2006, 21), and the *Searoad of Hyannis* (1956) was designed to transport three semi-trailers (Post 2021).

"RORO" stands for "roll on, roll off," which is an apt description of how RORO ships handle cargo. There are modern RORO ships with very high capacities. For example, the *Eco Livorno* (2021) "comprises seven decks capable of accommodating 7800 linear m of cargo, equalling 495 trailers and 182 cars" (Marine Digital 2024).

While RORO (roll on, roll off) ships may be seen as an alternative to container ships, there are hybrid ships ("ConRO") "where containers are carried on the upper deck and wheeled vehicles are carried under deck" and RORO ships carrying "trucks with trailers carrying loaded standard containers" (Misra 2015, 326).

Barges (lighters) can also be loaded on a "mother ship." In 1969, the first LASH (lighter aboard ship) carrier, the *Acadia Forest*, was commissioned. The lighters (barges) each had a capacity of 385 metric tons and 550 cubic meters, and the empty weight was 80 metric tons. The barges were towed to the carrier and lifted on board by a gantry crane. The carrier could carry 73 barges (Zehner and Scoggin 2020; Shaw 2019, 34).

LASH carriers had the advantage that they only needed an anchorage, not a berth, for loading and unloading. However, they were expensive to operate, and the lighters "were not easily incorporated into river tows." The LASH system was unable to compete with container ships, and the last LASH carrier ceased operation in 2007 (Zehner and Scoggin 2020).

Intermodal Transport: Container Ships

An 8 by 8 by 18-foot container for truck/rail/ship intermodal transport was used as early as 1906, but didn't catch on (National Research Council 1992, 18).

In the 1950s, Malcom McLean realized that if a ship carried only a trailer body, not the whole trailer, he would reduce the space occupied by one-third, and it would be possible to stack the bodies (Levinson 2006, 47). He proposed that a trailer would be "pulled alongside the ship, where the trailer body, filled with twenty tons of freight, would be detached from its steel chassis and lifted aboard ship." The process would be reversed at the destination (48).

However, there were three necessary ingredients: a suitable trailer, a suitable ship, and a suitable crane. The larger the trailer body, the more efficiently cargo could be moved, but the body had to be small enough to be highway legal, of a size that made efficient use of the ship's deck space, and light enough to be lifted on board. The ship had to have a strong enough deck, sufficient deck space, and means for locking the containers (trailer bodies) in place. Also, the crane had to be able to handle the load.

In 1956, McLean converted a T-2 tanker by putting a spar deck on top. His Brown containers were 33 feet long and chained to the trailer chassis. They had hooks on top, at each corner, that could be engaged by a "spreader bar" on the crane. They also had hollow tubes attached to the sides; the tubes would pass through holes in the spar deck and be locked into place by rods inserted from below. McLean's *Ideal-X* could carry 58 containers and a container could be transferred to it by a large dockside crane every 7 minutes. The estimated cost of loading the Ideal-X was 15.8 cents per ton, versus $5.83 for "loading loose cargo on a medium-size cargo ship" (Levinson 2006, 49–52, 55).

The next significant development was a container-only ship in which containers could be stacked in the hold. (The *Ideal-X* also carried oil.) In 1957, McLean converted a C-2 cargo vessel, the *Gateway City*, to carry 226 containers. The hatches were widened to make the holds fully accessible, and cell guides—vertical steel corners—were installed. These created cells 1.25 inches longer and 0.75 inches wider than the trailer bodies: smaller and it would be too hard to fit the body in the cell; larger, and the body would shift too much at sea. The trailer chassis was given "a new

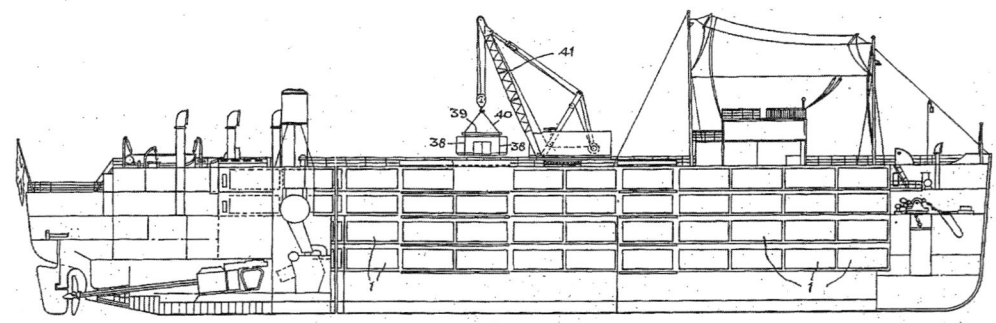

A profile view of an early proposed container ship, with a crane holding one of the containers (Kellett 1942, Fig. 1). The patent assumed that the containers were carried to the ship on railway rolling stock and were of half or quarter carload size. The patent discusses a rollers-and-trolley mechanism for moving the container over the deck, and means for locking the container in the desired position. Note that the patent predates McLean's entry into the field.

locking system" that could be engaged and disengaged faster than the old chains. The bodies were given "heavy steel corner posts," to make it possible to stack them safely. They also had castings at each of their eight corners that could accommodate a "twist lock," connecting stacked bodies. Finally, the ship had two gantry cranes, fore and aft, which could load or unload 15 bodies (containers) an hour, without any need for dockside cranes (Levinson 2006, 54–57).

"A conventional break-bulk cargo ship would typically required 150 or more longshoremen working for at least four full days to unload and load a vessel's cargo. With a container ship such as *Gateway City*, the same task could be accomplished by a crew of fourteen in a little over a single eight-hour shift" (Cudahy 2006, 35).

There were, of course, further developments, but covering them is outside the scope of this book. Containerization changed the nature of work at sea, but whether it improved it depends on your point of view. It definitely reduced the time that ships spent in port and consequently the cost of shipping. However, that also meant that ships reduced their crews and the crews got less shore leave.

4

"Haul Taut!"

(Lifting Machines, Part 1)

Until the early nineteenth century, ships were dependent on muscle power and, if equipped with sails, wind power. But even if a ship had sails, muscle power was needed for every operation. The sails hung from horizontal yards, in turn mounted on vertical masts. However, yards and masts were raised and lowered, and sails furled and unfurled, by muscle power. Likewise, muscle power raised and lowered the anchor and the ship's boats, turned the rudder, operated the ship's pumps, and embarked and disembarked the cargo. The ship's boats themselves were probably rowed. If the ship were armed with cannon, muscle power traversed and elevated them, and brought them back to firing position after recoiling.

However, even in ancient times, it was known that muscle power could be used more effectively with the aid of machines.

Machines

Many different machines were used by sailors. Note that until the nineteenth century, these machines were usually human-powered.

John Harris's *Lexicon Technicum* (1704) defined a "machine" as "whatsoever has force sufficient either to raise or stop the motion of a body." Renwick said that a "machine is an instrument, by means of which we change either the direction or the intensity of a force, or both" (1832, 126).

A machine is also "a device that transforms energy available in one form to another to do a certain type of desired useful work." Also, it "is not able to move itself and must get the motive power from some source" (Singh 2012, 11).

The connection between the two definitions is that a force, acting to move an object over some distance, is performing work, and in the process is transforming energy from one form to another.

Work equals force times distance, and some machines were designed that provided a "mechanical advantage": applying a lesser force, over a greater distance, in order to perform the same amount of work.

Machines themselves could be "simple machines," or combinations thereof. Renwick defined the simple machines as the lever, the wheel-and-axle, the pulley, the inclined plane, the wedge, and the screw. All were used on shipboard, either alone or in combination with other simple machines.

Lever. There are three classes of levers. The seesaw (with effort on one end and the load on the other) and the plier are levers of the first class; the fulcrum (the pivot point) lies between the point where effort is applied and the load.

The wheelbarrow and the nutcracker exemplify levers of the second class, in which the load is between the fulcrum and the effort.

With a second-class lever, the "effort arm" (effort-to-fulcrum) is necessarily greater than the "load arm" (load-to-fulcrum), and a first-class lever may be designed so this is the case. If so, since the effort is moving a greater distance than the load, less force is needed (work equals force times distance), and the lever is said to provide a mechanical advantage (the ratio of the force needed to the weight of the load).

Finally, the catapult is a lever of the third class, in which effort is between the fulcrum and the load. By definition, the effort arm is shorter, so there is no mechanical advantage. But it does mean that the load will move faster, which is helpful when hurling a projectile.

The principal use of machines was to raise and move weights that were too heavy to be carried, or which had to be moved to a height above a sailor's reach. You could attach one end of a line to the weight (or some sort of net, basket or cradle holding the weight), and the other end to a suitable machine.

Pulleys and the Block-and-Tackle

Overview

A pulley-block* consists of a wheel (the sheave), with a circumferential groove to receive the rope; a housing (the shell); and a pin, connected to the housing, around which the sheave rotates. It may also include a "coak," a bushing located between the sheave and pin.

There were a number of different methods of attaching it to the ship. The shell could have scores cut into its sides and ends to admit a strap ("strop") (Brady 1864, 32; Qualtrough 1881, 265). A "strop" was a ring of rope (cf. Falconer 1784, "Block"), and it would pass around the shell as well as a mast or spar. Alternatively, there would be attachment points (hook or eye) on the top and bottom of the housing.

Sailors called the simple fixed pulley a "single whip." It was used to change the direction of an applied force (Walton 1968, 46). The crew pulls on the line, which runs around part of the circumference of a grooved wheel (sheave) mounted on an axle and pulls directly on the load. The earliest shipboard application of the pulley was probably in hoisting a sail.

---

*Strictly speaking, all pulleys are blocks, but not all blocks are pulleys. "Dead eyes" were disc-shaped blocks with three holes but no rotating element. Nonetheless, a pair of dead eyes with a lanyard reeved through them could provide a mechanical advantage, much like the combination of a fixed and a movable pulley, although the frictional loss of power would have been greater.

Mechanical Advantage

A mechanical advantage (reduction of the input force needed to move the load) may be obtained if the pulley is movable and arranged so that the direction of pull is also the direction in which the load is to be moved. For this to happen, one end of

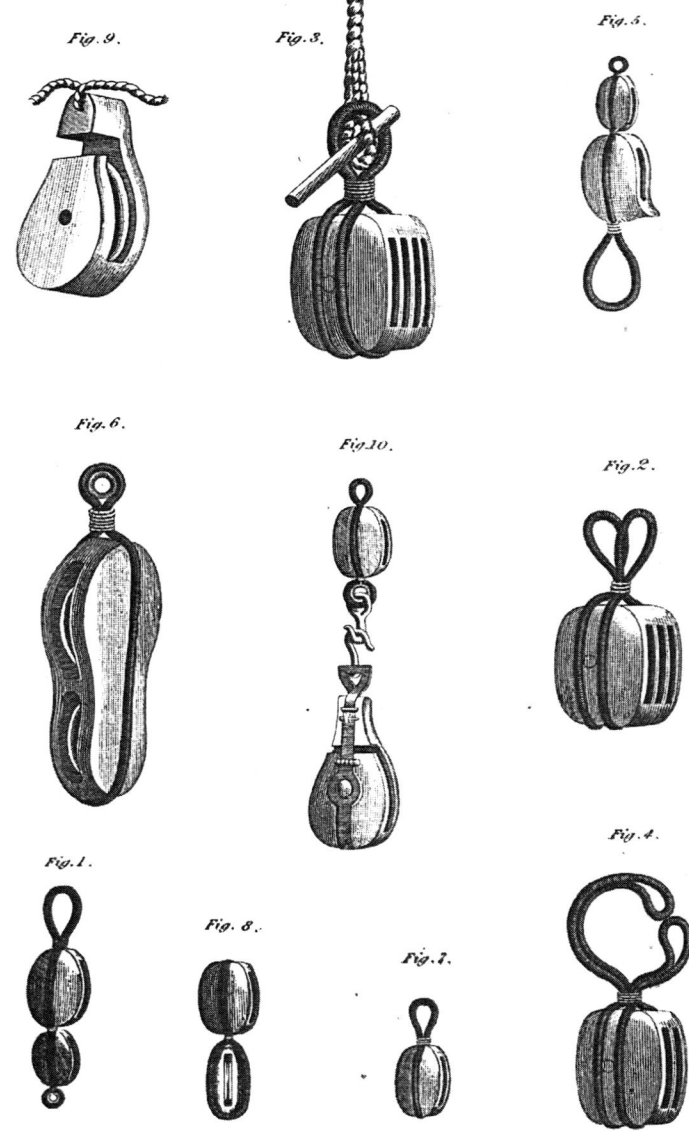

Blocks, from *London Encyclopedia* (1845). "Block Machinery," after 4:234. Fig. 3, Brunel's improved clew-line block. Fig. 5, heart. Fig. 6, long-tackle block. Fig. 7, nine-pin block. Fig. 8, shoulder block. Figs. 9 and 10, snatch blocks. The other blocks were not explicitly identified by the encyclopedia, but Fig. 1 appears to be a common single block; Fig. 2, a common treble block; and Fig. 4 remains unidentified, but it is two single blocks, one above the other. For additional depictions of blocks, purchases, and so on, see Brady (1964, 38).

the line running around the pulley is fixed at a point above the pulley and the load to be moved (assuming the load is to be lifted). A second line attaches the bottom of the pulley to the load below it. The weight of the load is divided equally between the two ends of the pull line as long as the load remains below the fixed end and the pulley (Walton 1968, 46). If the required direction of pull is inconvenient, than a second, fixed pulley can be added to change the direction (47).

A further mechanical advantage is obtained by means of a compound pulley system. This uses two or more moveable pulleys. In theory, one could use one moveable pulley to lift a second, and a second to lift a third. Each would have a mechanical advantage of 2, so the system would have a mechanical advantage equal to their product: 8. However, that would need three anchor points, one for each pulley, and four ropes, and since the pulleys are free-swinging, their ropes could get tangled (Walton 1968, 48–9).

It is more convenient to use a combination of fixed and moveable pulleys. For example, three fixed pulleys are connected by a frame, one above the other, to a single anchorage, and similarly three moveable pulleys are connected by a frame, one above the other. A single rope runs around all the pulleys, alternating between fixed and moveable ones, and ends at the bottom of the fixed pulley frame,* and a second rope attaches the moveable pulley frame to the load. The theoretical mechanical advantage equals the number of parts of the line ("fall") that support the load—in this example, six (Walton 1968, 49).

It is even more convenient if instead of "stacking" the pulleys, they are side by side. Continuing the example, suppose we have a fixed block with three sheaves mounted on a common axis (but turning independently of each other), and a moveable block with another three sheaves so mounted. A single rope runs around all the pulleys, alternating between fixed and moveable. The load is attached by a second rope to the moveable block. The theoretical mechanical advantage is again six (Walton 1968, 50).

Manwayring (1644) alluded to the possibility of a block having as many as five sheaves on it. Falconer (1784, "Block") indicated that single, double or triple blocks were the norm, and blocks with a larger number of sheaves (up to seven) were "not used about the yards or sails," but rather for heavier loads. Steel (1794, 148) says that blocks may have up to eight sheaves, but only those with one to four were in "general use."

When the pull on the rope is in the same direction that the load is to be moved, the tackle is said to be "rove to advantage." Both blocks are then moveable. When it is in the opposite direction, it is "rove to disadvantage." One block is moveable, and the other is fixed ("standing"), serving merely to reverse the direction of the "pull." In the first case, there is one more part of the line supporting the load so the mechanical advantage is greater by one.

The problem, obviously, with using a tackle that's "rove to advantage" to lift a load vertically is that you have to be above the load from beginning to end. If the

---

*If instead there were three fixed and two movable pulleys, the end of the rope would be attached to the movable pulley frame.

load is being lifted from the hold or from a boat alongside, to the deck the sailors are standing on, you will have to run the line over a fixed block, so you can pull down to lift up, and the tackle is then "rove to disadvantage." The most likely use of "rove to advantage" is to move a load horizontally.

Assuming that the tackle is rove to disadvantage, we have the following common arrangements: The gun tackle has one single-sheave movable block and a single fixed block, for a mechanical advantage of two. The luff tackle has a double sheave block and a single block, and an advantage of three. The double purchase has two doubles, for an advantage of four. And the triple purchase has two trebles, for an advantage of six.

The reader may wonder, why use anything other than a triple purchase? Remember, there is a trade-off between force and distance. With a triple purchase, the team exerts six times the force, but for a given amount of work moves the load only one-sixth as far as with a single whip. So the tackle is chosen with both the expected load and the available crew in mind.

Friction

So far we have only addressed the theoretical mechanical advantage. The practical mechanical advantage is smaller because of friction. There is "friction of the

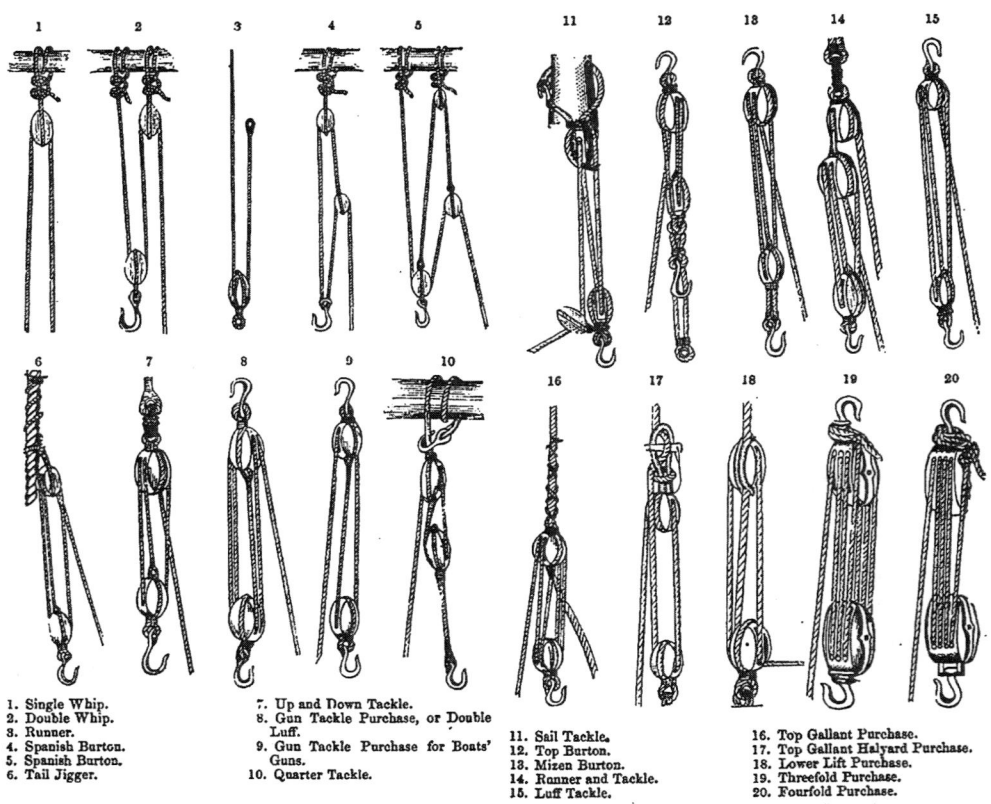

1. Single Whip.
2. Double Whip.
3. Runner.
4. Spanish Burton.
5. Spanish Burton.
6. Tail Jigger.
7. Up and Down Tackle.
8. Gun Tackle Purchase, or Double Luff.
9. Gun Tackle Purchase for Boats' Guns.
10. Quarter Tackle.
11. Sail Tackle.
12. Top Burton.
13. Mizen Burton.
14. Runner and Tackle.
15. Luff Tackle.
16. Top Gallant Purchase.
17. Top Gallant Halyard Purchase.
18. Lower Lift Purchase.
19. Threefold Purchase.
20. Fourfold Purchase.

Various purchases (Admiralty 1883, 212–217).

ropes against each other and against the shell of the block and between the sheave and the pin" and "power lost in bending the fall around the sheave" (War 1924, 21). The latter is due to "the internal friction between the individual wires and strands in the rope as they adjust themselves to the different lengths along the inner and outer curves" (Beckwith 1962, 3).

MacKenzie (1920, 36) suggested increasing the effective weight to be lifted by "1/8 of the weight for every sheave" in the purchase. Strictly speaking, the amount of friction per sheave would depend on the materials chosen for the sheave (and its pin and bushing) and the tackle.

An Army manual says to allow 1/10th of the load weight per sheave (including those on change-of-direction blocks) if the blocks are in "excellent condition," 1/8th if they are "good condition,"* and 1/5th if in "poor condition" (Army 1979, 2–242), and more if the block condition is worse.

A more modern method of expressing the effect of friction is to state it as the increased effort needed to move the load, as a percentage of either the actual effort or the theoretical (frictionless) one.

The effort needed to hoist and lower a Coast Guard motor launch was measured by dynamometer. With a 4,300-pound load on the lower block, and four parts supporting the load, the theoretical line pull was one-fourth (1,075 pounds). The standard rig used six sheaves, and the actual line pull needed was 1,750 pounds. However, it was noted that both the standing and hauling parts were chafing on the cheeks of the blocks. Two other rigs were tested, and with the final one, the required line pull was reduced to 1,450 pounds with six sheaves (USCG 1948, 15). (This shows that it is oversimplistic to estimate friction just by counting sheaves, because friction can occur in other places, too.) With the final arrangement, the frictional loss was 26 percent of the actual line pull and 35 percent of the frictionless line pull, and the increase in effective weight was 5.8 percent per sheave.

Another experiment showed that the average friction loss per sheave was less for sheaves with a "bronze roller bushing" than for those with a "graphite bronze bushing" (9.6% vs. 15.3% for hoisting, 10.8% vs. 14.0% for lowering) (USCG 1948, 16). Beckwith (1962, 3) suggested allowing 8 percent per sheave for "bushed plain bearings" and 5 percent for "anti-friction bearings (i.e., ball or roller bearings)."

Pulling Force Ergonomics

The use of block and tackle to move something requires the exertion of a pulling force. For horizontal pulling (moving a heavy weight across the deck perhaps), a standing person can exert 50 lbf if the whole body is involved, or 24 using just the arm and shoulder muscles. For vertical pulling (hoisting and lowering), the maximum force one person can exert depends on the height of the grip: 120 lbf it is above head height, 45 if it is at shoulder level (Canadian Centre for Occupational Health and Safety [CCOHS] 2022). Thus, for one person to lift 1,200 pounds by pulling on a

---

* A misprint in the manual gives the value "1/3." This was hand-corrected to "1/8."

tackle gripped above head height, you would need a *practical* mechanical advantage of 10:1.

In 1921, the War Department stated that men "can exert about half their weight on a horizontal pull," and with "average weight," this worked out to "60 pounds."

A 1979 Army manual says, "on a vertical pull, assume each man of average weight can pull 100 pounds. On a horizontal pull, assume 60 pounds." (The accompanying diagram shows that the "vertical" pull is actually oblique.) The manual notes that "usually you can use more men on a horizontal pull" (Army 1979, 2-241, 2-242). Bear in mind that a change-of-direction block can be used to convert a horizontal pull into a vertical one for hoisting a load below the level of the block.

Das and Wang (2004, 50) provides data for oblique pulls. For standing men (age 22–33 years), pulling 45° downward at normal reach (elbows near sides), the mean force is 142.04 N [31.93 lbf], at "maximum reach" (fully extended arm) 133.44 N [30.00], and at "extreme reach" (torso leaning, too) 115.29 N [25.02]. A vertical pull was 396.98 N [89.24] at maximum reach and 399.92 [89.91] at extreme—lower than the values reported by CCOHS.

A greater vertical pull can be exerted if one hangs from the rope—thus applying one's own weight as a pulling force—but there is less control in that situation.

Disposition on Ships

A single yard (a spar on which a sails is set) would be equipped with several blocks, named according to purpose. The lower, longer, heavier yards had more blocks. In the late nineteenth century, the lowest spar might carry seven pairs of blocks, one block of each pair on each side of the yard (Harland 1985, 34). Blocks would also be mounted at the masthead (Sephton 2011, 109, 114). For the names of the blocks on each yard, see *London Encyclopedia* (1845, 4: 232–3).

The block-and-tackle was ubiquitous on sailing ships, being used for both sail and cargo handling. By "the 1880s a schooner of 150 tons might use as many as 50 single blocks and 25 double blocks" (Sager 1996, 31). And that count was small compared to that on a Napoleonic warship: one of seventy-four guns required 622 "single blocks from 5 to 26 inches"; "130 double ... from 7 to 26"; 74 "other blocks, generally large, and several of them treble"; and 444 more blocks (six for each of the 74 guns), for a total of 1,270 blocks (*London Encyclopaedia* 1845, 4:234).

Blocks and Block Making

Blocks varied in size and shape, depending on their purpose. On HMS *Invincible*, the sheaves ranged in diameters from 330 mm down to 95 mm (Cousins 2022, 58). Steel (1794, 160) lists prices for blocks of lengths ranging from 6 to 38 inches. In the late-nineteenth-century American Navy, the shells "in general use" had lengths ranging "from 4 inches to 24 inches inclusive" (Qualtrough 1881, 265).

"Common blocks" had "simple oval shells and a single strop" (Cousins 2022). Steel (1794, 148) names 14 different blocks as "differing from the common shape." For example, "main sheet blocks" were "tear drop shaped" (Cousins 2022, 63). Overall, in 1815 there were "not less than 200 sorts and sizes" used in the Royal Navy (Burney 1815, 45).

Manwayring (1644) only mentions all-wood blocks. "Before the end of the 17th century rigging blocks tended to be handmade from local timbers," such as, in Britain, elm, ash, beech, and holly (Cousins 2022, 11).

Sutherland implies that elm alone was used for blocks 12–40 inches long, whereas either ash or elm could be use for those 3–11 inches long (1717, 222). A 19-inch-long double block might have a breadth of 12 inches, a pin of almost 3 inches diameter, and a "shiver" (sheave) of 8.5 inches diameter (232).

The shell could be "made" or morticed. A morticed shell was carved from a single piece of wood, whereas in a "made" one, "two or more pieces" were "pinned together" (Qualtrough 1881, 265).

The preferred wood (for the sheave and pin) came to be a tropical hardwood, lignum vitae ("tree of life"). The trees (*Guaiacum* genus) that produce this wood are native to the Caribbean, and the wood was first exported to Europe in the early sixteenth century as a supposed anti-syphilitic (Thompson 2022, 107) . However, "by 1677 the Royal Navy (RN) had begun to outfit its ships with the new timber" (Cousins 2022, 11).

Green lignum vitae has a specific gravity of 1.05, and with 12 percent moisture content, its compressive strength parallel to the grain is 78,600 kPa and its side hardness 20,000 N (FPL 2010, 5-19). For American elm, the comparable values are 0.46, 38,100, and 3,700 (5-4).Lignum vitae has "self-lubricating properties" attributable to its "high resin content" (FPL 2010, 2-28), and because of its "oiliness" its coefficient of friction is less than that typical of wood (4-17).

Cousins observes that while lignum vitae could be used to make a "stronger, longer lasting, self-lubricating block," it was "almost twice as heavy as ash," and he suggested that this would "increase the pressure on the masts and the rigging overall," and consequently "there would have to be an evolution in the rig and technology to cope with the excess weight" (2022, 11).

Without doubting that there were changes in the rig over the course of the eighteenth century, it is questionable that they were prompted by an increase in the weight of the blocks.

By 1794, the sheave (wheel) was "commonly made of lignum-vitae" (or, for "very laborious purposes," metal), and the pin (axle) of "lignum-vitae, cocus, green-heart, or iron," while the shell was still made of "elm or ash" (Steel 1794, "Block-Making"). Hence, the weight of the block was not doubled.

Moreover, while the weight of the blocks was not trivial, they were small compared to the rest of the "top hamper" (the wind-catching apparatus). I regret that I do not have any weights for the eighteenth century. However, in the 1840s, for a "first-class A three-decker," the total weight of masts and yards was 269,226 pounds; of the sails (including spares), 23,570 pounds; and of the rigging (including the blocks), 374,281 pounds. Of the latter, the blocks were 23,520 lbs (Chatham 1847, 119).

This table shows typical weights quoted in 1879 for 500- and 1,000-ton sailing ships and a 180-ton schooner.

Table 4-1: Weights of Top Hamper, Long Tons

	1,000-ton ship	500-ton ship	180-ton schooner
Lower masts and bowsprit	21.6	9.1	6.4
Topmasts and yards	18.6	8.8	1.9
Spare gear and booms	7.5	4.1	1.2
Standing rigging	11.0	9.4	1.9
Running rigging	11.4	6.5	1.4
Blocks	5.4	4.2	0.3
Ship's sails	3.8	2.1	1.3
Spare sails	2.3	1.5	0.8

Wooden masts and spars, wire rigging. Mackrow (1879, 350). A long ton is 2,240 pounds.

Cousins more plausibly proposes that "the change in materials from ash to lignum at the turn of the century [i.e., around 1700] also allowed for a stronger rig," as evidenced by the increase in the length of the ship's yards and the area of the sails that occurred over the first half of the eighteenth century (2022, 11–12).

Until the mid-eighteenth century, blocks were handmade. "The shells of blocks are first sawed to their length, breadth, and thickness; and the corners or angles are taken off." The block was fixed with wedges in a "clave," and sheave holes were gouged (with a chisel) and bored (with an auger) from either end, meeting (hopefully) in the middle. The scores for the straps were gouged out and knives (the "stock-shave" and the "spoke-shave") were used to finish the outside of the shell. The sheave was rough-cut with a saw, the hole for the pin was bored through and reamed with an auger, and the outside of the sheave was "turned smooth" on a lathe. The pin, except for its ends, was also turned on the lathe (Steel 1794, 153–5).

"Making blocks by hand was slow, laborious, and subject to errors of measurement and slips of the tool that could make the finished product irregular ... In handmade blocks, according to one account, the pinholes were 'so rough and so uneven ... that the blocks and shivers often caught fire through the violence of the friction'" (Cooper 1984, 184).

Edward Phillips's 1720 dictionary indicates that a pulley is a wheel fastened to "a piece of wood or iron." However, it also says that pulleys on ships are called blocks, and these are "wooden pulleys." This suggests that the transition to use of metal had begun on land, but not at sea.

In 1756, Walter Taylor* sought to improve the nautical block by "changing the pin from Lignum vitae to iron and adding a metal coak, a type of bearing to the sheaves." Since this would have "reduced the amount of friction on the blocks," it would have reduced the effort needed to lift weights (friction reduces the mechanical advantage from its theoretical value) (Cousins 2022, 12).

While one might expect that the use of metal rather than hardwood would have increased the weight, it came to be recognized that the mechanical advantage provided by the pulley was not dependent on the diameter of the sheave, and hence the

---

* There were actually two individuals, father and son, named Walter Taylor, who entered the block-making business. The elder was a ship's carpenter who died in 1762. The younger died in 1803.

blocks could be made smaller. "Captain Bentinck, therefore, assured of the strength of Mr. Taylor's shivers and pins, ordered his ship, the *Centaur* …, to be rigged with blocks and shivers of little more than half the usual dimensions, which proportionately reduced, first, the price, and secondly, the weight: the latter was found, by experiment, to be diminished twenty-six hundred weight, which was taken off from the mast only" (Quin 1814, 271). According to *Wikipedia*, this field trial took place in 1762. (Taylor also received a patent in 1786 for "coakes … made in two parts…, so that when riveted together a small space be left between them to contain grease.")

The wreck of the *Invincible* (1758) revealed that most of its blocks used lignum vitae for both the sheaves and the pin around which they rotated. However, there were a few blocks with iron pins (Bingeman et al. 2021, 37). There were also eighteenth-century blocks in which the wheel was made of lignum vitae but the groove lining was brass (Morton 2019, 267).

The Taylors, father and son, had also devised machinery for the mass production of blocks, and in 1761 this family won a Royal Navy contract. A fire in 1770 that destroyed the navy's entire block inventory forced it to adopt the Taylor blocks as standard (Cousins 2022, 12), and these were of the smaller size promoted by Bentinck (Cooper 1984, 186).

In 1762, Elizabeth Taylor, as executrix of the senior Taylor's estate, received a patent on a "set of engines, tools, instruments and other apparatus for making blocks, shivers, and pins." In 1775, her son (Walter Taylor junior) received one for "bushing cast-iron or other metal shivers for ships' blocks."

On February 23, 1776, a parliamentary committee reported testimony that the Taylor blocks reduced expenses by at least 25 percent, and that with them, "two men will perform as much with these as three men can with blocks on the old construction" (Parliament 1803, 794).

Taylor's original machines appear to have been hand-powered (Cooper 1984, 186). Taylor moved his operations in 1781, and in the new location he was able "to make ready use of both river and steam power to drive his specially designed machinery," including "his prime invention, the circular saw" (Bingeman et al. 2021, 41). However, Taylor only "briefly and unsatisfactorily tried out steam power" (Cooper 1984). Taylor may also have used horse power (Steel 1794), to run a lathe and a frame saw.

"The uniformity of the Taylors' blocks meant they could be stored in British colonial harbours along sea routes and, being uniform and interchangeable, slotted in to replace broken or defective ones" (Bingeman et al. 2021). In 1797–1805, production averaged 100,000 blocks a year (Bingeman et al. 2021).

Taylor's machines mechanized "boring, sawing and mortising," but they still left "turning shaping, completing mortises, scoring to handwork" (Cooper 1984, 189). In addition, they were not specialized; they needed to be set up differently depending on the block or sheave to be made (191–2).

In 1801, Marc Isambard Brunel attempted to interest the Taylors in his plans for more specialized blockmaking machines, for which he had received a patent, but the Taylors made the fatal error of sending him away. Brunel turned to Samuel Bentham, the inspector-general of naval works, who was something of an efficiency maven (Cooper 1984, 195). The new machines were built for the Portsmouth dockyard by Henry Maudsley. In 1805, the Taylors' Royal Navy contract was terminated (186).

Ultimately, "there were about forty-five machines altogether, of twenty-two different kinds" (Cooper 1984, 198). They had a number of interesting features. First, they were of all-metal construction (197). Second, they made use of steam power (197, 204), previously introduced to the dockyard by Bentham in 1799 (195). Third, they were organized into three assembly lines, for small, medium, and large blocks, respectively (206). The small (4–7 inch) blocks received wooden pins, and the medium (8–10) and large (11–18) blocks, iron pins (Burney 1815, 45). Labor requirements were reduced by a factor of 10 or more (Cooper 1984, 206).

The *Edinburgh Encyclopaedia* (1830, 3:607–12) provides a detailed description of all of the machines, and how they operated. Gilbert (1965, 11) shows in photographs the progression from raw shell to finished shell, and from raw sheave to finished sheave. A corresponding flowchart, naming the machine used at each step, is provided by Coad (2005, 79).

Later in the nineteenth century, the demand for blocks declined (as a result of the replacement of sail power with steam power), and wooden blocks were largely replaced by metal ones (Cooper 1984, 198). On HMS *Devastation* (1870), the Royal Navy's first steam-only capital ship, blocks were needed only for signal halyards, hoisting ship's boats, and other minor purposes (109). "By the early 1960s annual production [at Portsmouth] had fallen to around 2,000 blocks, the great majority made of metal" (110).

Ropes and the Like

The ropes used in the standing rigging, which support the masts, are called stays, shrouds, and backstays. Those used in the running rigging, which control the sails, are called ties, halyards, purchases, braces, lifts, tricing lines, and so on. Those used to tow or moor a ship were called hawsers or cable.*

They all have the same purpose: forming a flexible connection that is capable of withstanding a tensile (stretching) force exerted by a load (weight, wind, etc.) or applying such a force to another object.

The load determined the required breaking strength of the rope, which (for a given rope material and construction) in turn determined its minimum circumference (or diameter), which in turn determined its minimum bending radius and thus the minimum size of the sheave it was wrapped around.

Strictly speaking, a "rope" in nautical usage may be a "fiber rope" (made of natural or artificial fibers) or a "wire rope" (made of iron or steel wire) but would not include a chain (interlocked metal links). Nonetheless, I have decided to use "rope"

---

* This section quotes extensively from this author's "Tethered Balloons and Kites in the 1632 Universe, Part 2," and this statement is in lieu of the usual quotation marks.

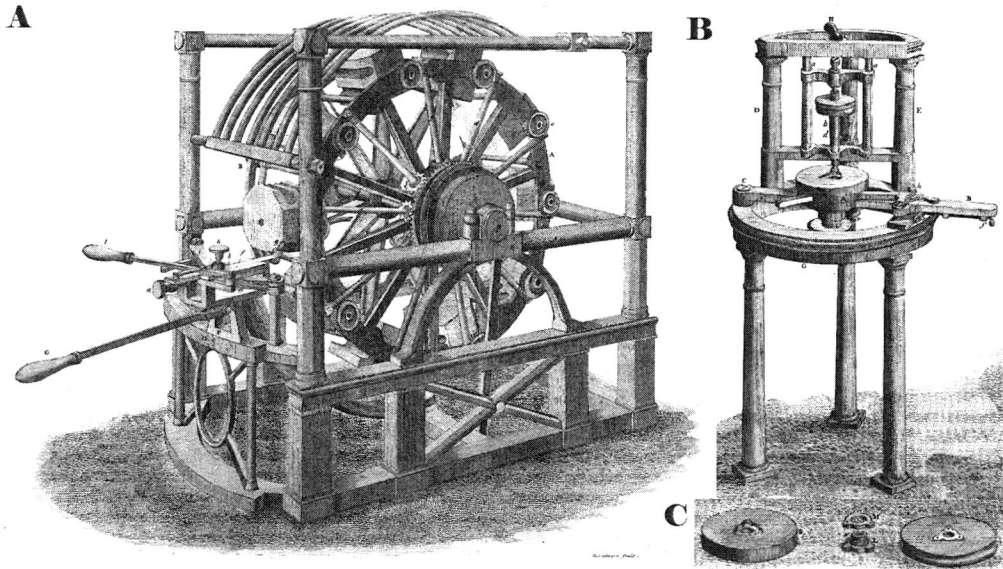

A. Perspective view of the shaping engine, "for forming the outside of the blocks to their proper figure." Note that it is handling 10 blocks simultaneously. The double wheel rotates, bringing them against a cam-guided gouge (*Edinburgh Encyclopedia* 1830, Vol. 3, Plate LIX). B. Perspective view of the coaking engine, "for forming a cavity in the form of three small semicircles in the center of the sheave, for the reception of the coak or metal bush" (Plate LXI, Fig. 1). C. From left to right, a sheave with the coak cavity formed; two coaks; and a sheave with the coak inserted (Plate LXI, unnumbered). For other illustrations of Portsmouth block-making machinery, see generally the *Edinburgh Encyclopedia* (1830, Vol. 3, Plates LVII-LXI), and also Cresy (1865, 1043).

as a generic term for flexible, tensioning elements (including chains) and elaborate as needed.

The most important characteristic for the ropes used on ships was a high strength-to-weight ratio. The simplest method of expressing the strength of a rope is in terms of the static load that, hanging from the rope, will break it.

In modern scientific literature, it is more common to express it in terms of the specific strength (breaking load per unit cross-sectional area), measured in megapascals (MPa) (1 megapascal is 1 million pascals; a pascal is one newton per square meter) or psi (pound force per square inch). This implicitly assumes that the strength of a rope is proportional to its cross-sectional area (CSA), a long-recognized phenomenon (Steel 1794, 62). However, with manually twisted hemp ropes, narrow ropes generally had higher MPa values than thick ones.

The working load is the expected load on the rope. The breaking load divided by the working load is the safety factor. Luce and Ward (1884, 29) recommended a safety factor of at least three for naval practice (with hemp ropes), but I doubt that would be considered adequate today, if hemp ropes were still used. Modern safety factors vary depending on the material: For cargo slings, they are manila (SF5), nylon (SF9), polyester (SF9), and polypropylene (SF6) (Kulweic 1985, 516). Higher safety factors may be used for more critical applications.

In theory, the strength of a rope is not dependent on its length. However, the rope material is not homogeneous; there can be local flaws. The longer the rope, the greater is the probability of a flaw. Still, this is a minor effect.

The principal "lays" are "plain-laid" (three strands, twisted right-handed), "shroud-laid" (four strands, twisted right handed around a core or "heart" strand), and "cable-laid" (three plain-laid ropes twisted left-handed, thus, nine strands) rope (Hines and Wilmer 1910, 460). The term "hawser-laid" is best avoided.*

Rope may instead be braided, that is, the fibers formed into a tight tube. The braid may have a hollow center. Braided ropes are more flexible, but have less stretch and are more difficult to splice (Kenninger 2019). There appears to be a difference of authority on the issue of strength.

Kenninger reports that 5/8-inch diameter nylon has a minimum break strength of 7,200 pounds braided, versus 9,350 twisted. But Moyer (Moyer and Everett 2013, 74) quotes comparative data for cotton, polyester, and polypropylene in which braided was stronger than twisted of the same diameter.

Fiber Ropes (Lines)

To make a fiber rope, fibers are first spun together into yarn. To make (lay) twisted rope, yarns are twisted together to form a strand, and strands are twisted together in the opposite direction to make a rope. The length was limited by the work area, which was called a ropewalk. A typical ropewalk length was 400 yards (Rees 1819, 30: "Rope-Making"; cf. Steel 1794, 60), more than enough for the standard eighteenth-century Royal Navy 120-fathom coil.

Natural fibers may be extracted from leaves, fruits, or plant stems ("bast"). Hemp rope, made from fibers from the stem of *Cannabis sativa*, was found at el-Amarna, the ruins of the capital city of the late Eighteenth Dynasty of Egypt (Robinson 1996, 111). It was used on the *Syracusia*, Hiero II's flagship (114), and Venetian galleys and ships were rigged with locally manufactured hemp rope (115). It was the principal rope used in European naval ropemaking until the second half of the nineteenth century. The USS *Constitution* used 8 miles of hemp in its running rigging; the *Cutty Sark*, 7; the *Preussen*, 9½ (Royce 1998).

Manila rope is made from fibers from the leaf sheaths of the abaca plant, *Musa testilus*. The plant is native to the Philippines and was cultivated prior to the Spanish colonization. There was both local use of the fiber (primarily in textiles) and sale of the fiber to Chinese traders (Spencer 1953). The Spanish used it to manufacture rope for the rigging of the Manila–Acapulco galleon, but it did not become an export commodity until 1812 (Hayase 2018). The first shipment to the United States was in 1818 (Spencer 1953).

Manila was weaker but more durable than white hemp, and stronger and less durable than tarred hemp (Murray 1916, 301). Hemp rigging stretched in dry

---

* Steel (1794, 55) defined "hawsers" as "ropes made of three or four single strands," thus encompassing both plain-laid and shroud-laid rope. But Kipping (1857, 70) insists that "hawser-laid and cable-laid rope is all the same; it is composed of nine strands."

weather and shrank when it was humid; manila was less susceptible to this (Royce 1998).

By a century after manila rope's American introduction, hemp rope was "but little used." Tarred hemp was "used on shipboard for such of the standing rigging as is not of wire and for the heaviest of the running rigging of sailing ships." The U.S. Navy, having abandoned sail for steam, only used hemp for "small stuff" (ropes up to 1¾ inch circumference) (Knight and Baldridge 1921, 34).

Other natural fibers that have been used in naval ropemaking include cotton, agave, jute, abaca, and sisal (Fillingane and Williams 1993, 3-1). Coir rope, "made from the husk of the coconut," was "one-third as heavy as hemp rope," and thus would "float on water" (Murray 1916, 301). Unfortunately, it was weak: 3-inch circumference coir had a breaking strength of 1,570 pounds, compared to 6,700 for tarred hemp, 7,500 for manila, and 9,400 for white hemp (Murray 1916, 372).

As of 1995, the only natural fiber ropes still used by the British navy were manila ("as a check stopper for towing operations") and sisal ("as a sliprope during replenishment at sea"). Their perceived advantages over artificial fibers were that they stretched less, surged more readily around a winch or capstan, and did not fuse when heated (by friction) (Admiralty 1995, 3–5). In the U.S. Navy, the natural fibers most often still used in 1993 were "marline (tarred hemp) and manila" (NAVEDTRA 1993).

Wire rope. This is composed of twisted metal wires, originally wrought iron but later steel. Leonardo da Vinci's "sketches include apparatus for drawing iron or copper wire and twisting it into rope." A crude wire rope was in use in the 1780s and the mining industry was one of the first to experiment with wire rope (Martin 2014, 152–3). "By 1857 three-quarters of all ships fitted out in Liverpool were rigged with wire rope" (Martin 2014, 153).

To inhibit corrosion, iron or steel wire rope may be galvanized (which reduces flexibility and strength). Phosphor bronze wire is used when "either noncorrosive or nonmagnetic properties are desirable" (Bureau of Naval Personnel [BNP] 1964, 29).

Wire rope is made of strands that are each in turn made of individual wires. For example, there may be six strands of 19 wires each. "The strands are laid up around a central core, which may be only a single wire, a single strand of wire, or hemp" (BNP 1964, 30). Wire rope is damaged by acid, and "many times" wire-rope that has given way has been found to have been acid-damaged (33).

The recommended safety factor in the early twentieth century was "six to seven for running rigging, and five for standing rigging" (Murray 1916, 301).

Artificial fibers. These are made of synthetic polymers. Nylon (polyamide) became available in cordage form in 1939. Later, the cordage industry made available polyester, polypropylene, and polyethylene. As of 1995, the British navy also made limited use of aramid (Kevlar, a polyamide derivative), high-modulus polyethylene, and a guard-wire rope with a polyester core and a polyethylene sheath. The latter replaced polyolefin (a polyethylene derivative) (Admiralty 1995, 3–7, 10).

In 1964, the U.S. Navy observed that "nylon has several advantage over manila. Size for size it is nearly three times as strong and lasts five times as long … [It] does not rot or age like natural fiber … It is less bulky, more flexible, and therefore, easier to handle and requires less stowage space" (BNP 1964, 28).

Nylon stretches under load, but if not stretched too much (40 percent is the critical point for rope- and cable-laid nylon) it will recover completely when released. It does not offer the "audible protests of natural fiber under heavy loads" and therefore sailors must note the visible signs of danger (elongation and thinning). It will part at 50 percent elongation and then there is a "decided," potentially dangerous, snap-back (BNP 1964, 29).

In 2010, the synthetic fibers in use for making line were "nylon, aramid, polyester (Dacron), polypropylene, and polyethylene, in descending order of strength" (NAVEDTRA 2010, 2-9). The advantage of polypropylene was lower cost, and the advantages of polyester, abrasion resistance and reduced stretch (2-12). Polypropylene "may lose as much as 40 percent of its strength in 3 months when exposed to tropical sunlight if the line is made without ultraviolet inhibitors" (2-13).

Synthetic fiber line has a lower coefficient of friction than the natural fiber lines, so care is needed "when a line is being payed out or eased from securing devices." It also has "poor knot-holding characteristics" (NAVEDTRA 2010, 2-9, 11).

For the material properties of ropes in nautical use, see Appendix 2.

Chain

Chain consists of interlocked links of wrought iron or steel. In the British Navy, "in 1812, just one chain-cable was issued per ship. In 1831 First Rates were issued with four hemp and four chain cables and by 1862 only one hemp cable was carried" (Harland 2013, 75).

Wrought iron is about five times as dense as hemp. However, iron also has greater tensile strength. Samuel Brown compared the strength and weight of chain to hemp, finding that "the 25-inch circumference hemp cable issued to the largest vessels and weighing about 140 lb per fathom, was comparable to chain made of 2.25-inch iron rod and weighing about 300 lb per fathom, both exhibiting a mean breaking strain around 110 tons" (Harland 2013, 75). The implication was that hemp had the superior strength-to-weight ratio.

The relative prices of hemp rope and wrought iron chain varied over the course of the nineteenth century, with the price of iron slowly declining and that of hemp rope oscillating. Ultimately, this gave chain a price advantage. In 1860, a 6-ton hemp cable might cost 240 pounds and a comparable 12-ton chain cable cost 196 pounds. Moreover, the chain cable would last far longer than the hemp one (Harland 2013).

Use of the Block-and-Tackle

Cargo Handling

Cargo had to be moved from an outboard location (dock or boat) to the hatch leading down to the cargo hold, and back again. If the ship were docked and the dock was equipped with cranes, this could be done with the dockside equipment. If the dock lacked cargo handling machines, or the ship was moored and the cargo brought to it by lighters, then the ship's crew had to handle the cargo themselves.

On sailing ships, this would have been done with block-and-tackle. Assuming that the hatch was just in front of the mast, one block would be attached to a pendant slung from the forestay so that it was directly over the hatch. The other would be on the outboard end of the lowest yard of that mast. The fall of the yard block is used to hoist the cargo above deck level. It is then hooked to the line running from the pendant block, too. Pulling on the fall of the pendant block while easing gradually on that of the yard block will slew the cargo transversely across the deck until it is over the hatch. It can then be unhooked from the yard line and lowered just by the pendant block. It is possible that a single line could be run on both blocks, with a short line from the cargo running to a hook fixed in position on the purchase/slewing line. (Tipping shows how these blocks could have been used for cargo handling on a medieval cog, with the hauling done by capstan and windlass rather than manually.)

Much of the cargo in a premodern vessel would have been in the form of barrels of various sizes, and the weight would depend on both the size and the contents. Cargo could also be in the form of boxes, sacks, and miscellaneous forms.

If the cargo container didn't have a convenient, sturdy point for attaching the tackle, it would have to be slung or netted. A single barrel could be slung with the barrel axis either horizontal or vertical (War 1921, 13).

Containers could also be stacked on pallets. While skids were used in ancient Egypt, the true pallet, designed for handling by a forklift, was introduced in the period between the world wars. The pallet load itself would be slung or netted for transfer to the ship's hold.

Sail and Yard Handling

Sails could be sent aloft to be bent (attached) to their yards, or "bent to their yards on deck and sent up or down with them." The latter procedure was followed with the upper sails ("the royals invariably, the topgallants usually") (Harland 1985, 96–7). Block-and-tackle (see below) would be used for these operations (and their reverse), as well as for bracing (turning) the yards.

The greater the weight of the sail or yard, the more crew must be assigned to the task of handling it, or the more efficient the machine used to handle it.

Hard data on the weight of even nineteenth-century sails and yards is hard to come by. On a 74-gun warship, the heaviest yard would be the main yard, with a length of perhaps 32 yd 8 in and a weight of about 9,400 pounds (Boyd 1857, 124).

I have two sources for the weight of the main course, and they are somewhat contradictory. Boyd says that on a 74-gun ship, it would need "892 yards" of Number 1 canvas (Boyd 284). A bolt (40 yards by 24 inches) weighed 46 pounds (Harland 1985, 29).* I believe Boyd is only referring to the total bolt length, so $46 \times 892/40 = 1025.8$ pounds.

Another source says that the mainsail of a "seventy-four" would have an area

---

* Without going into details, there appears to have been some variation in both the length and width of a bolt of canvas.

of 9,655 square feet (Fincham 1851, 271). Assuming no wastage, that would require 40.23 bolts, for a weight of 1,850 pounds.

Gun Handling

In the Napoleonic era, the main armament of a Royal Navy "seventy-four" was a gun firing 32-pound shot. Its weight varied depending on barrel length (and whether the metal was iron or bronze), and Goodwin refers to a 10-foot gun of 58 cwt (6,496 lbs) on a 1,299-pound carriage, and a 9.5-foot gun of 55 cwt (6,160 lbs) on a 1,176-pound carriage (Goodwin 2002, 304).

For the methods used to aim a gun, and to bring it back to firing position after it recoils, see Cooper (2024a). Here we talk about how guns were brought on board nineteenth-century warships. The gun barrel and the carriage were brought on board separately. The barrel, being much heavier, was the more difficult part of the task.

In essence, two block-and-tackles were used. One acted vertically, and was attached around the neck ring of the barrel, and the other acted horizontally and was attached by a "chain strop" to the button (cascabel) at the breech end of the barrel. The main yard was braced "immediately over the port through which the gun was to be taken in," and it held the vertical hoist. The horizontal hoist was on "the deck that guns are coming in upon." The gun was "hoisted muzzle upwards," and "when the neck ring is level with the port," the horizontal purchase was hauled upon. When the gun was "over the carriage," it was "lowered into its place" (Nares 1876, 109–110, Fig. 244).

5

"Rig the Capstan!"

(Lifting Machines, Part 2)

"Yo, heave ho! Round the capstan go! Round, men, with a will! Tramp, and tramp it still! The anchor must be heaved" (Smith 1888, 9). It was to such a song, a capstan shanty, that a gang of sailors pushed on capstan bars, slowly pulling in the slack on the anchor cable and ultimately, with a supreme effort, freeing the anchor from the sea bottom.

The windlass, capstan, and winch are all machines for raising weights. The load is attached to a rope that is secured to a barrel to which levers are attached. This creates a lever of the second class, that is, one in which the load is between the fulcrum (the axis of the axle) and the effort (the far end of the crank or inserted lever). The mechanical advantage gained is the ratio of the distance of the effort from the fulcrum (barrel axis) to that of the distance of the point of application of the load from the fulcrum. The latter distance is the radius of the axle plus half the thickness of the wound-up rope (Mackenzie 1920, 28).

The etymology of "windlass" is obscure, but appears to come from Old Norse *vindáss* (wind-pole). "Capstan" is derived from Provençal *cabestan*, itself derived from Latin *capistrare*, to bind, and originally referring to putting a halter on a horse.

The windlass has a thick horizontal axle (barrel), and the capstan a vertical one. Both the capstan and the windlass were known in ancient times. In discussing stone throwers, Vitruvius comments, "Some are worked by handspikes and windlasses, some by blocks and pulleys, others by capstans."

A second-century CE mosaic from Ostia depicting a Roman *codicaria* (riverboat) appears to show a capstan on the afterdeck, and Casson (1965, 38) suggests that this was used to tow (kedge) the *codicaria* upriver.

Harland says that "small craft weighed anchor by hand, while merchant vessels up to 400 tons or so, used a horizontal windlass … Large vessels used the main capstan to weigh anchor" (1985, 260).

According to Richard Henry Dana (1841, 96), warships used a capstan to weigh anchor, while merchant vessels used a windlass. The *Bremmer Koge* (Bremen cog) (1380) actually had both (Harland 2003, 39), and this was also true of eighteenth-century colliers, such as Captain Cook's HMS *Endeavour* (46).

While the principal use of the windlass was for raising anchor, on whaling ships it was used to roll a sperm whale carcass (Raupp 2015, 261) and lift "heavy blubber pieces onto the deck" (189).

The Load: Anchors and Anchor Cables

"A large English man-of-war of the mid–1800s had at least three, and usually four, heavy anchors." These were named according to where they were stowed: as bower (at the bow), sheet ("abreast the fore-chains"), or waist ("abaft … of the channels") anchors (Harland 1985, 235). The anchor was sized to the ship, based on tonnage, maximum beam, loaded draft, or, in a warship, the number of guns (232). The weight stated in a historical source might be just that of the rings, arms, and flukes, or it might also include the horizontal stock.

For a merchant ship, Boyd suggested a rule of four pounds per ton, and Rubin, that the weight in pounds was three times the squared maximum beam (feet). For warships, a typical rule was that the bower anchor weighed one hundredweight per gun. "A First Rate of 100 guns, like *Victory*, carried a best bower of just under 100cwt" (Boyd). *Victory*'s beam is 52 feet; the bower presently found on board weighs 84cwt, the stock about another 21cwt" (Harland 1985, 231–2). A British hundredweight (cwt) is 112 pounds.

To this one must add the weight of the anchor cable. A "widely used rule" was that there should be "half an inch of circumference per foot of breadth of beam." By this rule, the *Victory*'s cable should have been 26 inches circumference; it was in fact 24. "A fathom of such cable would weigh something above an English cwt" (Harland 1985, 231–2). The cable might be 100 to 120 fathoms long (Harland 1985).

Clearly, it's a lot of weight to lift. And sometimes additional effort was needed, to break the anchor and cable out of the mud, and once it is rising, to overcome their hydrodynamic drag.

Winch

The term "winch" is a bit problematic. Falconer (1784) defines a "winch" as "a cylindrical piece of timber, furnished with an axis, which extremities rest in two channels placed horizontally or vertically." If he had stopped there, it would make it a generic term covering both windlasses and capstans. However, Falconer adds that it is "turned about by means of a handle resembling that of a draw well, grind-stone, etc." The implication is that it is equipped with a crank, and the winch of a draw well has a horizontal barrel, as in Falconer's illustration. Young's *Nautical Dictionary* (1863) defines it even more narrowly as "a species of small windlass or crab, turned round by crank-handles." And *Knight's American Mechanical Dictionary* agrees that "the turning-power" is "a crank" (Knight 1881, 3:2776). Hedderwick also identifies a winch as a "small windlass" (1830, 122) turned by crank-handles (314). A crank, it should be noted, is simply a bent lever that is attached to the axis of a rotating element.

Nonetheless, Harland chose to elevate "winch" to a generic term covering both capstans and windlasses, without limitation to cranked machines. Harland then distinguishes between a reel-winch, in which "the end of the rope is firmly secured to the barrel," so there is a finite maximum number of turns possible, and a "traction-winch," in which "two or three turns of the rope are taken around the barrel, and the grip depends solely on the friction between rope and barrel." Provided that tension is applied to the "inboard end" ("tail"), the number of turns on the barrel "remains constant, and in theory there is no limit to the length of rope which can be handled by the device" (Harland 1991, 151).

Applying tension (tailing force) to the inboard end is called "holding off." Mainwayring (1644) wrote, "if the cable be very stiff and great, or else have lain in slimy, oozy ground, it surges and slips back, unless that part which is heaved in be still hauled away hard from the capstan, to keep the cable close and hard to the capstan whelps."

Note that this tailing force is what is needed to keep the load from pulling the rope off the barrel, and is separate from the force that must be applied to the levers in order to lift the load.

It seems to this author that there was a fine balance between having enough turns on the traction-winch so that the load wouldn't pull off the rope, and not so many that the inboard end couldn't be pulled in.

The Windlass

According to Glynn (1867, 4), on Chinese junks the windlass barrel extended the full breadth of the vessel, with its ends, which were "reduced and rounded to form pivots," resting on the bulwarks. Consequently, the barrel could be as much as thirty feet long. In contrast, on western ships, the barrel ends were supported by vertical supports called the windlass- or carrick-bitts. Each of these was formed of two pieces so that they could be separated in order to remove the cable from the windlass (Falconer 1769, "Knight-Head"). The bitts themselves were braced on their forward side by knees.

The windlass was positioned near the foremast, and "the lower part of the windlass is usually about a foot above the deck" (Falconer 1769, "Windlass").

Small windlasses, such as the winding mechanism on a crossbow, or those used to draw water buckets up from wells, generally featured a crank as the lever. This could be sized for one- or two-hand operation.

A shipboard windlass instead usually had a barrel with holes for the insertion of levers. This potentially increased the force that could be applied, as a long barrel could accommodate more sailors than a normal crank. But unlike the crank, which could be turned continuously if the crank arm weren't longer than the sailor's reach, these levers could only be pushed or pulled so far before they struck the deck or became too awkwardly positioned for further effort. Therefore, there needed to be either a series of levers around the circumference of the barrel, so as one moved to an inconvenient position, another became available, or a series of holes for receiving a lever, the lever being moved as needed. The latter solution was the one adopted. The lever used was called a handspike; it was a wooden bar which was round and tapering at one end and square at the other (Falconer 1769, "Handspec").

Harland (1991, 263) describes a windlass with an eight-sided barrel. Each face would have a series of horizontally spaced square holes, and the number of holes per face would determine the number of men abreast who could work the windlass simultaneously (assuming one per lever arm).

The holes could be placed on every face, or only on every other face. If on every face, half of them could be offset from the other half to avoid weakening the barrel.

The usual mode of operation was to place the handspike in the hole when it faced upward, then haul it down and back until it was horizontal, thus turning the barrel one-quarter. If the load were especially heavy, two men might be placed on a single handspike.

With this system, unlike with a crank, truly continuous motion was not possible, but rather a series of heaves. An experienced crew no doubt could minimize the time between heaves. Dana says (1841, 147, 156) that when a merchant ship weighed anchor, the second mate took a handspike at one end of the windlass.

To prevent slippage, each face of the barrel had a pawl notch that engaged a pawl. The pawl was simply a short lever, sloped downward from a structure called the pawl bitt, which engaged the pawl notch so the barrel could only turn in one direction. Pawls could be made of wood or iron, and the pawl notches could be lined with iron (Falconer 1769, "Windlass").

The windlass pawl is mentioned in Henry Manwayring's *Sea-Mans Dictionary* (written 1623, published 1644; "Windlasse"), and as he notes, if the sailors "cannot heave forward, or one should faile," the windlass will pawl itself (i.e., engage the pawl notch and halt the backslip).

With a pawl notch on each face of an octagonal barrel, there is an engagement every eighth of a turn. Falconer points out that "if the windlass is twenty inches in diameter, and purchases five feet of cable at every revolution of the machine, it will be prevented from turning back … at every seven inches nearly."

Unfortunately, even a seven-inch roll back could cause a sudden and unexpected recoil of the sailors' handspikes, causing injury. One early expedient was the use of "pawl and half pawl." In essence, this used two pawls of different lengths. As the barrel turned, first the long pawl and then the short one would engage, so there would be an engagement of one or the other every one-sixteenth turn, thus cutting the potential roll back in half (Homer 1895, 127; Glynn 1867, 5).

According to Harland, Hutchinson (1796) considered "the introduction of pawls that engaged every sixteenth of a turn" a "great improvement."

A later alternative to cutting the pawl notches into the barrel was to give the barrel an iron pawl "rim" (although not always at the end of the barrel, so perhaps better to call it a "wheel" or "ring") offering notches or teeth to engage the pawl (Glynn 1867, 6). That in turn meant that the number of "catches" per revolution wasn't limited by the number of faces imparted to the barrel.

If there were concern that a pawl might fail, one could use multiple pawls that engaged the pawl notches or teeth simultaneously. Harland (2015, 39) shows a "classic" windlass with two pawls extending from the pawl bit, simultaneously engaging separate notches in the same face of the barrel, while later (41) he shows another with three pawls simultaneously engaging different teeth of the same pawl wheel.

Because of the orientation of the pawls, the windlass was only turned one way. In contrast, the capstan could be and was rotated in either direction (Harland 2013, 55).

The windlass was referenced in these lines from "The Shipwreck," a 1762 poem by William Falconer (1822, 52):

> Roused from repose aloft the sailors swarm,
> And with their levers soon the windlass arm
> The order given, upspringing with a bound,
> They fix the bars, and heave the windlass round;
> At every turn the clanging pawls resound;
> Uptorn reluctant from its oozy cave,
> The ponderous anchor rises o'er the wave.

Less poetically, in his *An Universal Dictionary of the Marine* (1769), Falconer wrote, "As the windlass is heaved about in a vertical direction, it is evident that the effort of an equal number of men acting upon it will be much more powerful than on the capstan, because their whole weight and strength are applied more readily … It requires, however, some dexterity and address to manage the handspike to the greatest advantage; and to perform this the sailors must all rise at once upon the windlass, and, fixing their bars therein, give a sudden jerk at the same instant; in which movement they are regulated by a sort of song or howl pronounced by one of the number."

Consistent with Falconer's opinion, but providing a bit more information, Knight declared, "The windlass is more powerful than the capstan, in proportion to the men employed, as a man can exert a power of about 150 pounds on a windlass-handspike but only about 35 pounds on capstan bar. A greater number of men, however, can reach the capstan" (Knight 1881, 3: 2780).

So, in essence, the advantage of the windlass relative to the capstan is that in heaving down with the handspike, the sailor can use gravity—acting on his own body weight—to advantage. Indeed, human motion studies confirm that pulling is more forceful than pushing, and pulling is most effective when the lever is above the shoulder so the force is mainly due to the weight of the body (Garry 1930).

At least by Hedderwick's time (1830), the windlass barrel was not merely wood, but rather wood turning around on an iron spindle (122). Hedderwick credits Stephen Wright's 1786 patent for the first proposal of both iron spindles and iron pawls, and says that the latter engaged with every one-sixteenth turn. However, the only patent issued to "Stephen Wright" in the eighteenth century was No. 699 (April 2, 1755), which identified him as a North Shields (Newcastle) "master and mariner" (Woodcroft 1854, 1:129). It indeed referred to "iron spindles," but to "wooden pawls" (Abridgments 1872, 22).

Hedderwick (1830) gives a rule for proportioning the windlass pawl, based on the ship's extreme breadth and keel thickness. Likewise, he proportions the length and diameter of the windlass, and the thickness of the windlass bitts, to the breadth of the ship. The length between the bitts is to be "one-half of the breadth of the ship's main deck amidships," and the central diameter, "two inches for every 3 feet of the ship's extreme breadth" (158–9), that is, 1/18th thereof. The end diameter, inside the bitts, was five-sixths of the central diameter (304). As for the windlass spindle, its diameter was 1/96th of the ship's extreme breadth, unless the ship was deeper in hold than half the ship's breadth, in which case the diameter was increased (159). Henderson also gives proposed windlass dimensions for eight different ship types (242).

Obviously, there's no direct relationship between the breadth of the ship and the maximum strain of the cable on the ship at anchor. Rather, that maximum strain would probably be proportional to the area the ship presented to the wind when the wind was on the beam, assuming that current was not also a factor. That area, in turn, would be more closely related to the length of the ship, and the height of its main deck and superstructure above the waterline. As for the strain when attempting to break the anchor free of the sea floor, that would depend on the size of the anchor, how deeply it was embedded, and the characteristics of the "ground."

Hedderwick (1830, 304ff) also discusses the construction of the windlass. He warns that the timber intended to be shaped into the windlass barrel "must be very carefully examined" as, "should any defect in the piece be discovered after it is dressed," it will be condemned, as it is then "unfit ... for any other purpose without great waste of material" (303–4). While the British preferred to make the barrel from a single piece of timber, "in several parts of the United States, the body of the windlass is made in two pieces," which are subsequently shaped, and jointed to encompass the spindle. Hedderwick acknowledges that the carpenter would then see whether the piece is perfectly sound in the middle before proceeding further, and that the spindle may be put in "more centrically" than if the timber were bored. But Hedderwick complained about the "additional expense and weight" created by the two piece method. Another source on construction is Steel (1810, 391ff).

Patent Windlasses

In Britain, windlass patents have been issued since 1770 (Woodcroft 1854, 171, citing Stuard, Patent 954). In the United States, the first windlass patent was issued in 1804, and 105 such patents were issued through 1895 (Whitney 1896, 120). I do not have a count for British windlass patents.

Most patented inventions never are manufactured and sold, and of those, only a few truly achieve commercial success. Consequently, this chapter examines only a few examples of patent windlasses.

Pump-handle (pump-brake) windlass. This windlass dispensed with the insertion and removal of handspikes. Instead, it was actuated with a back-and-forth pumping action on a lever that in turn moved a driving pawl. While a retaining pawl merely prevents rotation in one direction, a driving pawl pushes against the tooth of a toothed wheel (gear), forcing it to rotate (Harland 2015, 41). In the early 1900s, these windlasses were sometimes called "Armstrong patent windlasses." This was "a jocular reference to the way it was powered ... rather than a surname" (Harland 2013, 158)—that is, by a "strong arm" rather than by steam.

The most primitive form is one Harland found depicted in Hutchinson's *Naval*

Architecture (1794). This shows a "seesaw" (two-handled) lever with a crossbar having a large claw at the bottom (Harland 2015, 42). The fulcrum of the crossbar was just above the top of the windlass barrel. The claw engages teeth on a sixteen-tooth ratchet wheel. A ratchet wheel is a type of gear with asymmetric teeth, shaped somewhat like the fin of a shark. In the power stroke, the trailing point of the claw engages the steep slope of a ratchet tooth, turning the wheel. On the return stroke, the leading point of the claw slides over the shallow slope of a ratchet tooth. Hutchinson's figure shows one handle almost touching the floor, and the other vertically above the sailor's head, so each action would have involved first a pull and then a push.

In 1832, Benjamin Cowle Tyzack, Thomas Storer Dobinson, and John Robinson patented a windlass with a double seesaw actuating mechanism. The two fore-and-aft seesaws were each formed by two bars, with handles, fitting into a double arm socket. The two double arm sockets were connected by the "warping shaft," held in the slot created by a bitt positioned behind the windlass. The shaft extended slightly beyond the windlass bitts, so the sailors working the windlass were not directly in front or behind it. The shaft served as the axis for both seesaws, and it was parallel to the axis of the windlass barrel.

In the center of the barrel is the conventional pawl wheel, with teeth shorter in radius than the barrel, which are engaged by the retaining pawl from the pawl bit. At each end of the barrel is another pawl wheel, also fixed to the barrel, but with exposed teeth. These are each engaged by three driving pawls held in a pawl box. The pawl box (creeper) has parallel "side bosses" and, viewed along the barrel axis, a profile somewhat like that of a light bulb without its socket. The circular part of the creeper has a large circular hole and fits around the side pawl wheel. The three driving pawls are depicted as being of different lengths, anchored irregularly within the creeper, and set at different angles, but Tyzack et al. don't give particulars of their geometry.

A lever is attached to each end of the warping shaft, and is connected by a rod to the narrow end of the corresponding creeper. Tyzack et al. explain that "the windlass will be driven round with great force at each up stroke of the creeper." The creeper moves with the seesaws, but the side pawl wheel turns with the barrel. Note that unlike an ordinary seesaw, the load is at the warping shaft lever, between the fulcrum (the warping shaft axis) and the effort (the handle of the seesaw)—thus, the seesaw bar, a lever of the second class. (The warping shaft lever is also a lever of the second class, albeit one with a rather small mechanical advantage.)

If the warping shaft levers point in opposite directions (one fore, one aft), as shown in Tyzack's Fig. 6, then the rotation of the barrel will be nearly continuous, as a down-stroke on the fore side will correspond to an up-stroke on the other, and vice versa. In the figure, the axis of the warping shaft is also set at a slight angle to that of the windlass barrel. The expectation is that this would smooth out the rotation by causing the port driving pawls to take effect at a slightly different point in the barrel rotation than the starboard ones.

Tyzack et al. note that the creepers (and, presumably, the corresponding pawl wheels and levers) may be positioned on some other part of the windlass barrel than the ends, thus anticipating a feature of the Taunton windlass discussed in the following.

The inventors refer to it as having a "compound lever" design, although that is debatable. A "compound lever" is one in which levers are connected so that the load arm of one acts upon the effort arm of the other. If two such component levers each have a mechanical advantage of 4:1, the compound lever combining them has one of 16:1 (Walton 1968, 17). The first lever action offers a mechanical advantage equal to the ratio of the distance of the seesaw handle (effort) to the warping shaft axis (fulcrum), to the distance of the rod–lever connection (load) to the same axis. The second possible lever action is one in which the short section of the creeper is the effort arm and the end of the pawl is the load arm. However, their relative distances from the barrel axis are unclear, so they might not provide an additional mechanical advantage.

Tyzack, Dobinson & Company was a manufacturer of chain cable, anchor and patent windlasses, and the anonymous *History of Tynemouth* (7) says that its patent windlass "has become in very general use."

A "pump-type windlass of European design" is depicted by zu Mondfeld (2005, 181). He does not provide attribution, but the figures are plainly derived from Taunton's second British patent specifications (Taunton 1845). Harland (2015, 47) reproduces the figures from Taunton's first patent (Taunton 1841). Harland comments that "in practical terms [Taunton] may be considered the father of the pump-brake windlass."

I begin with the common features of the two patents. The barrel itself was cylindrical in the center and tapered toward the ends. In the center, there was a pawl "wheel" that was engaged by three "retaining" pawls of different length (Taunton 1841, Fig. 2; Taunton 1845, Fig. 13). These extended from a "pawl bitt" positioned "behind" (that is, on the side away from the cable) the barrel. On either side of the pawl rim was a ratchet wheel.

The seesaw subsystem, mounted on the top of the pawl bitt and thus behind and higher than the barrel, consisted of two bars with "tee"-handles, a crosshead, and a crosshead pin. (Windlass bars are elsewhere called "brakes"—see Russell 2004, 62—and hence "pump-brake.") The crosshead pivoted back and forth around the crosshead pin, and it had square holes into which the bars were inserted. (The advantage of these arms being removable was that it increased the deck clearance when the windlass was not in use [Davis 2012, 239]). The pin axis was perpendicular to the barrel axis; that is, the seesaw was "athwartships."

On the bottom of the crosshead, near each of its two ends, there was a pivotably mounted vertical connecting rod, running down inside the interior of the pawl bitt. At the bottom end of each rod was a pawl mechanism, which I will describe in due course.

The key point is that unlike the "retaining" pawls already mentioned, which merely served to keep the barrel from rolling back the way it had come, this pawl mechanism was a *driving* pawl, intermittently engaging the teeth of one of the ratchet wheels and pushing against them, causing the barrel to rotate. Because of the seesawing of the bars, as one side disengaged, the other side engaged, rotating the barrel once more. The rotation of the barrel was thus almost continuous.

Harland (2015, 46) comments, "The novel feature was the simplification resulting from the way the driving ratchets were shifted toward the centre of the barrel,

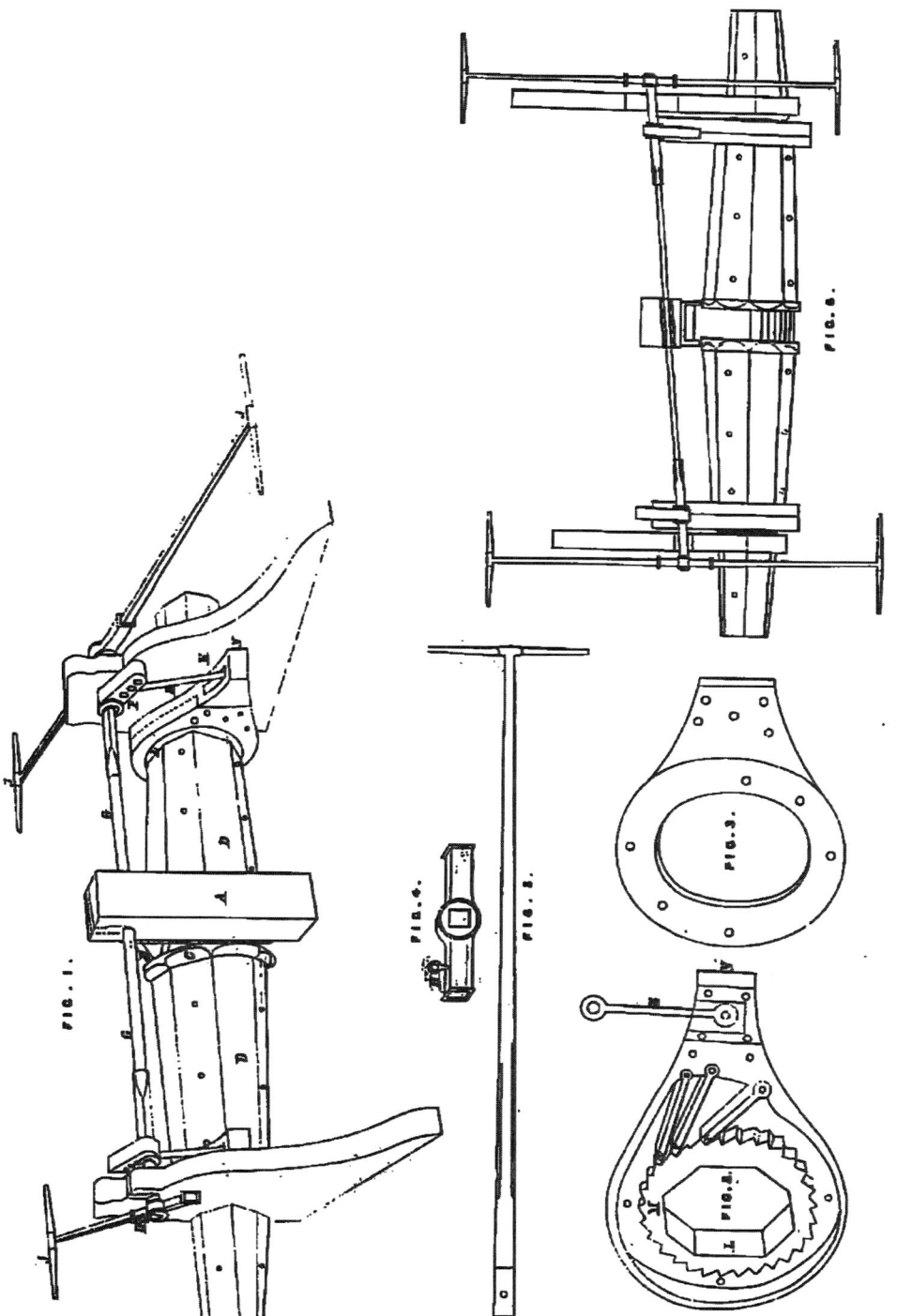

Figures 1–6 of the Tyzack patent (1832).

that allowed both to be engaged by a single transverse rocker and moved in a reciprocating fashion."

Neither Harland nor zu Mondfeld explains precisely how Taunton's pawl mechanism drives the windlass, and indeed Taunton's descriptions of the pawl mechanism leave something to be desired. Each ratchet wheel was sandwiched between two roughly teardrop-shaped plates. There was a pivot pin connecting the protruding ends of the plates. A bent lever was pivotably attached at one end to the bottom of the connecting rod, and at its elbow to the pivot pin of the plates. The bend was so the driving end was turned upward (Taunton 1841, Fig. 4).

In Taunton (1841), the other end of the bent lever acted as the driving pawl. Taunton writes, "It will be evident that by causing the lever to descend, the end of the driving lever, e, will enter below a tooth ... it follows that the raising of the lever will cause the barrel to revolve."

Unfortunately, there is an ambiguity here. Since the driving lever is pivoted at its center, if one end descends, the other rises. So it is unclear what Taunton means by the "lever" descending or raising, and thus whether Taunton intends that the barrel rotate on a downward stroke of the handle, or on an upward one. A downward stroke would permit more force to be applied. In that event, the rod end of the driving lever moves down, and the pawl end moves up. That would mean that the direction of rotation of the barrel is clockwise as viewed from the port (left) side of the windlass and pulled cable first passes underneath the barrel. But with hemp cable, "the rope came on and off the top" (Harland 1991, 154). If it were so wound on Taunton's windlass, then the power stroke would be the up stroke of the handle, which is weaker.

The pawl mechanism in Taunton (1845) is more complicated and puzzling in its own way. It had three major elements. First, as in 1841, there was a bent lever (R). The lever engaged a "recessed segment" (F). This was a piece with arms roughly in the shape of a "V" or "U" with one arm longer than the other. Each arm had a portion of its surface curved and toothed to engage the ratchet wheel, presumably like a meshing gear. The elbow of the "V" formed a recess, hence the name. A nearly triangular piece (V*), which also had a curved and toothed surface, fit loosely into the recess.

The basic idea was that the segments F and F* alternately engage the ratchet wheel and that this would somehow make the rotary motion more continuous. I believe the notion is that while the teeth of segment F are in the indentations of the ratchet wheel and thus ineffectual, those of F* will be forced against the teeth of the ratchet wheel, and vice versa, thus smoothing out the motion. Taunton teaches that F* is "somewhat smaller" than the opening in F, but I think that it more specifically must be smaller by half the distance between the teeth of the ratchet wheel. Also note that the only way that F* can engage the ratchet wheel is for the driving lever to force F against F*.

There are much simpler and more effective ways of achieving continuous rotation using a ratchet and pawl. Basically, these use a double pawl in which, alternately, one pawl pushes and the other pawl pulls, but in the same direction of rotation. One, lever-operated, is depicted as mechanism 999, "double-pawl ratchet," by Hiscox (1899, 254). When lever a is pushed down, pawl b advances the ratchet wheel.

5. "Rig the Capstan!"

And when lever a is pulled back up, pawl c advances the ratchet wheel. For videos, see the ratchet mechanisms 18–20 and 22–24 posted on mechanisms.club.

The *Comet*, a Pacific Coast lumber schooner, was built in 1886 and wrecked in 1911 (Russell 2004, 15, 17). The wreckage has been studied, and the starboard half of the windlass and the iron parts of the port half survived (31). The windlass, which was of the "pump-brake" type (56), was mounted in the bow section of the main deck (45).The iron purchase rims (ratchet wheels) were 23.5 inches diameter. Since no "pawl rim" was found, it is assumed that the retaining notches for the retaining pawl were cut into the barrel itself. Shafts and gearing have been found that archaeologists believe to have been part of a system "for driving the windlass by an endless

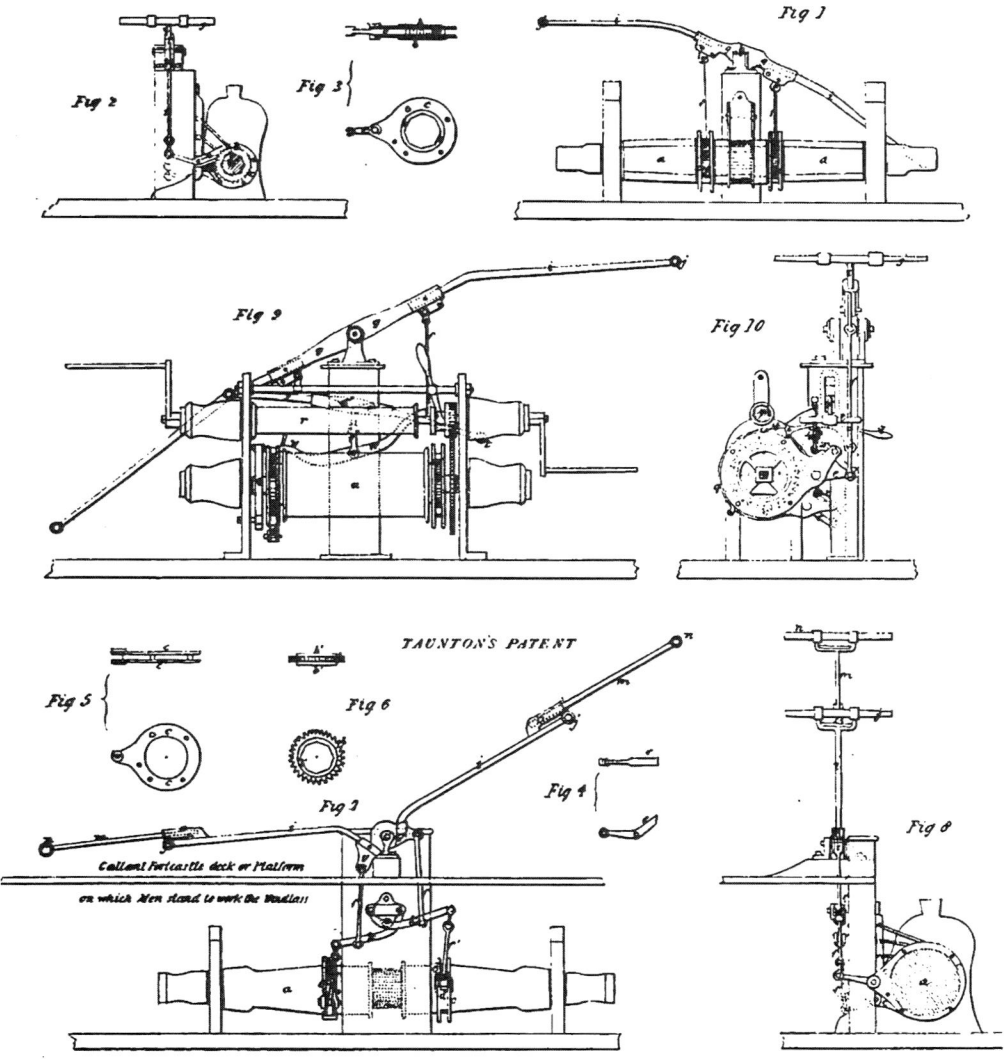

Figures from Taunton's 1841 patent.

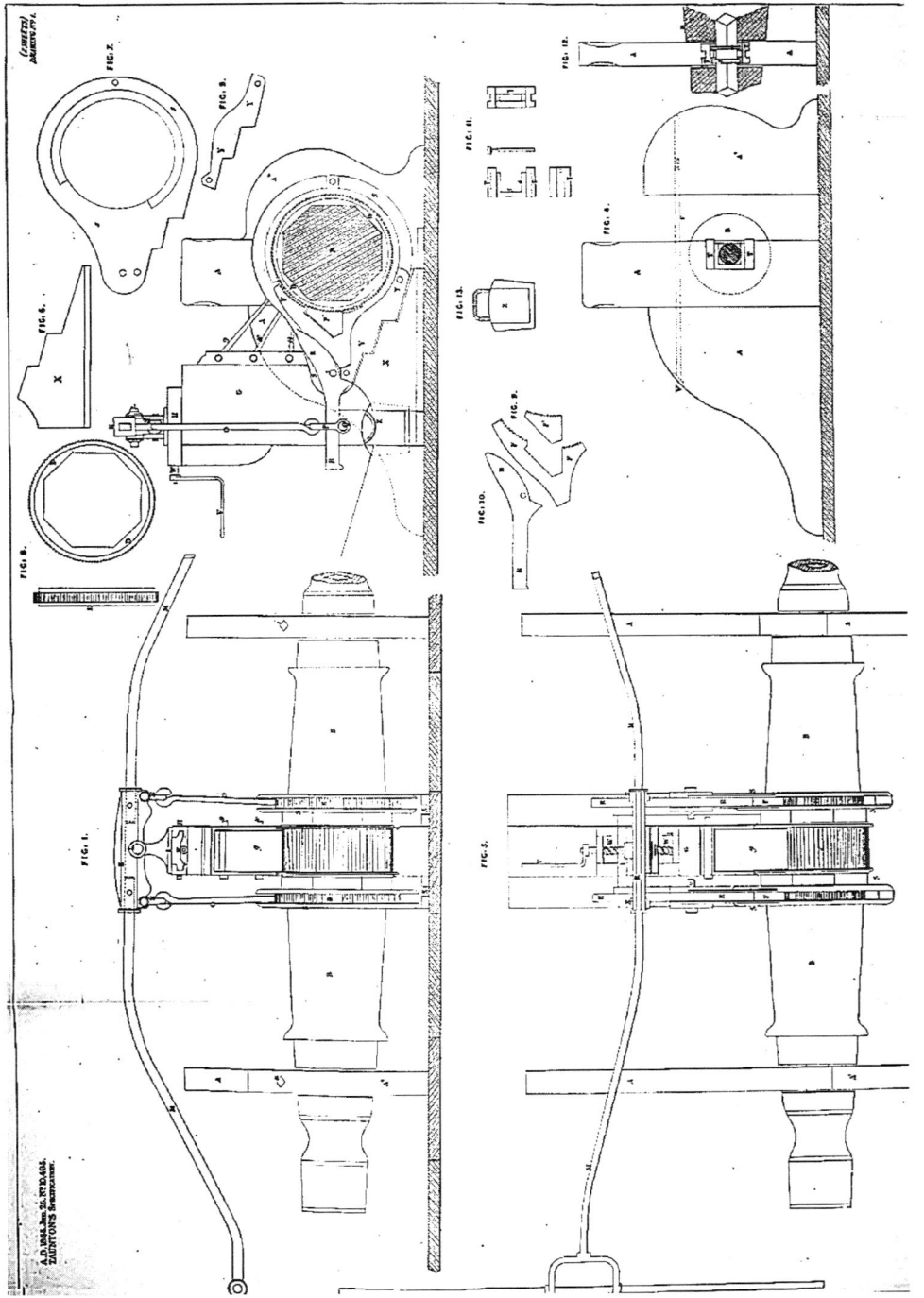

Figures from Taunton's 1845 patent.

Artist's conception of a split barrel pump-windlass. There are six men working the windlass and a seventh handling the chain; two loops are wrapped around the barrel and the tail of the chain is attached to a fixed hoop. The lever arms are bent up, rather than down as in the Taunton patents. Note the simple retaining pawl engaging the central ratchet wheel, and the tapering of the barrel. The linkage between the starboard lever and the barrel is visible (Hayes 1892, 18).

chain from a donkey [steam] engine," possibly added to the *Comet* in the 1890s (57).

The Russell study referenced and duplicated a series of six drawings of a "20 inch split barrel windlass" prepared in 1969 by the Smithsonian Institution, commenting that it "depicts details of a windlass very similar to Comet's" (Russell 2004, 58). The term "split barrel" refers to the placement of the ratchet wheels near the center of the barrel, as on the Taunton windlass.

Reel Winches

In general, reel-winches have horizontal barrels, like a windlass. The reason is that you don't want the turns of the rope to ride over each other in such a way that a knot is formed, jamming the winch. With a vertical barrel, the turns would slide down and tangle. Even with a horizontal barrel, care must be taken that the rope approaches the barrel close to perpendicularly to the barrel axis (Harland 2003, 3–5).

One of the nineteenth-century innovations was the use of a reel winch to manipulate the braces, the lines that controlled the angle the yard made with the centerline of the ship. If the ship was not sailing directly downwind, the sailors had to haul in the brace on the lee side and pay out the one on the weather side.

The braces were simple tackles, which did not provide any mechanical advantage (Villiers 1953, 124). Before the introduction of the brace winch, a whole watch might be needed to man the braces (Findlay 2005, 120). It was dangerous work—men went "over the side or they were spilled about the deck." And when the men lost control of the brace ends, the swinging yards "could damage the rigging, dismast the ship, or at least rip the sails" (Villiers 1953, 124–5).

The brace winch not only provided a mechanical advantage, it was able to manipulate both braces simultaneously.

Harland (2003, 16–17) points out that a brace winch designer was confronted by the "incompatibility between the arciform movement of the yardarm and the effectively linear movement of chain on and off his winch." When the yardarm is "braced sharp"—at an acute angle to the centerline—"the lee brace shortens more than the weather brace lengthens."

While Cunningham received an 1867 patent for a brace winch, the design that became dominant was that of John Jarvis, patented in 1890. The Jarvis winch featured two conical drums, one for each brace, with the narrow ends facing each other. The lines are fastened at the narrow ends of the drums, and when the yards are "square" (at right angles to the centerline), each drum is half-filled with line. A turn of the drums, which are on the

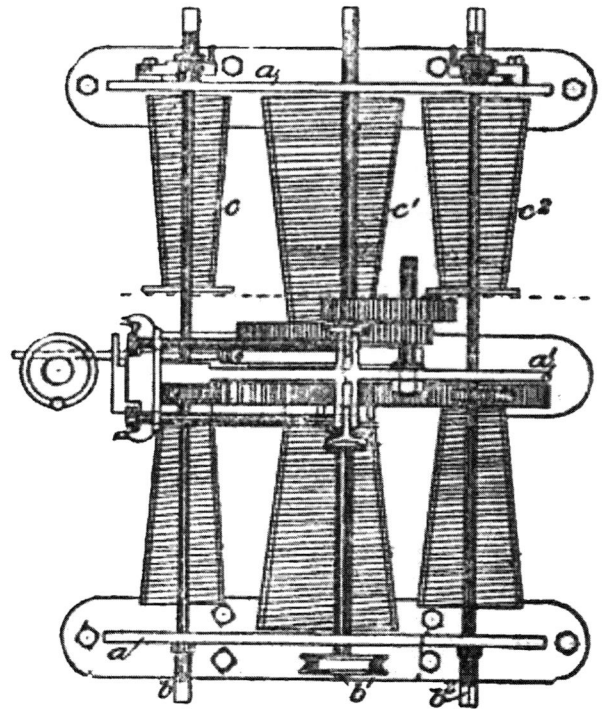

Jarvis patent winch for bracing ships' yards (Jarvis 1890, Fig. 1). The frames a, a1, a2 are bolted to the deck. The shafts b, b1, b2 each carry a pair of grooved conical barrels c, c1, c2. "The port brace of a yard is coiled round one and the starboard brace on the other of a pair." The barrels may be rotated simultaneously or independently, and by power or a hand crank. There is gearing for speed changes and pawls or brakes to hold the barrels.

same axle, will take up more of the line hauled in and give up less of the one played out.

It was typically necessary to customize the drums for a particular yard on a ship, and Jarvis suggested making "cast iron frames preferable [sic] with six, twelve or twenty four faces and piecing up with hard wood blocks which can be turned and bevilled [sic] to the required dimensions" (Jarvis 1890). Harland (2003, 19) describes a customizable drum featuring "seven or nine cast iron strips, which slide in slots on the periphery of two beveled disks." (If those slots were radial, then the two diameters defining the truncated conical shape would be adjustable.)

The Jarvis brace winch "allowed the yards of the largest vessels to be braced ... by only a few men, and, unlike any previously adopted technology, with almost no loss of life" (Ledyard 2012, 177). Indeed, on Jarvis's *Duntrune*, "with the brace-winches, one small apprentice could swing all the main yards in good weather" (Villiers 1953, 193). More typically, "what a dozen men were required for on the old-fashioned long-braces, two or three could do with the winch" (125).

While in the nineteenth century it appears to have been used primarily on large German schooners (Villiers 1953, 127), Harland (2003) asserted that "winches of Jarvis' pattern are still being built."

The Capstan

There are references to shipboard capstans in fourteenth-century literature (Tinniswood 1945, 98–9). The principal use of the capstan was in hoisting weights, especially those too great to be handled by the windlass. However, there were some oddball applications. Edward Shorter developed a propeller in 1800 and in 1802 it was mounted on the *Doncaster*. It was human-powered, by eight men, using "a capstan, and it propelled the ship at the rate of a knot and a half an hour" (Allan 2022; Fincham 1851, 341).

Mainwayring (writing in 1623) indicates that at least some English warships had two capstans. The main capstan was abaft (behind) the main mast, and the less powerful jeer capstan was between the main and the foremast. The main capstan was used to weigh anchor, to hoist or lower topmasts, or to hoist ordinance. The jeer capstan heaved on the jeers (the tackles used to hoist the main and fore yards) and could also be used to assist the main capstan.

The most primitive form of capstan is simply a log that has been pierced for insertion of lever bars. John McPhee (McPhee and Adney 1975, 79) has written about log drivers in Maine using a capstan "made from a segment of log" and set up on a raft to haul in a 300-pound kedge anchor.

However, I would expect that by the time that ships carried anchors heavy enough to require a capstan to manipulate them, more sophisticated designs were the norm.

The basic shipboard capstan was a vertical barrel with horizontal capstan bars. The sailors would walk in a circle, pushing on these bars. This has, as previously noted, the advantage of continuous effort and the disadvantage that pushing is less forceful than pulling down from above shoulder level. The bars were not permanent fixtures, but rather were pinned into place. That had the advantage that they wouldn't encumber movement on deck when the capstan was not in use.

The simplest form was the "crab" capstan, typically with two bars inserted so that they protruded from both sides, thus creating four levers. Necessarily, the two bars were at slightly different heights. Harland (1985, 260) describes an early seventeenth-century capstan as having three such "two-sided" bars, and thus six levers. The bar were typically "at breast height, say four and a half feet."

The crab capstan was still used by garrisons in the early twentieth century. Henderson (1907, 324) comments that "the ordinary permissible working stress on a crab capstan is from two to three tons, and this can be obtained with from four to eight men on the capstan bars" (four levers shown).

It was replaced, at least for larger ships, by the drumhead capstan invented in the 1670s, possibly by Samuel Morland. The bars, rather than being inserted into the capstan barrel itself, were embedded in a large wooden drumhead, on top of the barrel and of larger radius. "This enabled separate bars to be installed on each side," up to 12 such levers to a drumhead, all at the same height. Eight levers was the most common. The transition from crab to drumhead was facilitated by the ease with which the former could be retrofitted. For example, in 1676 the heads were cutoff both the main and jeer capstans of the fourth-rate *Diamond* and replaced by drumheads (Davies 2008, 73–4; Endsor 2020, 50–1; Hedderwick 1830, 311). Citing Lavery, Harland (2003, 40) says that in English warships, the drumhead was "relatively rare" in 1720 but "almost universally fitted in 1740."

Nineteenth-century illustrations show that the sailors push the bars with their chest, while holding the bar from above or below, rather than pushing merely with the hands (Harland 2003, 51–52). That might allow greater force to be applied to the bars, and also reduce the risk of fracturing hand or arm if the capstan slips back. But what if one broke a rib instead?

By the late seventeenth century, rather than having the heel of the spindle pivot directly on the step, it featured a narrower pin that fit into an iron saucer (Harland 2003, 47).

What is the hauling force on the capstan? Nominally, it is the number of sailors manning the capstan, times the force each exerts on his capstan bar, times the mechanical advantage provided by the capstan. But the actual force is harder to calculate.

With regard to the individual force exerted, it might take hours to haul in a long cable, and the force exerted may decline over the course of that period, due to fatigue, unless there are rest breaks or replacements.

The mechanical advantage obtained is not the same for all. With, say, 10 men on a bar, one cannot assume that the mechanical advantage is the radius of the bar divided by the radius of the whelp. Only the outermost man will enjoy that advantage; the advantage will be progressively less for the men closer to the barrel. Thus, the "effort arm" is their average distance from the capstan axis (Mackenzie 1920, 28).

Finally, there is the question of frictional losses. While the men aren't lifting the capstan, the frictional force it experiences is proportional to its weight.

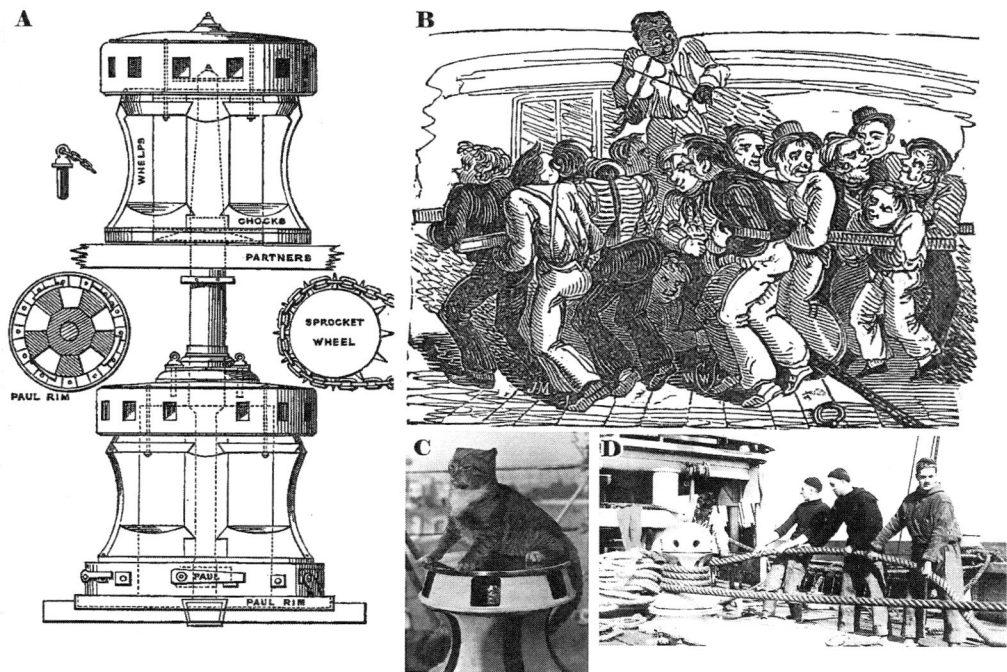

A. Profile view of a double capstan (Boyd 1857, 45). The upper and lower capstan could be used independently or locked together by drop bolts, one of which is depicted to the left of the whelps of the upper capstan. On the base of the lower capstan one can see three "pauls" (pawls); the rightmost one is in the locking position. These engage holes in the "paul rim" just below; a "paul rim" is shown in plan view to the left of the spindle connecting the two capstans. If the capstan were intended to be used with chain, it would have been fitted with a sprocket wheel (shown in plan view to the right of the spindle.) B. Artist's (J. Meadows) depiction of "Heaving at the capstan" ("Steerage Passenger" 1832, 222). Note the number of sailors per bar, and how they are pushing on the bars. Also note the fiddler standing on top of the capstan, playing a tune so they can time their movements to it, and the sailor crouched down, most likely trying to prevent the cable from riding too high and tangling.

On a double capstan, the capstans on two different gun decks shared a common spindle, increasing the number of men that could be devoted to a single hoisting effort. On the HMS *Victory*, the capstan on the middle gun deck had 14 bars, and the coaxial lower one 12. The head of the lower capstan was sometimes called a "trundlehead," and it was commonplace for it to have fewer holes than the upper capstan drumhead (Harlan 2003, 41). Assuming 10 men on each bar, that meant a maximum of 260 men could be working its double capstan simultaneously. Harland quotes an estimate that they could have lifted 10 tons (1985, 261–2).

While the double capstan permitted a cable on one barrel to be turned by sailors working both drumheads, some double capstans had removable pins so that the barrels could be turned (and used) independently (Harland 2003, 43).

Double jeer capstans were in use by the early seventeenth century and double main capstans by the 1730s (France) and 1740s (England) (Harland 2003, 44). Triple capstans were rare (46).

A single jeer capstan would most likely have its barrel on the main (upper) deck and its spindle stepped on the gundeck below. A double jeer capstan might have the upper barrel on the forecastle and the lower one on the main deck, or the upper barrel on the main deck and the lower one on the (upper) gundeck. A single main capstan would probably have its barrel on the lowest gun deck and its spindle stepped on the orlop deck. In a three-decker, a double main capstan would probably have its barrels on the middle and upper gun decks. In a frigate, the upper barrel might be on the quarterdeck (Harland 2003, 44–6).

The diameter of the capstan barrel determined the maximum thickness of cable that could be handled. The thicker the cable, the greater is its minimum bending radius. HMS *Victory*'s main cable was 24 inches in circumference, whereas its main capstan was two feet in diameter. Hence, the main cable was connected by "nippers" to a thinner "messenger" of 15 inches circumference, and the messenger was secured (by friction) to the cable (Harland 1991, 154).

The capstan barrel itself was polygonal in plan, with a slight upward taper, and it also was reinforced at the base by upwardly tapered "whelps." The whelps typically had a "step" (an outward jog), perhaps two feet high, "to prevent the messenger from riding too high" (Harland 1991, 198, 260). It couldn't be higher because the men were stepping over the already-hauled-in cable.

"The whelps increased the [effective] circumference of the barrel without a proportionate increase in weight. The effective diameter was about five times that of the cable handled" (Harland 1991, 198, 260). This would have allowed the capstan to handle a thicker (stiffer) cable, with a larger bending radius, than would otherwise be the case, but a price would be paid in that it would also reduce the mechanical advantage of the capstan.

Hedderwick (1830) provides suggested proportions for the capstan. The diameter of the barrel is to be 1/28th of the extreme breadth of the hull, and the diameter of the drumhead is 1/10th of the same breadth. The whelps are one-half or five-ninths the diameter of the barrel. The diameter of the iron spindle was 1/144th the sum of the extreme breadth and the depth of hold (159). Hedderwick also describes the construction of the capstan (312ff), as does Steel (1810, 388ff).

Hedderwick's proposed proportions for the windlass and capstan barrels are 1/18th and 1/28th, respectively, of the extreme breadth of the hull, which would make the windlass barrel thicker for a given vessel. But Harland (2013, 55) says that the diameter of the capstan was up to four feet, and that of the windlass, only up to two feet. The two are reconciled by the observations that windlasses only appeared on small merchantmen, which were of relatively small breadth, thus placing an upper limit on the size of a windlass.

For example, the Ronson Ship, an eighteenth-century merchantman known to have had a windlass, had a maximum breadth of 27 feet (Riess 1987, 34). The windlass dimensions aren't known to me, but by Hedderwick's guideline, its barrel diameter would have been 18 inches.

That said, this author wonders whether any shipboard capstan, at least prior to the twentieth century, was four feet in diameter. On the *Sovereign of the Seas* (1637), the main capstan was 29 inches diameter and the jeer, 22 (McKay 2020, 52). Steel

(1805, 310) said that the capstan of a 74-gun warship should have a barrel diameter of 28.25 inches, whereas the windlass on a 330-ton merchantman would have a central barrel diameter of 22 inches. In 1921, an informercial in the *Marine Journal* reported that the SS *America* had the largest capstan ever built in the port of New York, with a 29-inch diameter barrel, weighing over two and a half tons (*Marine Journal* 1921, 2).

In 1867, Peake (272) set forth the number and length of capstan bars for various sizes of ships. They ranged from eight bars, 8.5 feet long, for a ship of 10–14 guns, to 14 bars, 14 feet long, for a 120-gun first-rate.

Pawls

Until the eighteenth century, the capstan did not have a self-acting pawl (to prevent backslip). No doubt this was because of the greater ease of maintaining continuous motion. When the effort had to be halted to adjust the position of the cable on the capstan, a "sliding" (and horizontally pivoting) pawl, near the base of the capstan, was kicked into place, blocking a whelp (Harland 2015, 40).

Hence, at the capstan, when the pawl was not in place, if there were a failure of effort or a sudden strain on the cable, the sailors working it "may be throwne from the cap-staine, and their braines beaten out against the ship sides, if they waigh in a sea-gate" (Manwayring 1644, "Windlasse"). The capstan pawls were iron ("Capstaine").

In the late eighteenth century, hanging pawls that "depended from the deck beams" could arrest the lower capstan of the double capstan (Harland 2015, 40). There were two of these, one for each direction of rotation, and presumably only one was in the "down" position when the capstan was in use (40).

Harland credits Antoine (Anthony) George Eckhardt (1740–1810) with the invention in 1772 of a "hinged reversible drop pawl." It was "secured to the lower rim of the capstan barrel and engaged a circular ratchet plate fastened to the deck." Eckhardt was a prolific inventor (Davids 2022, 191), elected a Fellow of the Royal Society in 1774.

Harland indicates that by about 1800, Eckhardt's device was universally adopted. However, there were still capstan mishaps. Thus, while the HMS *Agincourt* (1865), a broadside ironclad screw frigate, was being unmoored one morning in 1869, "the capstan overpowered the men at the bars; and three of the men were severely hurt on their heads and arms. One has been sent to the hospital at Haulbowline with his arm broken and a severe gash in his head" (Davis 2024).

"Power" (Geared) Capstans

It was possible to increase the force exerted by the capstan, at the cost of moving around it a greater distance (and thus of speed), without actually increasing the size of the capstan. In Eckhardt's 1771 capstan, this was accomplished using a gear train in which there are a central "sun" gear, three intermediate "planet" gears, and an outer ring gear, the last with inward facing teeth (Harland 1991, 156–7). The planet gears mesh with both the sun gear and the ring gear. Depending on the relative sizes of the gear, and which is held stationary and which move, you get different ratios of turning rate and consequently different mechanical advantages.

A model of the Eckhardt geared capstan, which was patented in the Netherlands in 1770, is in the Rijksmuseum. "The capstan has a double head, each for twelve bars, and two pawls in the heel to prevent it from turning back." The barrel only turns clockwise, and the setting of internal pawls determines whether it is controlled by the upper head and associated gearing, or by the lower head. With one setting, the bars are placed in the lower head, and the heads and barrel turn clockwise as a unit. With the other setting, the bars are placed in the upper head, and that head is rotated counterclockwise, rotating the sun gear the same way. This causes the planet gears to rotate clockwise, and the barrel does so, too, but one turn for every three of the upper head. Thus, it provides increased "power" at the cost of less "velocity." The gear train is inside the capstan, above the deck, but the spindle extends to the lower deck, resting in a "step" (Rijksmuseum 1771; Harland 2013, 99).

The Eckhardt capstan was placed on HMS *Defiance* (1772), a 74-gun two-decker. A plan describes the *Defiance* as equipped with a "triple-purchase capstan" (RMG 1772), which implies that it offered a 3:1 mechanical advantage.

One problem with the Eckhardt design is that it could only turn in one direction. Harland notes that "the turns of the messenger are taken with the sun [clockwise] in the case of a starboard cable and against the sun [counter-clockwise] with the port cable" (2013, 99). That, and the cumbersomeness of the mode change, limited its utility.

The Phillips capstan likewise has sun–planet–ring gearing. There were two modes of operation. In one, a vertical bolt locked the drumhead to the barrel (called a "single purchase"), so both turned the same amount. In the other, the bolt was removed. The drumhead then just acted upon the spindle, which turned "wheel-work" (gearing) such that "the spindle makes 3 turns and the barrel makes 1, and they revolve in opposite directions" (Knight 1881, 1: 456). This increased the force threefold, but reduced the speed by the same factor.

Charles Phillips obtained two geared capstan patents (1819, 1827). His 1819 invention may have played a role in his promotion to captain in 1821, and the certificate of his election as a Fellow of the Royal Society in 1829 described him as "a gentleman devoted to science and inventor of the Patent double capstan and many other highly important mechanical contrivances." His election was unsuccessfully opposed by John Knowles, on the grounds of plagiarism of Eckhardt's invention (Soane 1829). A committee investigating the matter apparently concluded that Phillips's invention was similar to but distinct from Eckhardt's (Soane 1829a).

The drawings from Phillips's 1827 patent are duplicated, albeit fuzzily and at reduced size, in Harland (2013, 101), and it appears that he did not need a second drumhead. Harland says that the Phillips design also lacked the internal gear-controlling pawls and that it used vertical bolts rather than horizontal ones.

A contemporary expressed skepticism as to the market for a geared (power-multiplying) capstan, since in warships, the number of men who could be put on the capstan bars was "sufficient for any force that its spindle could bear," whereas merchant vessels used windlasses rather than capstans. Nonetheless, according to Harland (1991, 156), "in 1832 no fewer than 230 English ships were fitted with the Phillips capstan." It is possible that his success was attributable not just to his design changes, but also to improvements in gear design and gear-cutting machinery (Harland 2013, 101).

James Brown, a ship-rigger, devised a capstan (patented 1833) that was turned by means of a "winch" (crank) acting through a right-angle gear drive; "two men at the winches will do the same work as many men with capstan bars" (Ure 1843, 253 and Fig. 257).

The nineteenth century witnessed two technological developments that affected capstan design: chain cable and steam power.

Chain-Handling Capstans

When chain cable was first introduced, the chain cable was secured to a hemp messenger. When the hemp messenger was replaced with chain, the wood whelps of the capstan were damaged. The solution to that problem was to armor the whelps. But there was a second problem: The whelps were shaped to permit a hemp messenger to surge and the grip of the chain messenger on the capstan was less than ideal (Harland 2013, 76).

In 1835, Benoit Barbotin proposed "to fit the capstan with a shaped sprocket capable of grasping the links of chain" (Harland 2013, 77). While Barbotin popularized the sprocket capstan, he was not the first to invent one, with that honor going to Gordon and Company in 1828 (Harland 2013a, 339).

The next step was to modify the capstan to have holes or indentations to receive the links of the chain. The casting was called a wildcat in the United States and a gipsy in Britain (Harland 2013a). A popular capstan of this type, designed by Thomas Brown, "at one time was in use with slight modifications on nearly all the vessels of the U.S. Navy" (Whitney 1896, 117). There was something of a design disagreement with respect to both sprocket and shaped holder capstans as to whether to aim for "one-size-fits-all" or to match a specific link size (Harland 2013a, 340).

Combination Geared Capstan and Windlass

This was in some sense a hybrid of the windlass and the capstan. It was a windlass in the sense that the barrel on which the cable was wound had a horizontal axis. However, it was driven by pressing on horizontal levers attached to a capstan-like structure mounted above the windlass. The vertical shaft of this structure features a spur bevel gear, that is, a gear with an inverted conical profile. That gear meshes with another spur bevel gear, mounted on the end of the windlass barrel. Thus, a vertical axis rotation was converted into one with a horizontal axis.

What advantage did this apparatus possess over a more conventional windlass or geared capstan?" One possibility is that it provides for continuous effort, which a windlass does not. Another possibility is safety. The windlass barrel could be secured by automatically operating windlass-style dropping pawls. A third is that a cable coiling on a horizontal barrel is less likely to have one turn ride over another and perhaps cause a knot.

An early example of the combination apparatus was Hawke's Patent Capstan.

An 1836 drawing (Hebert 1836, 315) shows the windlass positioned alongside the capstan, which would prevent the capstan gang from making an uninterrupted circle around it unless the capstan bars cleared the top of the windlass and extended beyond it horizontally, and even then this would limit how many men could be placed on each bar.

A more practical example was that of James (Joseph?) Andrews, described by *Scientific American* (1849). This positioned the capstan above a two-barrel windlass (*Scientific American* 1849). It received a silver medal at the 1850 Massachusetts Charitable Mechanic Association exhibition, and the judges commented that "either half or end of it may be used without interfering with the remaining part; and all may be used in connexion with or independent of the capstan; or the capstan by itself" (Hooper 1850, 114).

James Emerson obtained several patents on windlasses of this type in 1855–56. All three taught the use of the windlass with chain. The last of these taught that the windlass barrel bore a grooved wheel, and the groove was "of V-form and is provided with teeth or projections ... between which the links of the cable catch." However, persuading owners and captains to install his apparatus on their ship proved difficult. The stumbling block, apparently, was that it used cast iron gears. One said, "you might as well have a glass windlass." Another called the apparatus a "d___d cast iron jimcrack." Another concern was that "the links of cables varied in length so much that it would be impossible to handle them in the way proposed."

Nonetheless, Emerson eventually secured a sea trial and by 1885 was able to assert that his apparatus was installed on 6,000 vessels (Emerson 1894, 138–145).

The three-masted schooner *Lucerne* (1873), which foundered in 1886, was one of the ships equipped with Emerson's windlass (or a knockoff) The capstan was on the top of the forecastle, and the linked windlass on the forecastle deck. The capstan shaft had two bevel pinion gears, one engaging a large gear on the starboard side of the windlass, and the other a smaller gear on the port side, for power and speed, respectively. The gears could be disengaged from the windlass barrel (Cooper et al. 1991, 47). It is also known that on some ships, such as the *Formby* (1863), there were means for driving the Emerson apparatus by auxiliary steam (49).

Steam Capstans and Windlasses

Steam capstans. The most obvious advantage of the steam capstan is that it requires few men to operate. That is not its only advantage. An 1888 report quoted Captain Lefevre, of the Ocean Steamship Company (Savannah, Georgia): "It takes up but little room, and can be worked effectually where men could not exert their force with bars or cranks on decks covered with ice or snow" (Trowbridge 1888, 40).

John Schaffer's '894 patent (1857) claims "The drum (C) on the shaft of the capstan (B) as arranged, the capstan being steam driven by geared shafting." His '935 patent (1857) clarifies that the source of the steam is an auxiliary steam engine. However, in the '935 patent, the capstan may be either steam-driven by torque applied to the capstan shaft, or manually driven by handspikes inserted in sockets in the shaft-head.

I have found an 1860 reference to the presence of a steam capstan on board the steamer *Peytona* (Harper's 1860, 156). but I doubt that was the first one to be installed.

Skipping over Schaffer's later patents, the next patent of interest was that of John S. McMillin (1866). This shows more clearly the transmission of power from the steam engine to the capstan. It also notes that it was conventional in the art to use the auxiliary engine in "handling freight." Not only does it allow for both steam-powered and manual operation, but it allows the gearing to be adjusted depending on which is more desirable, power or speed.

The 1860s and 1870s appear to have been a period of transition to the steam capstan. In 1862, *The Artizan* reported (230) that the HMS *Black Prince* (1861) was equipped with a steam capstan (which experienced a problem during its sea trial on August 30, 1862) , and Cunningham (1862, 106) says that the HMS *Warrior*, at least of his time of writing (1862), also had one. An 1864 report by Union Rear Admiral Porter stated that the gunboat *Fort Hindman* succeeded, with her steam capstan, in moving the bow of the *Eastport*, which had run aground, but without freeing her (Moore 1862, 137).

It is relatively unusual for the nautical literature to identify which make and model of steam capstan was on a particular ship, but there are exceptions. We know that in 1882–3, the Navy contracted for Providence steam capstans for the *Omaha* and *Kearsarge*, and J.P. Manton models for the *Nipsic*, *Shenandoah*, *Ossipee*, and *Marion*. The Providence capstan, taking 2-inch chain, weighed 13,823 pounds, occupied 40 cubic feet, and cost $6,000. It was driven by "three trains of spur-gear." The equivalent Manton version weighed just 10,000 pounds, occupied only 18 cubic feet, and cost a mere $3,200. It used "the worm and wheel" (Navy 1883, 283–4).

Steam windlasses. Beginning in the nineteenth century, some windlasses were steam-powered. The American Ship Windlass Company "began to introduce steam windlasses" in 1863. "The steamship *City of Tokio* (1874), carrying 7000 pound anchors run on 2 5/16-inch chain, could, with a man and a boy operating a Providence steam windlass, haul in sixty fathoms of chain in six minutes, whereas the same task, with a manual windlass, would have forty men an hour" (Fitch 1881, 40). The first American navy vessel with a steam windlass was the screw frigate USS *Trenton* (1877) (Whitney 1896, 121).

Electric and Hydraulic Power

Electrically powered winches, capstans, and windlasses became available, at least on land, in the 1880s or 1890s. As dynamos and generators were installed on ships for other purposes, such as lighting, it was natural to increase their power enough to support hoisting devices, too.

Derricks

According to the *Oxford English Dictionary*, "derrick" was "the surname of a noted hangman at Tyburn c1600." It quickly became a term used to refer to hangmen in general and, by extension, the gallows. In the eighteenth century, it took on the meaning of a machine for hoisting or moving heavy weights (other than condemned prisoners). The first reported usage in this new sense was in connection with a tackle on the mizzen yard (1738).

Steel (1806, 6) says that it may also mean a raisable boom, with a head hanging a tackle over the hatchway and supported by guys, and a heel in a socket fastened to the deck. Burney (1815) describes it as a "temporary crane" and indicates that the guys ran to a ship's mast, and the head carried a luff tackle. A winch may be used instead (121).

Usage was much the same a century later. Tait (1907, 272) said that the heel of the derrick should rest on a shoe large enough to cover two or three of the deck beams (planks?), and "the heel of the derrick is rounded to fit in the hollow of the shoe." The deck below is shored up, and "preventive stays" are attached to the mast to bear the extra strain. The derrick is equipped with two "treble purchases" (tackles), one as a "topping stay" ("topping lift") running from the head of the derrick to the masthead, to control the derrick's elevation (luffing), and the other for lifting the weight (hoisting). The derrick must "swing freely," but the horizontal swing (slewing) is controlled by "guy purchases."

Baldwin, describing the stern wheel steamer *Queen City*, notes that it has derricks on either side of the mast, "seated in hemispherical deck sockets" (1897, 19). Luce and Ward (1884, 112) say that the "heel of the derrick pivots in a shoe, which travels on a fore and aft guide track and is fitted with stout fore and aft tackles."

The Tait Patent Derrick apparently had the heel travel on an athwartships guide

Emerson ship's windlass. From Emerson (1894, 140).

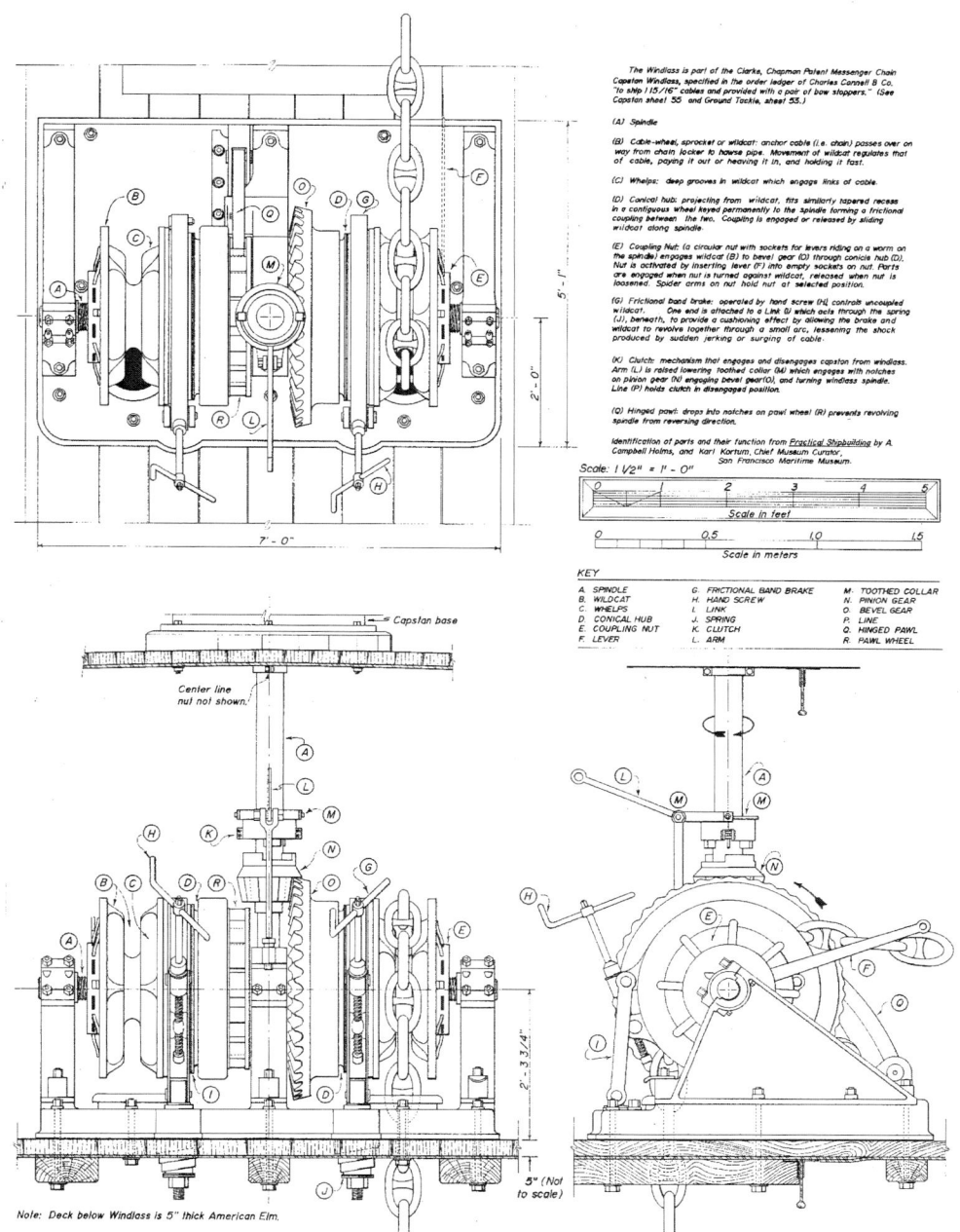

An engineering drawing of a Clark, Chapman Patent Windlass, from the three-master *Balclutha* (1886), now a National Historic Landmark preserved in the San Francisco Maritime Museum National Historic Park (Library of Congress Prints and Photographs Division. LOC 1988); drawing is from the Historic American Engineering Record, National Park Service. Note that the windlass was mounted on the deck below the capstan. Turning the capstan would, through gearing, turn the windlass. An engineering drawing of the *Balclutha*'s capstan is available online but not included in this book's figure (Library of Congress 1989).

bar, by the action of a screw. Shifting the heel to port would cause the head to swing to starboard, "the weight of the load supplying the only power required." It was claimed that with a winch that could handle a one-ton load, "the derrick would land easily 100 tons per hour" (*Marine Engineer* 1893, 285–6).

On steamships, the masts were no longer needed to carry sail, but masts might still be provided to support the derrick. "The number of booms is often two per mast; more generally in important vessels, four, or even six, and in extreme cases, ten" (Cunningham 1924, 23–24).*

The winches were likely to be steam powered, and wire ropes used instead of fiber rope. In 1899, the cargo steamer *Monarch*, carrying 18,000 measurement tons, had nine hatchways, 12 winches, and 18 derricks (De Rusett 1899, 456).

Wooden derricks, "which generally lifted only up to three tons," were replaced with steel structures. There were also changes in the method of pivoting the boom. There was a "gooseneck" (a notch) near the base of the mast, and the heel of the boom had a flange by which it could be bolted to the gooseneck (House 2018, 124).

Derricks were not permanently erected. When at sea, the booms would be stowed in "crutches"; the rigging might or might not also be stowed away (House 2018, 125).

Tait suggests that a steamer needs one derrick, and a sailing ship two: "One purchase plumbs the hatch, and the other is positioned over the side." Such use of two derricks (or cranes) in combination, with their "falls" attached ["married"] to the same "cargo hook," is called a "union purchase" or "burtoning," and the booms are "fixed in the overboard and inboard positions" (Buxton et al. 2012, 34; Eyres 2001, 270–2).†

Ideally, the cargo is held by a Seattle-type triple swivel hook. The three swivels form a "Y"; each cargo runner is attached to one of the flanking "double-eye" swivels; these are attached to the center "jaw-and-eye" swivel, which holds the hook. It is "used so that no wire is forced to have foul turns in it" (House 2018, 127; Army 1979a, 2-10). An unconfirmed source states that the "Seattle cargo hook" was invented by Lewis Thertrup (*Pacific Builder & Engineer* 1918).

Once the outboard derrick lifts the cargo above deck level, the cargo is moved inboard by winching in the inboard derrick's cargo runner and letting out the outboard one's (House 2018, 127). "The angle between the married cargo runners should not normally exceed 90°, and an angle of 120° should never be exceeded" (130). "A pair of derricks, each having safe working loads of 5 tonnes when rigged singly, would be able to lift no more than 1½ tonnes in union purchase because of the forces which this arrangement sets up in the standing guys" (Buxton et al. 2012).

The alternative to union purchase is a swinging derrick. Some modern ones can lift 20–25 tons and may have "two widely spaced, powered topping lifts" (Buxton et al. 2012, 35).

---

* This is perhaps a good place to note that strictly speaking, a mast is a vertical spar on the centerline of the ship; one may instead place them in pairs on either side, in which case they are called Sampson posts. Those, too, can be used for support.

† It is also called "married falls" or "yard and stay."

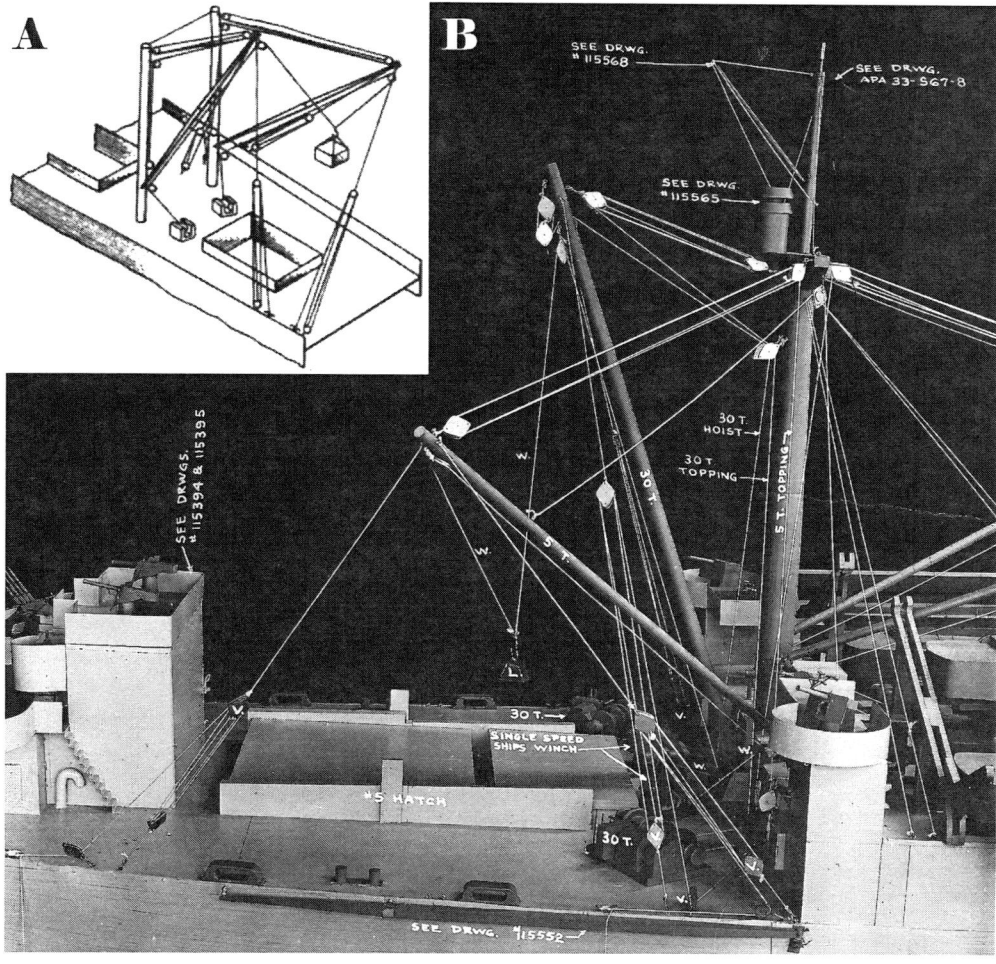

A. The burtoning system with a conventional vang (guide rope), "commonly used for loading and discharging medium weight cargo of 1 to 1½ tons between ship and shore" (Bureau of Naval Personnel 1962, 193, 195). See also the typical Union Purchase rig depicted in Fig. 2.4.1 of IMO Rules. B. Burtoning cargo from #5 hatch to starboard on a scale model of a Bayfield Class APA (attack transport) conversion, February 20, 1943 (Official U.S. Navy photograph).

In 1923, it was reported that at Southampton and Antwerp, an "average gang of 14 men," working "8 hours," could achieve the rates of loading and unloading (Cunningham 1924, 149–150) shown in this table.

Table 5–1: Tons of General Cargo Loaded or Discharged Per Hatch

	Southampton		Antwerp	
	Loading	Discharging	Loading	Discharging
Ship's winches only	140	163	56	70
Shore cranes only	163	186	70	112
Combination	230	291	77	122

Heavy Lift Derricks

Why would you need a heavy lift derrick? Well, suppose you were transporting a railway locomotive, or a giant electrical transformer, or tubing for a tunnel.

House (2018, 131) characterizes the Hallen Universal Derrick as "one of the most successful advances in lifting gear over the last fifty years." In its most typical form, it has a Y-shaped mast, with a crossbar near the top of the mat, forming outriggers. It can be "operated by one man," and "the lifting capacity may be up to 200 tonnes through a working radius of 170°, being topped up to 85°" (131). The "topping lifts" are on the arms of the Y and thus can slew the boom (when used in opposition), as well as lift/lower it (when used in unison). Only a single boom is depicted in Hallen's 1965 U.S. patent, but there are two, moved in parallel, in his 1969 one, which was intended for handling containers.

The Velle Derrick has a T-shaped mast, with the derrick heeled in a gooseneck near the base of the T, and the distal end held by topping lifts running from the arms of the T. "Each of the topping lift winches has a split or divided barrel on to which the ends of falls are secured. On the luffing winch the falls are laid on to the split barrels in the same direction. Thus, both falls will hoist or lower the derrick simultaneously. On the slewing winch the falls are laid on to the split barrels in opposite directions. Thus, when the barrels rotate, one fall pays out while the other heaves in and the derrick slews to port or starboard" (Lavery 2013). This "derrick can engage in luffing and slewing operations at the same time while under full load" (House 2018, 132).

In 1959, Köhnenkamp received a patent on what came to be known as a Stülcken heavy lift derrick. This had two free-standing "hollow uprights, sloping obliquely outwards." These were "arranged athwartships ..., between two cargo hatches." Between the uprights, "a hollow boom is articulately mounted in a derrick socket." This boom can be swung to overhang either hatch.

The distal end of the boom is forked. The heads of both uprights and both "tines" of the fork carry guide pulleys and are capable of 360° rotation. A pair of tackles, one from each upright, are connected to the boom, just below the origin of the fork. There are two hanger-winches (for luffing and slewing) and two load winches (for hoisting).

The patent did not discuss the foundation, but House (2018, 141) states that "the heel is set on a tabernacle that has a roller bearing for derrick movement."

Some Stülcken derricks could lift "more than 500 tonnes" (House 2024, 35). Danton (1996, 387) reported that Stülcken derricks had been fitted to over 200 ships. However, "their use has diminished" (House 2018, 141).

Derricks Versus Cranes

The distinction between a derrick and a crane is rather blurry. Towne (1883, 3) wrote, "A hoist is a machine for raising and lowering weights. A crane is a hoist with the added capacity of moving the load in a lateral or horizontal direction." By that definition, a derrick is a type of crane.

The Occupational Safety and Health Administration (OSHA) defines a derrick as "an apparatus consisting of a mast or equivalent member held at the head by guys or braces, with or without a boom, for use with a hoisting mechanism and operating ropes" (29 CFR 1910.181[a] [1]). It defines a crane as "a machine for lifting and lowering a load and moving it horizontally, with the hoisting mechanism an integral part of the machine" (1910.179[a] [1]).

Deck Cranes

"Cranes were first installed in ships around the beginning of the [twentieth] century" (Buxton et al. 2012, 36). In the 1920s, the *Montgomeryshire* was equipped with 22 winches, 21 derricks, and no cranes, while the *Madoera* had 9 winches, 10 cranes, and 9 derricks (Cunningham 1924, 147). At the time, a common deck crane was a "2-ton crane with a radius of about 20 feet," and they were "usually located at or near the hatchway corners" (26).

"Early cranes were expensive compared with derricks; they were slower to operate and required more highly skilled drivers. Moreover, their capacity and outreach was limited ... For many ship operators, their ability to spot loads was insufficient advantage to outweigh these drawbacks" (Buxton et al. 2012, 37). It was not "until the late 1940s ... that cranes were extensively fitted in general cargo vessels" (36). A typical modern deck crane has "a capacity of 25 tonnes" (37). "In 1962, about 20% of ships entering service had at least one deck crane" (Hammond 1968).

A modern pedestal crane, used for general cargo handling, has a rotating pedestal, with an operator cab and hoisting, luffing, and slewing motors, on top of a fixed pedestal (the latter giving the operator a more elevated view). The rotating pedestal ("turret") is the functional equivalent of the mast of a derrick, but the mast of a derrick does not rotate. The jib (boom) is hinged at the base of the rotating pedestal and supported by a cable running from the pulley at the top of the rotating pedestal to the end of the jib (Del Rosso 2024, 85). A suitable foundation is also needed (Datawave Marine Systems 2021).

There are also "knuckle boom" deck cranes; these have booms that articulate at the "knuckle." Photographs show these as hinged at the top of the pedestal, but with an additional connecting rod running from the base of the rotating pedestal to the underside of the boom, away from the main hinge but short of the knuckle.

Both types have the advantage of 360° rotation capability (unless limited to avoid striking superstructure) (House 2018, 143). A disadvantage of a pivoting boom is that luffing changes the height of the hook as well as its radial distance from the pivot point. Level-luffing mechanisms were devised, beginning in 1845, so that the hoist winch would compensate for the action of the luffing winch, keeping the hook at a constant level (Chilton 1923, 42–53).

This problem is also avoided by the jib crane, which has a fixed horizontal boom, and the hoist hangs from a trolley that moves along the boom. While jib cranes are common in urban construction work, I could not find any example of their use at sea.

Some ships are equipped with gantry cranes. These have an inverted–U structure (the gantry proper), an overhead bar with supports at either end. The hoist can travel transversely on trolleys on the overhead bar, and the gantry travels longitudinally on tracks (House 2024).

Gantry cranes as such have been known for centuries (Zhang and Yang 2020, 80), but their use on shipboard is much more recent. In 1946, experimental gantry cranes were installed over each hold on the forward section of a C-3, the SS *Sea Hawk*. These were oriented so the tracks ran athwartships (and "out over the ship's sides on cantilever trusses") and the bridge girder was longitudinal, so the winches

Top: A. Deck crane of Figee type (Cunningham 1924, Fig. 8). B. Deck crane discharging overside (Fig. 9). *Bottom:* "Starboard bow view of the chartered merchant container ship SS AMERICA VETERAN underway" in the Indian Ocean. Note the two gantry cranes. Photograph dated January 1, 1995 (National Archives, NAID 6392689, local ID: 330-CFD-DN-SC-85-08585.jpeg).

ran fore-and-aft. The "conventional boom-type loading gear was retained in the after section." The "overhead crane gear was approximately 15 percent faster than the boom gear in the average for all cargo handled … The principal advantage to the crane gear was in the case of large drafts or those which required accurate spotting, while the disadvantage was for that cargo requiring dragging into the hatch wings" (Merchant Marine Council 1947, 171; Rohn 1946).

Shipboard gantry cranes of more modern design—rails for fore-and-aft movement of the gantry, a trolley for athwartship movement of the hoist—were obviously desirable for container ships, and there is a 1964 Maritime Subsidy Board decision relating to the delay in delivery of such a crane for a ship undergoing an experimental "containerization modification" under a 1958 contract amendment (Maritime Subsidy Board 1965).

Ships may also be equipped with light-duty gantry cranes, whose sole purpose is opening and closing steel hatch covers (Pellett 2018).

6

"Helm a-Starboard!"

(Steering Gear)

Logically, this chapter would have immediately followed the ones on navigation, since they tell you where you are and steering is needed to get there. However, we first needed to understand some of the machines used at sea, as they came to form a part of steering gear.

An obvious concern with steering gear is the maximum rudder angle that can be practically achieved, as it may be necessary to turn hard to port or starboard to avoid an imminent navigational hazard or to gain an advantageous position relative to an enemy warship. In the mid-nineteenth century, Peake stated that in the Royal Navy, the ability to turn the rudder 42° was considered necessary for "efficiency" (Peake 1851, 163).

The Steering Oar

The first steering gear for boats was the steering oar. It is depicted in Egyptian Old Kingdom art. A fulcrum was provided by having the oar rest on a notch or pass through a hole cut in the gunwale, or pass through a wooden bracket or rope loop (Edgerton 1927). "Models with single steering oars mounted on the stern have been found in Egyptian tombs dating to 2000 B.C." (Mott 1999, 125).

Another common steering arrangement, with examples from the ancient world (Mott 1999, 3) up to the modern *pinisi* of Indonesia (23), was the use of a pair of steering oars, positioned at the ship's "quarters" (aft, but off the centerline). This was the dominant system until the thirteenth century (3).

On the Nydam boat (circa 400 CE), "at the top of the steering oar a tiller was led off at right angles to the oar shaft, to pass transversely across the boat" (Marshall 1990, 19). Using the tiller would not only have been more convenient than leaning over the side to grab the steering oar, it could have been pushed and pulled by several sailors simultaneously, which might be needed in rough weather.

The Rudder

The modern sternpost rudder is, essentially, a combination of a relatively flat control surface (corresponding to the blade of the steering oar) attached to a more or less vertical steering post so that it pivots around it.

Early rudder blades were wooden and usually flat, as on the *Mary Rose* (1545), but the rudder of the *Vasa* was "wedge-shaped, with the thick edge aft," and some shipwrights used wedged-shaped rudders with the thin edge aft, instead (Pipping 2000, 23). The wooden blade was ultimately replaced, first with a single plate, and later with double plates. The plates were later shaped to streamline the rudder, thus reducing the resistance it created (Muckle 2013, 229), although such a rudder produces "little lift ... at small rudder angles" (Mott 1999, 58).

Roman rudders had a low aspect ratio (mean underwater height/mean underwater length parallel keel); high-aspect-ratio rudders were introduced in the medieval period. For equal area, the high-aspect-ratio rudder has a higher lift/drag ratio (Mott 1999, 61), but stalls (loses lift) at a smaller rudder angle (47). Beginning in the Roman period, rudders also tapered upward to reduce drag-inducing aeration (60).

On the *Vasa* (1628), the rudder was tall (the length along the forward edge of the rudder was 9.48 meters) and narrow, and raked aftward. It was widest at the base (1.12 meters) (Pipping 2000, 20–1, 27).

The stern (axial, center-hung) rudder was invented by the Chinese in the first century CE (Molland and Turnock 2011, 7). They also used the tiller (see next section) to control it. However, the Chinese vessels did not have a sternpost and their rudder was "suspended and held in place by an elaborate series of tackles" (Mott 1999, 120). A sternpost rudder was used by Arab mariners in the tenth century, but their rudder was then controlled by lines rather than a tiller (121).

Stern rudders did have the disadvantage, at least during the Middle Ages, that the rather bluff stern of a cog created turbulence that decreased stern rudder effectiveness (Mott 1999, 114–115).

Having two quarter rudders inhibited yawing and provided double the turning moment of a single rudder (Mott 1999, 132). Quarter rudders continued to be used on some European vessels into the seventeenth century (and in Asia, to the present day), and some ships had both types (9, 145).

The Arab stern rudder was lashed to the sternpost (Mott 1999, 124); lashings were also used with the quarter rudders (101). A more secure attachment method for stern rudders was pintle-and-gudgeon. The pintle was an L-shaped piece composed of a horizontal bracket, attached to the side of the forward end of the rudder blade, and a downward-pointing iron spike. The gudgeon was a horizontal iron ring attached to the sternpost; the spike engaged the ring. The pintle-and-gudgeon is depicted in late-twelfth-century European artwork (106), and at least three evenly-spaced pairs were used (111). *Vasa*'s rudder was attached by seven pintles and gudgeons to the sternpost (Pipping 2000).

By the fifteenth century, "scores" (notches) were placed in the forward edge of the rudder to accommodate the pintles, and thus to reduce the clearance needed between that edge and the sternpost (Mott 1999, 112).

The advantages of pintle-and-gudgeon included less chance of rudder breakage as a result of hydrodynamic forces on the rudder and invulnerability to the time-honored battle tactic of disabling an enemy rubber by cutting its lashings (Mott 1999, 111–13). However, if the mount were damaged, it would have been difficult to repair or replace at sea (115–16, 130–31).

In early practice, the pintles were on the rudder and the gudgeon on the stern post, as the pintles were more likely to break and it would have been easier to replace the rudder than to fiddle with the sternpost. However, on modern plate rudders, the gudgeons may be on the rudder (at the end of the arms holding the rudder blade to the rudder stock) and the upward-pointing pintles on a rudder post attached to the stern (Army 1966, 220).

While the pintles defined the axis of rotation of the rudder, additional elements were needed to actually rotate it. One was the tiller, a horizontal lever (see next section). The other was a vertical rudder stock. This could be an integral extension of the rudder (Harland 1985, 71) or, later, a separate shaft (D'Antonio 2018). In either case, it was vertically aligned with the pintles (if any). It was convenient to divide the shaft into upper and lower parts, coupled just above the top of the rudder; this permitted the rudder to be removed without removing the entire rudder stock, too (Walton 1907, 185).

The pivot line of the rudder could be at its leading edge ("hinged" or "unbalanced" rudder) or "about one-third of the way aft" ("balanced rudder"). The hinged rudder was self-centering; that is, "left to itself, it would swing into the fore-and-aft line." However, the balanced rudder was "easier to turn than a hinged rudder of equal size" (Harland 1985, 70–71).

Attaching a balanced rudder to the stern post by means of the conventional pintle-and-gudgeon was not possible; the pintles would not have aligned with the desired pivot axis. Early forms were mounted on a rudder stock supported only from above (Harland 1985); this is nowadays called a "spade" or "hanging" rudder. However, by the nineteenth century, they could be supported from below by a skeg, an extension of the stern. Either the skeg had an upward-pointing pintle that engaged a socket in the heel of the rudder post (Meade 1869, 405), or the heel had a downward-pointing pintle engaging a socket on the skeg (Reed 1869, 252). In a variant form, the skeg fits into a notch in the middle of the rudder (Army 1966, 221).

The Tiller

A tiller was usually attached to the top of the rudder stock, the head. On Elizabethan ships, the rudder stock was completely outside the hull, and the tiller passed through an aft opening to engage the rudder head. In a later design, the rudder stock passed through a hole in the bottom of the overhanging part of the stern and the tiller was completely inboard. On a large ship, relieving tackles could be run between the tiller and either side of the vessel, to increase the force that could be applied to the tiller (Harland 1985, 173).

The tiller is a lever of the first class. The effort is at the end of the tiller; the fulcrum is the rudderhead, and the load is the center of pressure on the rudder blade. As usual, the mechanical advantage gained is the ratio of the effort arm to the load arm. Note that with a balanced rudder, the load arm is shorter than for an unbalanced rudder with the same rudder blade geometry.

In the sixteenth century, to steer larger ships, the size of the rudder and hence the tiller was increased. To support the weight of the tiller, the sweep was introduced. This was a timber beam sheathed in metal and coated in grease in order to minimize friction (zu Mondfeld 2005, 148). The tiller (helm) could have a roller on its underside that engaged the sweep, as on the *Vasa* (Hocker 2023, 404).

The Whipstaff

John Smith (1580–1631), in *The Sea-Mans Grammar*, Chapter 2, says that "the tiller playeth in the Gun-room over the Ordnances by the Whip staff; whereby the Rudder is so turned to and fro as the Helmesman pleaseth."

The whipstaff, a "vertical lever" (and "push stick") for manipulating the tiller from the deck above, most likely first came into use sometime in the early sixteenth century; "a fragmentary whipstaff" was found on the Cais do Sodré wreck of circa 1500 in Lisbon (Hocker 2023, 391).

The evidence with regard to the arrangement and operation of the whipstaff is actually quite limited, and the *Vasa* (1628) wreck is by far the most informative, On the *Vasa*, the "rudder …, helm,* whipstaff, sweep, rowle, rowle bearings and wooden pieces covering the rowle" were preserved, as were "a spare helm and a spare whipstaff" (Pipping 2000, 20–21). Moreover, they were still in more or less their proper positions in the ship's structure. (This is the equivalent to a paleontologist finding a fully articulated dinosaur skeleton at an excavation site.) The arrangement is discussed in more detail later.

We begin with an idealized explanation of how whipstaff steering works, and then address how the gear on the *Vasa* differed. The helmsman, holding the whipstaff, faces forward. The whipstaff passes through a hole in a swiveling bearing (the "rowle") fitted into the deck; the bearing has a fore–aft axis and thus permits the top of the whipstaff to be moved to port or starboard. The bottom of the whipstaff is attached to the fore end of the tiller (helm), which turns in the compartment below the helmsman (in the gunroom in the *Vasa*'s case).

To turn the rudder to port, the whipstaff is pivoted to port and also pushed downward. The tiller's fore end would thereby move to starboard, and thus take the rudder to port, initiating a turn to port (zu Mondfeld 2005, 148; McKay 2020, 48; Anderson and Anderson 2012, 107).

The shape of the rowle is sometimes compared to that of a rolling pin. However, on the *Vasa*'s rowle, the spigots' diameters are 120–125 mm, compared to 255–280 mm for the barrel. The ends (spigots) "pivot in bearings cut into the planking and underlying timbers" (Hocker 2023, 407), and the spigots are held down by caps (408).

The first complication is that the circular motion of the fore end of the tiller (helm) causes it to move aft when displaced from the neutral position. To

---

* Hocker prefers to use the term "helm" rather than "tiller" when referring to the horizontal lever that turns the rudder of a large ship.

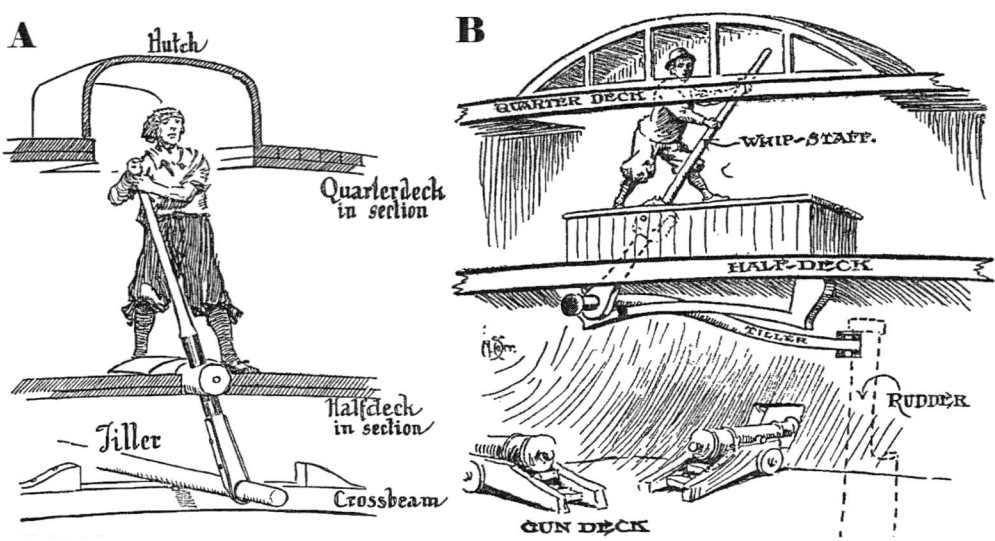

Two early-twentieth-century artistic conceptions of whipstaff steering. They show how the whipstaff acts as a vertical lever upon the tiller, and that the whipstaff fits in a hole in a rowle with a longitudinal axis. They do not make it clear that the whipstaff slides in the hole. Also, both drawings suggest that the helmsman had a better view of the outside than was likely to be the case. What contemporary "artists portrayed was a sort of skylight, well above the helmsman's head, rather than a … structure out of which he peered" (Harland 2011, 114). A. Culver's depiction of a steersman with a whipstaff (Culver 1924, 297, drawing by Gordon Grant). This shows the whipstaff as passing through a "bearing" that rotates on a horizontal longitudinal axis. His whipstaff is reinforced at the lower end by an iron strap, and also has a narrowed waist. If the sliding of the whipstaff would be limited by this waist, as he probably expected, the range of rudder angle would have been quite limited. His tiller is straight and rides on a straight sweep, with two stops. B. Chatterton's depiction (Chatterton 1913, 189, artist unknown). It shows the helmsman on a raised platform (much taller than the bench on the *Vasa*). Since this was "sketched on board the replica of the *Revenge*," it was possibly a feature of that vessel. The raised platform would have given the helmsman a better view, but reduced the extent to which the whipstaff could have been thrust downward. His helmsman's chest faces the side of the ship, but his head is turned to look forward. His tiller is riding on a ceiling-suspended curved sweep, and has an elongated S-curve shape.

accommodate this, either the top of the whipstaff has to "tilt forward" (causing the lower end to tilt backward) "or the lower end of he whipstaff has to slide forward on the end of the helm" (Hocker 2023, 393). On the *Vasa*, the hole through the rowle is flared at both ends, permitting some pitching motion.

While the original rowle is in such poor condition that the degree of motion it permitted can't be quantified, we do know that when the *Vasa* was found, the helm was touching the port knee timber in the gunroom, corresponding to a rudder angle of 23 degrees, so it must have allowed the corresponding forward tilt of the whipstaff.

The rowle was "covered by a decorative carving in the form of a monster's face, with the whipstaff coming out of the mouth as if it were the tongue … The carving is oriented with the chin aft, and the mouth is carved in a smile which accommodates the tilting motion of the whipstaff as it is pushed hard over" (Hocker 2023, 410).

6. "Helm a-Starboard!"

According to Hocker (2024, private communication), "At neutral, the whip actually tilts aft about 1.5 degrees ... At full travel, the tilt is 9.5 degrees forward ... The angle between the helm and the whipstaff changes as well. It is also not square at rest, about 1.5 degrees aft, and becomes more acute, to 7 degrees, or a total change of only about 5.5 degrees."

Some literary depictions of the rowle make it appear to bear a cylindrical hole; if so, the fore-and-aft play therein would be constrained by the relative diameters of the hole and the whipstaff, and the length of the hole. However, Hocker suspects that this was "simply a convention of small illustrations rather than a realistic portrayal" (private communication). He also points out that the whipstaff would repeatedly rub against the edges of the hole and "if the hole was not flared to begin with, it would quickly become so in use through excessive wear ... The only other preserved rowle, a spare from *Kronan* (1676) also has a flared hole" (private communication).

Another complication is

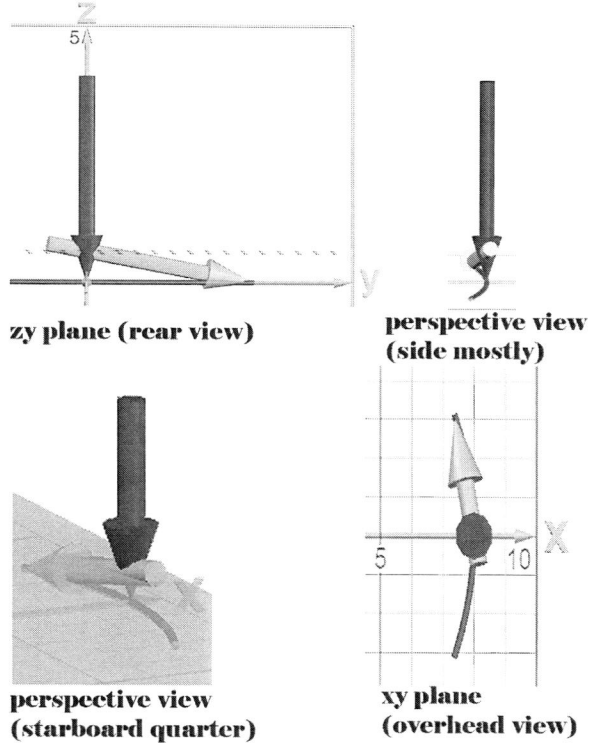

A schematic of a whipstaff in the neutral (dark 3d arrow) and extreme (light 3d arrow) positions, from different views. The schematic was created by the author in the Desmos 3d calculator, assuming a tiller (helm) length of 8 meters, a whipstaff length of 4 meters, and a distance of 0.6 meters between the rowle and the whipstaff-tiller junction with the tiller at neutral. It is assumed that the whipstaff is vertical at neutral and that the extreme tiller angle is 23°. Also, the plane of the deck holding the rowle and the plane in which the tiller rotates are assumed to be parallel. The axes are x-axis (fore-aft), y-axis (port-starboard), and z-axis (up-down). Top left: a rear view, so the whipstaff is projected onto the zy plane. The "striped line" shows the deck plane in which the rowle is located. Note how the whipstaff is not only tilted, but slid deeply downward. Top right: a perspective view, from the side but also from a little above and aft of the plane of tiller movement. The curved line is the path of the tiller arm from -23 to +23°. Note how small the fore-and-aft movement of the tiller arm is. Bottom left: a perspective view, roughly from the starboard quarter and above the deck plane. The deck plane is shown as translucent, hence medium gray. Again, it is evident that in the extreme position, only a small part of the whipstaff is above the deck. Bottom right: overhead view (xy plane). One sees the shallowness of the arc made by the end of the tiller arm in moving from extreme port to extreme starboard. For a more elaborate rear view representation of the *Vasa*'s whipstaff in multiple positions, see Hocker (2023, fig. 10.38).

that the *Vasa*, like many contemporary vessels, had a raked stern.* Thus, the rudder stock, about which the rudder and tiller pivoted, was not vertical, and this would cause a drop in the fore end of the tiller as the rudder angle increased. That, in turn, would increase how far the whipstaff had to be slid down for a given rudder angle.

However, on the *Vasa*, the tiller (helm) is not straight; it has a curved segment. The curvature is such that the forward part of the tiller is essentially parallel to the deck above,† as can be seen in Pipping (2000). In other words, it compensates for the rake. At 23°, the helm end only drops "approximately 100 mm from its height at neutral" (Hocker, private communication).

It has been suggested that in the earliest implementations of the whipstaff, it passed through a simple hole (possibly gasketed), rather than through a rowle. That would, of course, have granted much more freedom for tilting it forward or backward. But Hocker argues that "a gasketed hole would still need to have enough play to allow the whipstaff to slide and tilt, which would mean that the hole would have to be larger than the hole through the rowle, so more overall slop. It would be enough play that the whipstaff would rattle in the hole, imposing impact loads on both the whip and the helmsman (no fun). Most of the tilt is in the transverse plane, with a smaller amount fore and aft, so a rotating bearing makes sense here. It limits slop in the direction of greatest travel, while still allowing enough freedom of movement longitudinally. I think it is a brilliant solution" (private communication).

The whipstaff is a lever of the first class, with the effort at the point where the helmsman pushes on the whipstaff, the fulcrum at the rowle, and the load where it meets the tiller. The total mechanical advantage of the whipstaff–tiller combination in turning the rudder would be the product of the mechanical advantage of the whipstaff and that of the tiller.

The effort arm is the distance from where the upper end of the whipstaff is gripped to the center point of the rowle; the load arm is the distance from the latter to where the lower end engages the fore end of the tiller. The mechanical advantage provided by the whipstaff is the ratio of the effort arm to the load arm. It is thus variable, as the whipstaff must be slid down further as the tiller angle is increased (cf. Martin 2017). "At [tiller] angles of less than a degree, the whipstaff's advantage, if gripped at shoulder height of an average Vasa crew member (1.35 m above the deck) is almost exactly 2:1" (Hocker 2023, 413). (This would be about 2 m below the very top of the whipstaff.)

This mechanical advantage (or disadvantage) applies only to the turning force exerted by the helmsman perpendicularly to the end of the whipstaff. There is no mechanical advantage (or disadvantage) associated with a pushing force exerted parallel to (along) the whipstaff.

---

* Bruzelius (1995) says 11.5 degrees; Rose (2014) implies that the rake was 9.28 degrees before the sternpost extension and 8.95 degrees after it; and Pipping (2000, 27) implies that the rake is 6.78 degreees. Hocker (private communication) told me 10.5 degrees, but "some distortion in the sternpost rake was introduced during conservation, so the rake could vary by up 1.5 degrees."

† The deck that the helmsman stood upon had sheer—that is, it sloped downward toward midships, by about six degrees according to Hocker (private communication). Hence, we must be careful to distinguish between the true vertical established by gravity and the "helmsman's vertical" (perpendicular to the deck at the rowle). When calculating the whipstaff angle, the latter is used.

The overall mechanical advantage provided by the steering gear is the product of the mechanical advantage provided by the tiller (see previous section)* and the mechanical advantage provided by the whipstaff used as a lever.

As previously noted, an alternative method of accommodating the circular motion of the tiller would have been to allow the lower end of the whipstaff to slide outward along the tiller as the tiller angle was increased. We do not have any specific historical evidence for this approach, and there are two reasons it might have been disfavored. First, with the tiller in neutral, the attachment point would have been closer to the rudder post, and hence the effort arm would have been shorter (and the mechanical advantage of the tiller, smaller). Admittedly, both would have increased as the angle was increased. But if steering was done mostly at small rudder angles, it might have been considered best to maximize the mechanical advantage at zero rudder. Second, the frictional losses from the sliding action might have been significant. That said, Harland believes that there could have been at least a few inches of fore-and-aft play of the ring on the gooseneck (2011, 109, 117, 124).

There were several different methods of attaching the lower end of the whipstaff to the tiller. On the *Vasa*, the whipstaff is "a single piece of oak shaped approximately like a tennis racket with a very long handle" (Hocker 2023, 405). At the "racket" (lower) end, it has a circular 202-mm-diameter hole, with a rounded edge.† This engaged a tenon (156–163 mm diameter) at the forward end of the helm, and this tenon features a hole for a retaining bolt or linchpin (404). In contrast, in the whipstaffs found in the Scheurrak SO1 wreck (built c. 1580), "the shaft and the eye at the lower end are separate pieces bolted together" (Hocker 2023, 406).

Blanckley (1750) says that the helmsman holds the whipstaff in his hand, and that it "goes through the Rowle, and is made fast to Tiller with a Ring" (186; cf. 74). The "Rowle" is "a round piece of Wood wherein the Whipstaff goes, being made to turn about, that it may carry over the Staff the easier from Side to Side" (138). Blanckley says (66) that the "lanyard [Ring?] of the Whipstaff" may be fixed to a "Gooseneck," a "Piece of Iron fixed on the End of the Tiller," which he depicts as L-shaped.

Similarly, Sutherland (1727, 87) contemplates that the whipstaff is made of ash and is round; it differs in diameter at the two ends; and it has an iron hoop at the great end, attached by a ragged bolt. The eye of this hoop engages an "iron gooseneck" (consisting of a "hoop," a "shoulder," and a part that extends from the tiller) at the "foremost end" of the tiller, secured by straps and nails.

The *Vasa*'s "working" whipstaff "is now 4.16 m long, although the upper end is broken off … For most of the length commonly handled by the helmsman, the diameter was 70–90 mm." The spare whipstaff is 4.198 m long, and also has a hole 70 mm below the upper end (Hocker 2023, 405–6); the working whipstaff broke at

---

* For the *Vasa*, about 13:1. Hocker (2023, 413) says 6.3:1, equating the rudder width with the "load arm." However, the center of pressure of the lift force would be at the half-width if the rudder were rectangular, and it is in fact tapered, moving it closer to the rudder stock.

† This is reinforced: "The perimeter of the lower end was originally covered by an iron strap nearly as wide as the edge …, held in place by short nails" (Hocker 2023, 405).

the corresponding point. It is believed that this hole was for attaching a lanyard by which the whipstaff could be pulled back from an extreme position (Pipping 2000, 25–26), or even used to apply body weight so as to increase the force along the whipstaff and cause it to slide deeper into the hole (Hocker, private communication).

The longer the whipstaff, the further it could be slid down and thus the greater the angle it could push the tiller to (Hocker 2023, 393). However, with a long whipstaff, at small rudder angles the top end may be out of easy reach, reducing the effective force to be applied (Hocker private communication).

The length of the whipstaff was in turn constrained by the height of the ceiling of the steering room, since at zero rudder angle it would be close to vertical and only a small portion (0.6 m on the *Vasa*) would be below the rowle. On the *Vasa*, the steerage compartment, where the helmsman stood, is 3.46 m tall (Hocker 2023, 396).

There was a low (288 mm) bench that the helmsman could stand on (Hocker 2023, 410), and this would have given him a view of the "sails through a grating … in the forward bulkhead above the upper deck." He would have had "no view of the horizon or what was in front of the ship" (398). However, it is believed that the binnacle holding the ship's compasses would have been placed directly in front of that bulkhead (398). Also, the helmsman would have been protected from musket fire (Pipping 2000, 28).

An alternative to raising the height of the ceiling of the steering room was to cut a slot in the deck above the helmsman, for the whipstaff to pass through, and there is pictorial evidence that this was done on some French and Dutch ships (Harland 2011, 115). While the length thus above the helmsman's grip in the amidships position would have made it possible (when slid down) to reach a more extreme helm angle than would otherwise be possible, it would not increase the mechanical advantage of the whipstaff, the force being applied at the point of grip.

The usual explanation for the introduction of the whipstaff was that it was because ships got taller, and "the helmsman at the tiller found himself one or two decks below the level of the commanding officer," and so was "working blind, and had to steer by command only" (zu Mondfeld 2004, 148). However, on the *Vasa*, the helmsman still had to steer by command or by compass, because the view was so limited. It is true that the whipstaff allowed the helmsman to stand one deck closer to the open deck from which the ship was conned, and this would have made it easier to receive commands.

Vaughan gave another explanation for the whipstaff: so the helmsman "would have a means of controlling the tiller without following its movements from side to side of the ship" (1913, 232), like a tennis player being forced by an opponent to run back and forth to return the shots.

A third reason is that at least at lower rudder angles, the mechanical advantage offered by the whipstaff reduced the force needed for steering without having to increase the length of the tiller arm (and thus widen the hull).

There was some controversy as to how much the whipstaff could turn the tiller, and how easily. Early pronouncements on the subject were pessimistic. Zu Mondfeld said that a 40–50° "angular throw" of the whipstaff would "correspond to a rudder movement of only 5 to 10 degrees" (2005, 148). In 1985, Harland said that it could turn the tiller "less than 15° to each side" (176), and "functioned well" only when the tiller was within 5° of "amidships" (175).

This changed in response to Pipping (2000, 25, 27), who showed that on the *Vasa*, it was at least geometrically possible to achieve a tiller angle of 22°, and Harland (2011, 119) accepted this. (Hocking [2023, 397] states the maximum as 23°.)

We know that the whipstaff was used on several large ships besides the *Vasa*, including the 126-gun *Kronan* (1668) (Pipping 2000, 21), the 108-gun *Le Royal Louis* (1692) (Vaughan 1913, 236), and the *Hampton Court* (1678 Third Rate) and the *Coronation* (1685 Second Rate) (Bruzelius 1995). Curiously, Sir Henry Manwayring insisted that "in great ships they are not fitted, for, by reason of the weight of the rudder and the water which lies upon it in foule weather, they are not able to govern the helme with a Whipp, because conveniently there can stand but one man at the Whipp" (Vaughan 1913).

The rudder force is of course greater on a "great ship." Pipping (2000) includes as Appendix A an analysis of the rudder and whipstaff geometry and forces by Peter Trägårdh. He makes several interesting points. First, because the "trailing edge [of the rudder] is thicker than the leading edge," the "effective rudder angle" is increased, by as much as 1.5°, "at the cost of some increase in resistance." Second, while to achieve a rudder angle of 5° the whipstaff is 2.5 meters above the rowle, to reach one of 20°, the "whole whipstaff has to be pushed down through the rowle." Third, Trägårdh provides a formula for the whipstaff angle as a function of rudder angle, and this is plotted in Pipping (2000, Fig. A4). For a 12° rudder angle, the whipstaff is 70° from the vertical.

Finally, to achieve a particular tiller angle, the helmsman must apply forces both along and normal to the whipstaff. At low angles it is primarily a lever, and a high angles, a push stick.

The required forces increase as the square of the ship velocity (Pipping 2000, 35). At a 5° rudder angle and 5 knots, the force normal to the whipstaff is then about 30 N (Fig. A2), and that along the whipstaff is about 125 N (Fig. A5). The whipstaff is about 50° from the vertical (Fig. A4).

At a rudder angle of 20°, the required pushing force along the whipstaff dominates, and with Trägårdh's hydrodynamic assumptions, it is 400 N at a ship speed of 5 knots, and thus 1,600 N at the *Vasa*'s estimated maximum speed of 10 knots (Fig. A3, 35). The latter, thought Trägårdh, seemed "far above the capability of one man or even two."

At that rudder angle, the whipstaff angle is almost 80° from the vertical (Fig. A2), and only a small portion of the whipstaff is exposed above the rowle. Hocker (2023, Fig. 10.37) shows the displacement (slide of the whipstaff) at different helm angles. At 20°, it is about 2.25 meters. Since the distance of the lower end from the rowle is 0.6 meters at neutral, if we assume that the effective length of the whipstaff (from grip point to center of the lower hole) is 4 meters, that leaves 1.15 meters above the rowle.*

---

* Except at angles less than 5°, the displacement varies linearly with helm angle. At 23°, it appears that the displacement would be about 2.7 m.

However, Hocker points out that it would be very unusual to need to adopt a rudder angle more than 14 degrees, as the rudder would "stall" (lose lift) between 14 and 22 degrees (2023, 412). At 14 degrees, by Trägårdh's formula, the whipstaff angle would be about 73 degrees from the vertical, and the displacement about 1.5 m (2023, Fig. 10.37), and of course the force would be primarily along the whipstaff. Hocker believes it would be possible to achieve this by crouching and pulling on the lanyard (private communication).

If 1.9 m were exposed, the upper end of the whipstaff would be 0.56 m (22 inches) above the floor: essentially knee level. It is certainly possible to crouch low enough, but it would not be good for the helmsman's back!

A Canadian government ergonomic standard calls for limiting the horizontal pushing (or pulling) force to 225 N when standing with the whole body involved; 110 N when standing but just using the arm muscles; 188 N when kneeling; and 130 N when seated. It notes that pushing and pulling are more difficult "when the hands must be above the shoulder or below the waist level." If the worker can brace body or feet against a firm structure, a force of up to 675 N is possible (Canadian Centre for Occupational Health and Safety 2024).

The replica of the *Kalmar Nyckel*, a smaller vessel, is also equipped with a whipstaff, and it has been sailed extensively. Hocker says that on the *Kalmar Nyckel* replica, at a whipstaff angle beyond 45 degrees, "steering is accomplished by pushing the end toward the rowle … In heavy seas or at high speeds … it is usually necessary to brace against the bulkhead of the deck head (starboard) or chart table (port) do be able to exercise enough force … Once hard over, the helm can be held there by standing on the end of the whipstaff, making it bind in the rowle" (Hocker 2023, 418).

The Steering Wheel

The next major step was the introduction of a steering wheel, linked to the rudder via a windlass and other mechanisms.

Harland points to a Greenwich model featuring a "steering windlass" as a transitional form—the windlass, for which the barrel axis is perpendicular to the keel, is turned by a crank. "By rotating the barrel through 90 degrees and replacing the handles with wheels, we have the apparatus as it appears in models of a slightly later date" (Harland 1972, 44).

The date Harland ascribed to the model was 1703–5, but it is currently believed to date to 1702. It is a 90-gun "first rate" (National Maritime Museum, Greenwich 1702; Franklin 1989, 121–6) with a "steering windlass on the quarter deck" (Bruzelius 1998). Ball and Stephens (2018) say that "this is the earliest known evidence of a rope-operated steering mechanism"—ropes ran down from the windlass to the tiller—and he suggests that it "was probably an experimental fitting prior to the introduction of the steering wheel in about 1703."

The evidence for the 1703 date is a full hull model of a 50-gun "fourth rate" that features an eight-spoke steering wheel. However, it also features the "remains of a whipstaff," so it is not certain that the steering wheel was an original feature (National Maritime Museum, Greenwich 1703).

The whipstaff was not immediately displaced. Sutherland in 1729, presenting a specimen shipbuilding contract, refers both to the inclusion of a whipstaff and to having "a Wheel or Reel fitted on the Quarter Deck, with Block in the Gun Room, and Bolts drove for traversing the Tillar, without the help of a Whipstaff" (87). However, in 1750 Blanckley wrote, "Lately they are left off, and Steering Wheels are made use of" (74).

"Chapman's drawings show vessels up to 90 feet or so, between perpendiculars, with a tiller only," and Harland considered the "simple tiller" to be "much more practical than a wheel in small vessels, i.e., those a tiller shorter than eight or ten feet long." Nonetheless, wheels were "installed in smaller vessels, even down to smacks" (Harland 1972, 52).

Sixteen-spoke steering wheels became customary and the spoke that was on top when the wheel was in "neutral"—the "king spoke"—was shaped so it could be recognized by feel alone. "A full turn of the wheel changed the tiller angle about eighteen to twenty degrees" (Harland 1985, 174)—which implies an overall mechanical advantage between 20:1 and 18:1.

The helmsman stood on the weather side and could be assisted by another sailor standing on the lee side. On a large ship, two men might not be enough to control the wheel, so the "double wheel," which could be turned by four men, was introduced (Harland 1985). (On modern cruising sailboats, there may be "dual wheels"—two wheels, on separate shafts, but engaging the rudder stock via a common chain [American Sailing Association 2014, 50].)

The two wheels were connected by a drum, around which rope was wound (forming the windlass). The midpoint of the rope was fastened to the drum, and the free ends ran around blocks (pulleys) to engage the tiller, one (or more) on each side (Harland 1985, 174). When the wheel is turned, the rope on one end of the drum unwinds (so the free portion lengthens), and that on the other end winds up (so the free portion shortens).

Harland has pointed out that without further mechanical elements, there is a kinematic problem. On the one hand, the action of the drum is such that the free portion of one end lengthens by exactly the same amount as the free portion of the other end shortens. On the other hand, to accommodate the arcuate motion of the end of the tiller, the total length for the port block–tiller and tiller–starboard block segments, which are straight lines, will change as the tiller moves away from the neutral position. The segment length will be shortest when the tiller points directly at the block. The tiller angle and the block positions (closer or further from the rudder head than the end of the tiller) will determine whether the rope is stretched or slackened (Harland 2003, Chapter 2; Harland 1972, 41–45). But Harland acknowledges that "if the tiller rope is very long ... or if the tiller is very short, the location of the leading blocks is less critical" (Harland 1972, 52).

The Quadrant Solution

While Dunn did not invent the quadrant, his 1889 patent offers an unusually clear illustration of the quadrant and its relationship to the rest of the steering gear. His steering wheel did not operate on a windlass, but rather pneumatic or hydraulic cylinders, causing one piston rod to extend and another to contract. These piston rods were connected to ropes or chains, and the rest of the apparatus was deemed part of the "prior art"—and that is the part with which we are presently concerned. The ropes ran from each cylinder to port or starboard, passed around a pulley, and then ran aft and further inboard to the "quadrant," to which they were "secured." The quadrant as depicted was a metal frame in the shape of a quarter-circle (or close to it), with a round socket at the point corresponding to the center of the circle defined by the curved end, and a center bar running from the center of the curved side to the socket. The socket would fit over the rudder post. It can be seen from the figure that the lines running from the pulley are tangent to the ends of the arc.

Unfortunately, Dunn did not show how the ropes were secured to the quadrant, because his claims being related to the hydraulic or pneumatic control. For the sake of simplicity, let us assume a simple hookup, in which the ropes run along the arc of the quadrant to its midpoint, where they were fastened. When the steering wheel is turned, the guide rope is wound up more on one side of the windlass and less on the other. This causes the "shortened" end to pull on the center of the arc of the quadrant, causing it to pivot. This pivoting would take up the slack on the other end of the guide rope. In both cases, the change in length is along the arc of the quadrant, and by the same angle, and thus there is no change in the total length required by the new orientation of the quadrant. It is thus apparent that as long as the quadrant does not rotate more than 45°, there would be no kinematic problem.

There were other ways of securing the rope or chain to the quadrant. Harland (1972, Fig. 19) depicts "the quadrant of the rudder head in the *Joseph Conrad* of 1883." This was a double grooved quadrant, which had an "anchor" shape—that is, it omitted the diagonal arms of the Dunn patent. A chain ran from the port block along one groove to the starboard side of the quadrant socket head, and from the starboard block along the other groove to the opposite side of that head.

For a modern illustration of the steering quadrant's connection to steering wheel (using a chain sprocket drive instead of a windlass), see American Sailing Association (2014, 50).

The quadrant itself was typically shorter than the tiller it replaced, and thus provided a smaller mechanical advantage. However, "all necessary additional leverage can be obtained by using proper sheaves, by leading the lines properly, and by the leverage obtained by the use of drum and steering wheel" (Desmond 1915, 80). Use of a balanced rudder would have substantially reduced the lever arm needed for a given rudder angle and force exerted by the helmsman (Desmond 1915), and thus might have encouraged the replacement of the tiller with a quadrant.

It is rather unclear when the quadrant actually came into use. *Oxford English Dictionary*'s first documented literary use of "quadrant" in this nautical sense is

6. "Helm a-Starboard!"

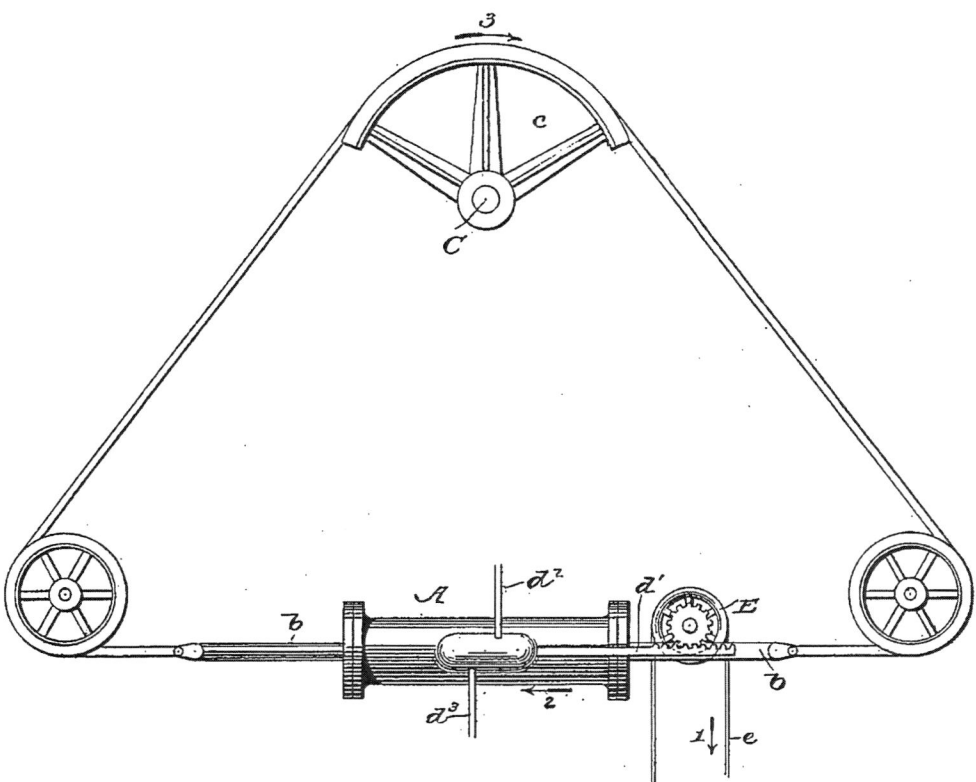

Figure 1 of Dunn (1889), U.S. Patent 404,472 (June 4, 1889). The quadrant e pivots around rudder stock C. We are not concerned with the rest of his apparatus; imagine that in place of the cylinder A we have a windlass.

from 1779,* in Thomas William Jolly and Robert Beatty's "Description of the New Patent Steering Machine" (British patent 1220 of April 22, 1779†). While the Jolly–Beatty machine indeed features a quadrant in place of a tiller, it does not use a windlass or steering ropes, and hence is discussed later in this chapter.

Steel (1794, 234) speaks of the tiller rope running from the "steering-wheel-barrel" through various blocks, with "each end then passed through an eye, on each side of the upper part of a hoop, that is bolted on the fore end of the tiller; the ends are then passed under the sweep, through an eye on each side, in the middle of a hoop, driven on the tiller farther aft." Conceivably this hoop, being arcuate, could be acting as the quadrant. However, the hoop might just be a ring on the fore end of the tiller, for attaching the tiller ropes.

---

* "The pinion Wheel and Spindle are placed parallel to the Stern-post, and made to act on the Quadrant by a Wheel fix'd on the Top of the Spindle, parallel to the Plane of the Quadrant, and turn'd by the Helmsman." Citing T. W. Jolly, Descr. New Patent Steering Machine, 5.

† Per Woodcroft, Bennet (1854), *Titles of Patents of Invention Chronologically Arranged rom March 2, 1617 (14 James I.) to October 1, 1852 (16 Victoria).* 224. Queen's Printing Office.

Moreover, even after the quadrant was invented, its use was not necessarily universal. As late as 1899, a "steering gear" was depicted with a drum (windlass) acting through pulleys on a tiller, with no quadrant or other obvious solution to Harland's "kinematic problem" (Hiscox 1899, 29).

There is some documentary evidence that the circular sweep, long used to support the tiller, was used to achieve some of the advantages of the true quadrant. Thus, Stalkartt (1781, 230) refers to the sweep, "a circular plank fitted to support the foremost end of the tiller, much improved lately by conveying the tillar-rope round it, keeping it always tight, which in the former method was never tight but when the tillar was in midships." For that to work, the leading blocks would have to be aft of the ends of the sweep, and the ropes running from those blocks attached beneath the tiller, just past where it overlies the sweep, so that they took up an arcuate shape around the fore edge of the sweep.

The Rapson Slide Solution

In 1839, John Rapson obtained a patent on an invention for which the object was "to obviate the mischievous consequences caused by the slackness of the tiller ropes which results from the end of the tiller moving in an arc, while the ropes move in a straight line" (Rapson 1839). (Since the quadrant would have eliminated those "mischievous consequences," the implication is that it had not yet been universally adopted.)

In the "Rapson slide," the "tiller ropes, instead of being fixed to the tiller, are attached to a slider which works between guides fixed across the ship tangentially to the arc described by the tiller end. The end of the tiller passes through a slot in the slider, or the tiller is slotted, and embraces a pin on the slider, so that the circular motion of the tiller end is converted into rectilinear motion in the slider" (Rapson 1839). Actually, it's the other way around: The motion in the slider is converted to tiller motion. As the *Inventor's Advocate* explained (1840, 195), the "frame [slider] constantly moves in a right [straight] line, and the ropes are therefore always tight; whilst the tiller, sliding through the frame, moves in an arc of a circle."

We thus have what Goodeve (1880, 351) called a "double slider crank" mechanism. Turning the steering wheel causes the tiller ropes to pull a sliding block to port or starboard within an athwartship guide, that is, a linear motion. But this block carries a pin running through a second sliding block, which slides within a slot in the tiller arm.

Harland (1972, 53) states that the Rapson slide was used on the USS *Constellation* (presumably he meant the 1854 sloop-of-war and not the 1797 frigate). He also indicates that it was practical only with an iron tiller (54).

The Rapson slide may be implemented so the sliding block is closest to the pivot point when the rudder is in neutral and furthest away when the tiller is hard over.[*] If so, the mechanical advantage is greatest at the high rudder angle.

---

[*] As can be seen in this video: YOYO, "SOLIDWORKS Rapson's Slide," https://www.youtube.com/watch?v=g9vlkvhtBBw.

The Rapson slide is used on modern vessels, but with, for example, hydraulic rams taking the place of tiller ropes (Russell 2021, 291).

Ameliorative Measures

While the quadrant and the Rapson slide offered exact solutions to the kinematic problem, they were not the only adaptations made.

Peake suggested "having two sets of leading blocks, so placed that the inequality generated by the forward pair is exactly balanced by the opposite inequality due to the after pair." Peake claimed that "it worked up to 45 degrees in *North Star*" (a sixth rate of 1824), and Harland conceded that this was possible "if the tiller makes less than a right angle with the sternpost," so "the tiller end comes closer and closer to the deck as the tiller is put hard over" (Harland 1972, 55–56).

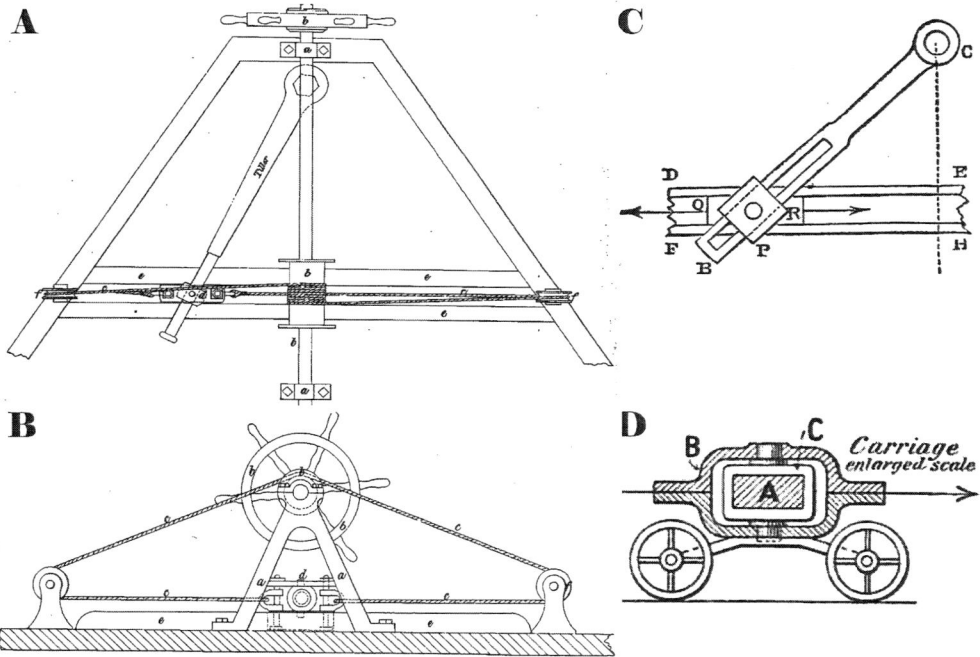

A (Upper left). Plan view of Rapson slide steering gear (Rapson 1839, Fig. 1). Turning the steering wheel turns axis a and thus barrel b. An endless rope or chain c is wrapped around barrel b and guide pulleys f, and is attached to the sliding block d, which travels within the slide guides e. This causes d to also slide radially along the tiller. Unfortunately, the slot in the tiller is not shown. B (Lower left) Rear view of the same gear (Rapson 1839, Fig. 2). C (Upper right). The generic double slider crank mechanism (Goodeve 1880, 351) used in the steering mechanism (Rapson slide) patented by Rapson in 1839. As shown by Goodeve, "block QR slides in the [linear] guides DEFH and carries a pin which runs through another sliding block P, which can move along the tiller CB [sliding in the unlabeled slot]," causing the arm to pivot around axis C. "The direction of the pull on the tiller ropes is shown by the arrows." D (Lower right). Rather than have the block travel within a slide, the block C may be mounted on a wheeled carriage (B) that runs on guide rails. A is the end-on view of the tiller (Dunkerley 1905, Fig. 224).

Another "empirical solution" was the "Wesselink" barrel. This was a windlass with a "differential barrel," that is, "wider at the ends." Harland says that the "leading blocks" (those closest to the tiller) must be "rather well aft." The barrel may be smooth or grooved. While Mossel (1859) (Harland 1972, n6) attributed the invention to Wesselink (born 1808), Harland believes that there are possible antecedents in the steering gear of the Danish frigate *Najaden* (1796) and the British fifth rates "*Foudroyant* (ex–*Trincomalee*) of 1817, and the *Unicorn* of 1824" (Harland 1972, 55).

Gears

Gears are essentially toothed wheels. A single gear by itself isn't very useful. However, if two gears mesh at an angle, you change the axis of rotation. Usually, they meet at right angles, to change vertical rotation to horizontal, or vice versa (Walton 1968, 93).

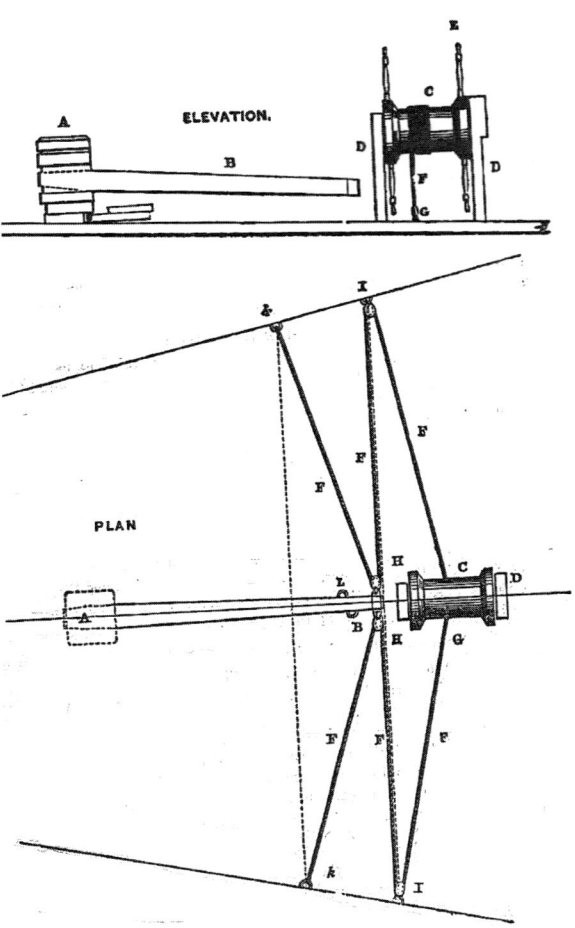

Peake's sketch (elevation and plan view) of the steering gear of the *North Star* (Peake 1851, 162).

If two gears that are on parallel axes mesh, but the gears have a different number of teeth, then you change the speed of rotation. For example, if a 10-tooth gear engages a 20-tooth gear, the former gear rotates twice for every rotation of the latter ("reduction ratio" of 2:1).

The relative diameter of the gears is also important. For engagement, the tooth spacing of the meshed gears must match. Thus, the 20-tooth gear is twice the diameter of the 10-tooth gear. When a force is exerted on the perimeter of the 10-tooth gear, it creates a torque equal to the force times the gear diameter. The same force is transmitted to the meshed tooth of the 20-tooth gear, but the "lever arm" is doubled, so the torque is, too. Thus, a reduction gear reduces speed but increases torque (Walton 1968, 95).

Note that with two gears, the output and input gears will turn in opposite directions. You can get around this by interposing an "idler" gear between them (Walton 1968, 99).

Early gears had wooden teeth, which were sometimes mere pegs, and had a tendency to slip out of mesh. One early solution was to have a toothed gear engage a "cage gear," essentially rods around the gear perimeter, parallel to the gear axis. A tooth of the toothed gear would engage the gap between a pair of rods. Ultimately, metal teeth replaced wooden teeth, and teeth of a shape better suited to turning the output gear at a uniform velocity were devised (Walton 1968, 96–7).

"Steam turbines must operate in a relatively high rpm range for greatest efficiency while propellers operate most efficiently in a relatively low rpm range" (Bureau of Naval Personnel 1958, 134). The most common solution to this conundrum was to use reduction gearing so that the rapid rotation of the crankshaft was converted to a relatively slow rotation of the propeller. In the mid-twentieth-century American navy, the reduction gears used for this purpose were of the double helical type; if you look at the rim of the gear, there are two sets of teeth, forming a "vee"; this "produces a smoother action of the reduction gearing and avoids tooth shock" (135–6). Its "ships built since 1935 have DOUBLE REDUCTION PROPULSION GEARS," which means that the reduction is performed in two steps (by a train of three gears) (136).

Chain and Belt Drives

Sometimes we need to transfer rotational motion from one axis to another. If they are widely separated, the use of gears alone would be inefficient and costly; one would need a gear train that spanned the entire distance. Instead, we use a belt or chain drive.

In a chain drive, the chain wraps around two gears, and the radial teeth of the gears (sprockets) engage the links of the chain rather than each other. In a belt drive, the belt (leather, fabric, rubber, plastic) wraps around untoothed wheels, and is driven (and turns the output wheel) by friction.

In either case, if the input wheel is larger in diameter than the output wheel, you get a mechanical advantage, just as with gears.

Chain and belt drives are used on ships in the propulsion and steering gear.

Axiometer (Helm Indicator)

One application of gearing was in the axiometer (helm indicator), the device used to indicate the current rudder angle to the helmsman. According to Lavery (2013, 27), "the first helm indicators in the British Navy appear to have been copied from the French 74-gun ship the *Invincible*, captured in 1747."

There were two basic types in the nineteenth century, those with a circular dial and those with a linear scale (Harland 1985, 175). In either case, these were geared to the steering wheel, and bear in mind that even with a circular dial, a reduction was needed, as the steering wheel was turned far more than the rudder. Harland depicts a circular dial axiometer that is on the same axis as that of the steering wheel, and a linear scale one that is mounted on the top of the post to which the steering wheel is attached. In both cases, the scale appears to face away from the helmsman, but it would have been visible to the pilot or the officer of the watch (Lavery 2013).

In 1905, some Royal Navy ships were equipped with an Evershead helm indicator. This had an angular position sensor and a radio transmitter at the rudder, and radio receivers and mechanical readouts at the bridge and the conning tower (Dreadnought Project 2013).

Gear Train-Type Steering Gear

The ultimate goal of the steering gear is to convert the rotational motion of the steering wheel, about a horizontal axis, into the rotational motion of the rudder, about a vertical axis.

In 1779, Jolly and Beatty chose to avoid this conversion altogether by use of a "Horizontal Wheel," that is, one turning on a vertical axis. This wheel had vertical handles spaced about on its rim, so it could be turned easily, and turning the wheel, via a pair of sheaves, moved a horizontal index below it, to indicate the rudder angle. The wheel turned on a spindle, causing the rotation of a pinion wheel on the spindle axis.

Their quadrant combined the arc frame with a square socket that would fit over the rudder head. Unlike the quadrants discussed previously, theirs was toothed, and the teeth of the pinion wheel meshed with those of the quadrant.

If the sternpost was raked, and therefore "the space abaft the rudder is not sufficient for the Quadrant to traverse clear of the Stern, we use the Quadrant (B) with the Teeth on the inside of the arch" (Jolly and Beatty 1779). If, instead, the sternpost was nearly perpendicular, the quadrant's teeth would be "on the outside of the arch."

Their quadrant "only extends two and a half, or three feet from the Rudder, even in Ships of 500 Tons burthen; and proportionately less in smaller vessels," and they suggested that it be placed in the cabin.

They asserted that it conferred a "superior power" to that of the conventional tiller. The mechanical advantage would be the product of (1) the ratio of the steering wheel radius to the pinion wheel radius and (2) the ratio of the quadrant radius to the moment arm from the rudder center of pressure to the axis of the rudder stock. They claimed that it was "used with success in a Ship of Four hundred Tons burthen, in the West India Trade, for four Years," until it encountered a hurricane in 1776.

There are two obvious problems with this admittedly innovative approach. The first is that given the proximity of the horizontal wheel to the sternpost, it would appear that the helmsman would have to be facing backward. The second is that the index of rudder angle is below the wheel, and could hardly have been easy to read.

The first problem might have been addressed by moving the steering wheel forward and using a chain drive to transmit the rotation of the pinion wheel under the steering wheel aft to another toothed wheel, meshing with the quadrant. As for the second problem, a better position for the index would have been on the other side of the wheel from the helmsman, and at a height above it, so it was in the helmsman's field of view.

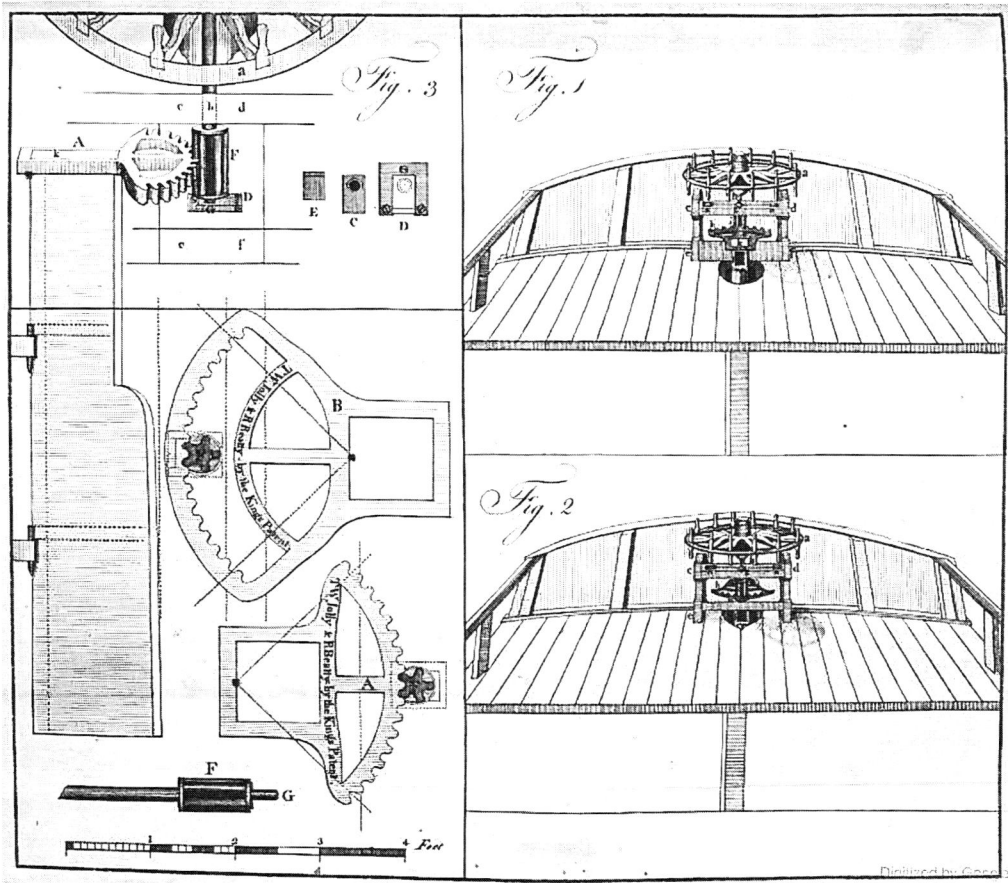

Figures 1-4 from Jolly and Beatty (1779).

If one were to keep the "vertical wheel" (horizontal axis), the simplest way of changing the axis of rotation would be by means of a pair of gears meshing at right angles to each other. Of course, the transmission system must have additional mechanical elements, as the steering wheel is above and forward of the rudder head, but the right-angle drive is the key.

The Godsoe patent (1860) indeed used a right angle drive. The output (vertical axis) gear's rotation was brought down to deck level by use of a vertical shaft, with a second gear at the bottom. This meshed with a "toothed segment ... which slides on a curved way"—essentially, a toothed quadrant, as the toothed segment is attached by a pivoting tiller to the rudder post. What I don't understand is where the helmsman is supposed to stand—in the figure it would seem that the tiller would be in his way. It would make more sense to place the lower gear, the toothed segment, and the tiller just below deck, or at least below a raised platform on which the helmsman stands.

North (1865) came up with an interesting variation. Referring to his patent drawings, his right-angle drive rotated an eccentric "cam" E (a circular gear

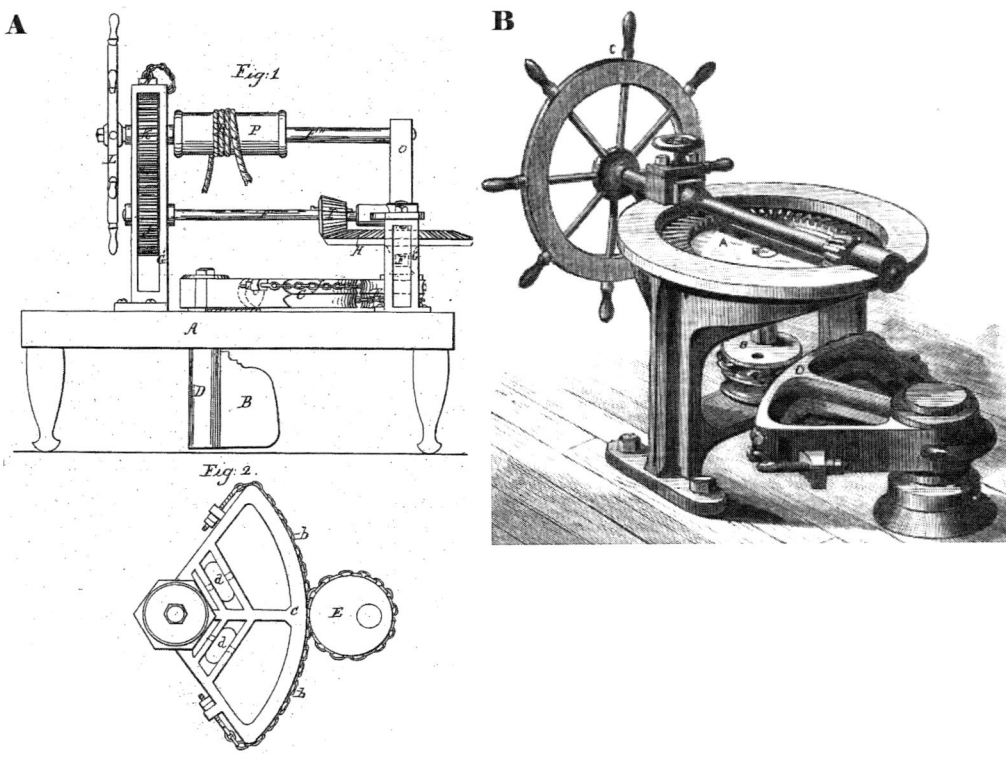

A. North's steering apparatus patent drawings, Figs. 1 and 2 (North 1865). B. Perspective view of a somewhat different embodiment. Here, the eccentrically mounted cam is B and the quadrant is D, but there is a difference in the implementation of the right-hand drive. (*English Mechanic and Mirror of Science* 1868).

mounted off-center), which meshed with an "eccentric-lever segment" C (a quadrant offering a quarter-elliptical rather than a quarter-circular arc). When the rudder was in neutral, "the longest extremity of wheel E and the shortest extremity of the wheel C are placed so to be in line with their shafts." That would be the position of least mechanical advantage, but the power would be increased as E and C rotated. According to Knight (1881, 2371), "When the rudder is 'hard up' the quadrant presents its longest radius to the shortest radius of the cam, thereby increasing the power on the rudder at the time when it is most required."

Screw-Type Steering Gear

The first patented device of this type appears to be that of John Rapson (1834). He explains, " A screw is formed on the shaft carrying a steering-wheel; a pinion on the end of this shaft gives motion to another pinion on the end of a similar screwed shaft, parallel to the first and beside it. On each of these screws works a nut, and each nut is joined by a connecting rod to an arm on the rudder-head. As the steering wheel is turned, the nuts are caused to travel in opposite directions, and motion is thus imparted through the connecting rods to the rudder" (Rapson 1834). The operation is best understood by review of the accompanying figures.

It is apparent that the helmsman would be forward of the wheel, facing the stern, as the screws block the opposite orientation. Hence, the helmsman would be completely dependent for proper steering on receipt of helm orders from an officer looking forward.

Nonetheless, according to *Mechanics' Magazine* (1836, 450), this steering gear was "adopted with success in ten or twelve vessels of considerable tonnage." However, at some point the apparatus was simplified by eliminating "one of the endless screws and its appendages." It was found that it was "wholly unnecessary, and only calculated to increase friction, and cause a considerable strain, whenever both screws happened not to be threaded exactly alike, or when both

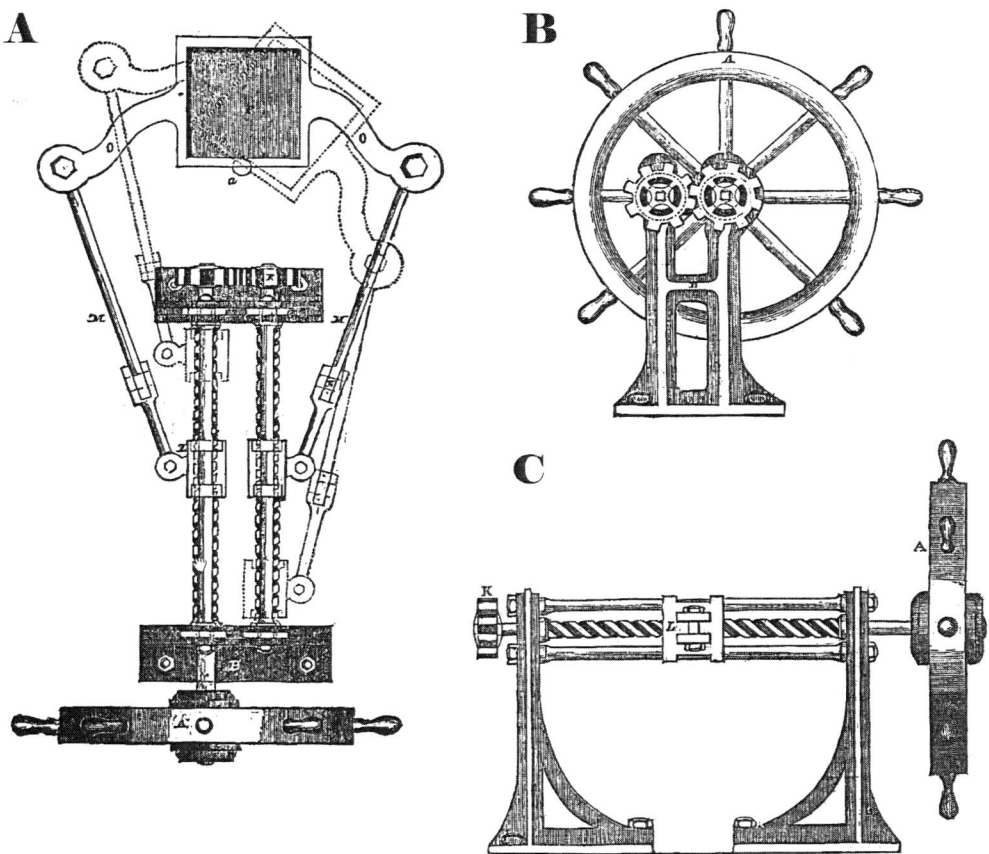

A. A plan view of the Rapson two screw steering gear patented 1835, showing the two screws, and the nuts, connecting rods and rudder head arms in the neutral and turned positions (*Mechanics' Magazine* 1836, Fig. 1). B. A frontal view, showing the steering wheel, the two vertical supports, and the interlocking pinions (gears) (*Mechanics' Magazine* 1836, Fig. 2). C. A side view, showing the steering wheel, a screw and nut, and the supports for the screw (*Mechanics' Magazine* 1836, Fig. 3).

connecting rods were not precisely of the same length, or when one was bent by accident."

Notwithstanding this objection, two-screw steering gears continued to be proposed, as evidenced by the Buffington 1873 U.S. patent, which seems identical in all important aspects to Rapson's.

I assume that Rapson's single screw derivative was single-threaded. In contrast, Arthur Murray employed a "double-threaded screw," that is, one "cut both right and left handed," on the American corvette *Niagara* in 1857. This was traversed by two nuts (sometimes called "half-nuts"), one right- and the other left-handed. The wheel was "placed directly over the axis of the rudder," and the wheel was connected to the nuts, and the nuts to a ring on the top of the rudder post, in such a manner that "when the wheel is moved the nuts travel in opposite directions and turn the rudder by a couple of nearly balanced forces" (Victoria and Albert Museum 1899, 274, entry 819).

A similar concept was used in the mail SS *Ville de Paris*, except that its screw was right-handed "for half its length, while the other half is left-handed." The nuts traversing this screw were "connected by links to a double armed lever on the head of the rudder post" (Victoria and Albert Museum 1899, 274, entry 820).

Powered Versus Manual Steering

As with other shipboard devices, there were considerable advantages gleaned by replacing muscle power with steam power. HMS *Minotaur* was conventionally steered using steam-powered gear, but had manual power as a backup. "With manual power ... seventy-eight men were required to steer her, and the helm was only got over to 23 degrees in 1½ minutes, while turning the circle occupied about 7½ minutes. With steam steering-gear two men put the helm over to 35 degrees in 16 seconds, the circle was turned in about 5½ minutes, and its diameter was less than two-thirds that" obtained manually (White 1894, 668).

The Flettner Rudder

The Flettner rudder was developed as an alternative to the powered steering engine. "The main rudder swings freely through 360° and is turned by the action of the small auxiliary rudder fitted in the after part of the blade. The area of this small rudder is about 1/20th of the area of the main rudder and it is controlled by horizontal rods connecting it with a yoke on a vertical shaft inside the hollow rudder post. When the auxiliary rudder is turned to port it swings the main rudder to starboard, which swings the stern to port" (Skene 2001, 105).

"Control of the Flettner rudder requires only about five percent of the power needed to manipulate an equally large rudder of the old type. Even a ship of 1000 tons may be steered by hand" (Martin 1926, 39). A further advantage, noted by Flettner in his 1928 U.S. patent (based on applications filed in 1917 in Germany and in 1920 in the United States), was that the "main rudder may be more rapidly thrown

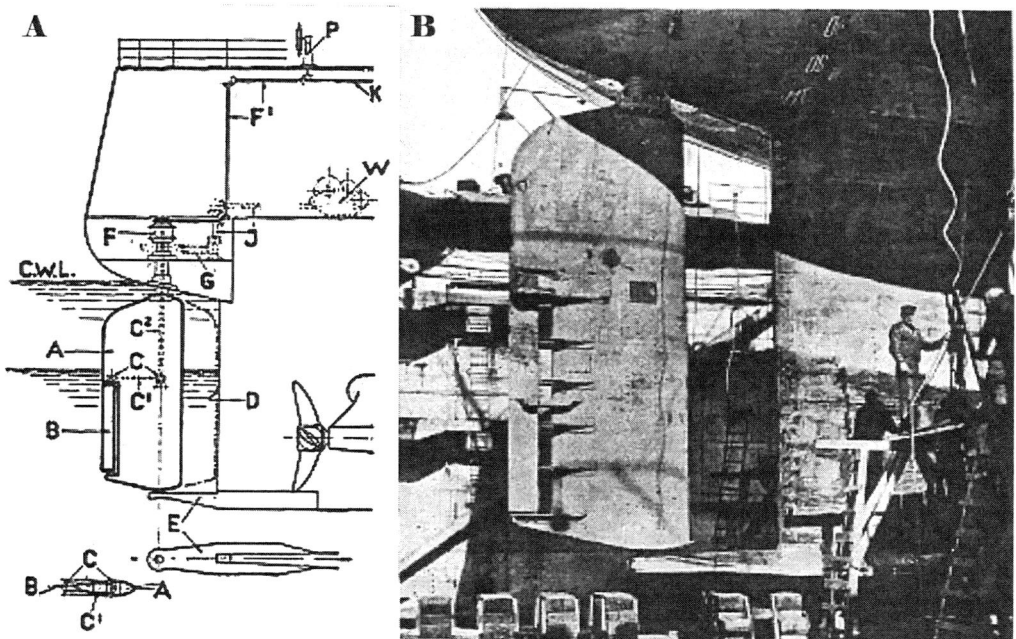

Photograph and profile drawing of the Flettner rudder system. Author reversed the photograph to correspond to the profile drawing. In the latter, we have hollow main rudder A attached to hollow rudder post F, and pilot rudder B aft the after end of A. It is mounted on yoke C (Foerster 1923, 237). Photograph is of the motorship *Odenwald* in dry dock (Flettner 1926, 32); "rudder is in mid-position while backing."

over" by the action of his auxiliary rudder than by a steering engine acting on the main rudder directly.

The Flettner rudder was first installed on the 200-ton cargo steamer *Frigido*, and then on the 5,000-register ton *Odenwald*. In the case of the *Odenwald*, the auxiliary rudder was 12 square feet, and the main rudder 140, both "partially balanced" (Foerster 1923, 237).

The auxiliary rudder, and hence the main rudder, could be steered by hand. However, on the *Odenwald*, an automatic steering system was also introduced, as discussed next.

Automatic Piloting

The earliest auto pilot was probably as old as the rudder itself: tying down the rudder or tiller to give the steersman a break. However, while this fixed the rudder angle, it had no capability to react if there was a change in the winds or currents.

In 1866, John McFarlane Gray patented a servomechanism that sensed the rudder angle and used it to control a valve on a steam steering engine. "The steam valve reduced power as the rudder approached the desired angle," which would have reduced "hunting" for the correct angle. This steering system was installed on SS *Great Eastern* in 1867 (Ma 2024, 125).

In 1906, Nelson applied for a patent on a less sophisticated steering apparatus in which a magnetic compass was used to sense deviations in course and actuate an electric motor controlling the rudder (Nelson 1912). The sort of control envisioned by Nelson—called proportional control because the applied rudder angle was proportional to the heading error—was old in the mechanical art, having been used in Watt's 1788 centrifugal governor for the steam engine (Mayr 1970, 2–3).

The magnetic compass, of course, suffers from the problems of magnetic variation and deviation that we discussed in Chapter 1, and since only the error in the course was sensed, there would likely be the problem of overshooting. Ideally, as the ship approached the correct heading, the controller would ease off on the rudder, or even apply "counter-rudder."

Continuing the saga of the *Odenwald*, the Flettner rudder was combined with a gyroscopic compass. If the ship yawed off course, it closed an electrical circuit, energizing a 0.5-hp electric motor "connected by a Gall chain to the axis of the steering wheel," and causing the steering wheel to be turned to bring the ship back on course (Foerster 1923, 237). From the description, it does not appear that *Odenwald* circuitry dealt with the problem of overshooting, either.

Elmer Sperry's 1911 experimental autopilot design was also gyroscope-based, but it sensed "heading error rate" as well as "heading error" (Roberts and Sutton 2006, 3; Sperry 1914; Sperry 1920). It could thus sense when the heading error was being rapidly reduced and "ease off," and Sperry referred to the corresponding mechanism as an "anticipatory device."

In 1922, Sperry Gyroscope introduced a commercial product, the "Gyro-Pilot," perhaps better known by its nickname, "Metal Mike." This had several adjustments to tweak its performance to a particular ship or even the instant weather condition and ship speed, and the autopilot could be disengaged entirely when in the presence of a navigational hazard (Mindell 2002, 72–75).

In 1922, Nicolas Minorsky analyzed the problem of automatic steering mathematically. He defined a class 1 controller as one that controlled the rudder angle, and a class 2 controller as one that controlled the rate of change of rudder angle. In both cases, these could take up to three different inputs: the heading error, the rate of change of the heading error (I'll call it "heading error velocity"), and/or the rate of change of the heading error velocity (I'll call it "heading error acceleration"). In calculus terms, these are the first and second derivatives of the heading error.

He showed that use of just the "heading error" would not work well on large ships (Bennett 1984, 11). Moreover, while a three-input, class 1 controller would work fine for damping out a transient disturbance in heading, it would not damp out a steady disturbing torque, such as one from a steady wind (12).

On the other hand, with a class 2 controller, that is, one for which *rate of change* of rudder angle was a suitable linear function of the heading error, the heading error velocity, and the heading error acceleration, a steady disturbance could also be compensated for. However, since it is the rudder angle that is controlled directly, Minorsky performed some mathematical magic and expressed the rudder angle as a linear function of the heading error, the heading error velocity (derivative), and the cumulative past heading errors (in calculus terms, the integral of the heading error with respect to time). This came to be known as a "PID" controller, because one term was proportional to the heading error, a second was its integral, and a third was its derivative (Bennett 1984, 12). The integral term looked at the "past" error in heading, the proportional term at the "present" error, and the derivative term at the "anticipated future" error if no correction was made. The integral term was the key to correcting for the steady disturbance.

Proportional control, by itself, responds to a transient disturbance by overshooting: an oscillation about the desired heading. Derivative control, by itself, responds with an "offset" error; the ship's course will be parallel to the one originally intended. Together, the two can compensate for a transient disturbance, but not a permanent one. For that, integral control must be added (Tetley and Calcutt 2007, 322–4).

In the past, others had empirically devised controllers that sensed the integral of the deviation of the desired output in some way. The earliest example may be the speed regulator on Périer brothers' steam engine (1790s) that powered a grist mill on the Isle des Cygnes; the speed was proportional to the flow rate, but that was controlled by a float, for which the level would be the integral of the net flow into the regulator vessel (Mayr 1970, 117). But Minorsky created a control *theory*, and advocated "continuously variable control action" (Bennett 1984, 10).

Minorsky tested an experimental steering controller on the USS *New Mexico* in 1923. That warship was "fitted with the Waterbury electrohydraulic steering gear" and therefore "permitted continuous control action" (Bennett 1984, 12).

Bennett has provided mixed teachings as to exactly how this controller worked. In his 1984 paper he said it had a class 1 controller (Bennett 1984, 12). But in his 1993 book he asserted that it controlled the "angular velocity of the rudder, not the angle," making it a class 2 controller. Initially, it was just a "PI" controller, and the resulting yaw was ±2°. Then Minorsky added the third input, making it a "PID" controller, and "the yaw" was now reduced to ±1/6° (Bennett 1993, 145–6).

In any event, "the work was abandoned in 1930 when the US Navy withdrew its support because '… the operating personnel at sea were very definitely and strenuously opposed to automatic steering'" (Bennett 1993, 18).

That did not, however, prevent the Navy from putting a Sperry gyro pilot onto the old battleship USS *Utah* (BB-31, commissioned 1911). In 1931–32, it was turned into an unmanned, remote-controlled target ship (as AG-16) (Leniman 2001). Nor did it keep Sperry from installing more than 400 gyropilot systems on civilian ships by 1932 (Åström 2012).

A PID controller was disclosed in Henderson's 1922 British patent: "A steering mechanism for sea and air craft comprises an element, such as a time-deviation integrator, controlled automatically by a deviation from the prescribed course and adapted to displace the rudder slowly until the deviation is corrected and to maintain the displacement thereafter to balance external forces tending to produce deviation. The element may be combined with other elements controlled in accordance with and producing displacement of the rudder proportional to the deviation and to the angular velocity of deviation of the ship."

Even the 1911 Sperry gyropilot is sometimes referred to in modern literature as a "PID controller" (O'Dwyer 2005; Ma 2024, 125), but Minorsky stated in 1937 that Sperry's commercial system used "proportional + acceleration control of the rudder angle" (Bennett 1984, 13; Mindell 2000, 141). An integrator was not disclosed in the foundational Sperry patent applications filed in 1922 (Sperry 1929; Mills 1928). Sperry acknowledged that as a result of various factors, "the vessel is often continually forced off the course in one direction," and the control system would result in "a continuous yawing movement of the ship mostly to one side of the true course." The solution disclosed in Sperry's 1929 patent was to define a "virtual center" for the rudder, differing from the "true center," that was "sufficient to counteract the effects of the extraneous forces and keep the ship on course." (This may be compared to a sailing ship helmsman giving the rudder "weather helm" or "lee helm" to compensate for the ship tending to pull to one side or another [Harland 1985, 56–7].)

If an autopilot maintains a fixed heading, the ship will follow a rhumb line course, a straight line on a map using a Mercator projection. However, we know from Chapter 1 that this will not be the shortest distance between two points on the spherical earth. Whitaker's 1926 patent related to a Sperry-type autopilot modified so that the desired heading was changed automatically in such a manner that the ship would follow a great circle route.

Steering Propulsors

Propellers

The ship's propeller may be used for steering as well as propulsion if it is pivotably mounted, like the modern motorboat's "outboard motor." Alternatively, a ship may be equipped with an auxiliary propeller dedicated to steering.

Shorter (1800), Dallery (1803), Trevithick (1815), Millington (1816), and Hunt (1839) apparently all proposed making it possible to swivel the propeller (Bourne 1855, 36). Bourne said that the arrangement "has not come into use, and there is little probability that as a propeller, the arrangement will meet with any considerable adoption" (37). Bourne did not explain his reasoning, but it likely that the increased maneuverability came at too high a cost.

The first U.S. patent in this line of development appears to have been that of Fessel (1858). His propeller was mounted in a frame "capable or rotating to some extent around a vertical driving shaft … and which is so geared with a steering apparatus that the propeller shaft may be set at any required angle to the center line of the vessel and the propeller thereby made to perform the duty of a rudder without interfering with its action as a propeller."

A vertical shaft C passed through the center of the frame A, and, via "a pair of bevel gears G and H" (i.e., a right-angle drive), rotated the propeller shaft D. Chains or ropes ran from the steering wheel to a chain wheel I. This geared with a pinion h, which in turn geared with circular spur-toothed rack g, which turned frame A, and thus swiveled the axis of the propeller.

Fessel disclaimed "applying a screw propeller in such a manner that its position can be changed to make it operate as a rudder," acknowledging that this was already known in the art, or at least was not considered "inventive" ("obvious" in modern patent law parlance). Rather, his claimed invention was the disclosed "arrangement and combination of the slotted frame A, propeller F driving shaft C, and chain wheel (I)."

Mallory obtained a steering propeller patent in 1874. (The differences from previous ones seem rather minor to me.) In 1878, Mallory demonstrated his steering propeller to the British. "*The Engineer* magazine was impressed by the maneuverability of the boat, noting that 'the only thing with which the evolutions of Colonel Mallory's craft can be compared are those of a perfectly trained circus horse' and went on 'for any purpose where extreme ease of handling is required, the Mallory system is simply perfection; for example, for tug boats or torpedo boats nothing can be imagined more suitable.'" Some concern was expressed with regard to the durability of the bevel gears, but the seawater apparently acted as a lubricant (Riviera 2008).

A Mallory steering propeller was retrofitted onto the early American torpedo boat *Alarm* (commissioned 1874). It "allows the whole energy of the main or driving engines to be instantly used for steering purposes, and then be instantly restored for driving purposes" (Matthews and Brown 1879, 503). But Caiella (2019) considered the Mallory propeller to have been a failure, as, in July 1881 tests, "the *Alarm*'s speed remained far short of what was desired," and the addition of the Mallory propeller added 3.5 feet to its draft, a serious concern for a warship intended to operate in shallow waters.

This criticism was not entirely fair. Because it was a retrofit, the draft was greater than if the vessel had been designed to bear the propeller (Isherwood et al. 1882, 6). And during the trials, the vessel's bottom was "entirely covered with barnacles nearly one-quarter of an inch high and so closely packed that no iron was visible between them. In addition to the barnacles, there was a slimy green vegetable coating, interspersed at short intervals with tufts of coarse sea-grass from four to five inches in length." A "diminution in speed" could be expected under these circumstances (53). The review board concluded, "the results of the experiments show the durability, reliability, and practicability of the Mallory propelling and steering screw" (53).

The board conceded that the Mallory propeller, as tested, turned the boat more easily to one side than another. This was the result of the torque exerted by the rotation of the propeller on the ship's hull. However, Mallory's 1881 patent proposed a solution, which was to provide "two oppositely-revolving propellers," so the torques would cancel each other out (Isherwood 1882, 55).

The real problem was that the raison d'être for the *Alarm* was flawed. It was an armored gunboat as well as a torpedo boat, and its large-caliber gun was aimed by pointing the bow at the enemy. It was thought that the vulnerability of the boat was greatly reduced by this orientation, as opposed to a broadside one. In order to be able to keep its bow so pointed, it needed to be able to change heading quickly, even when moving at a speed too low for the normal rudder to be effective (Isherwood 1882, 3–4). However, speed was more important than maneuverability for successfully evading enemy fire.

The use of a second, auxiliary propeller for steering was disclosed by Liburn (1866). His patent figure shows a "small propeller-wheel, which is used for steering the vessel," forward of the main propeller. However, the axis of this small propeller was not fixed. Rather, the axis shaft of the steering propeller is attached to a vertical shaft, and a pulley acts to rotate it around the latter.

The auxiliary steering propeller could no doubt still be used for steering, and yet be of simpler and cheaper construction, if it were non-swiveling, but fixed with its propeller axis perpendicular to the ship's centerline. But this would lose the advantage of serving as a backup to the main propeller. Liburn taught that "by unshipping the main propeller B the steering-propeller C can be used either for propelling or for giving direction to the vessel."

Steering propellers may be mounted at the stern, the bow, or even on the flank, although the last position is more for shifting the ship's position to one side or another, rather than for changing its heading.

The modern form of the steering propeller is the azimuthal thruster. The propeller's blades may be of fixed or controllable pitch, and the propeller itself may be fixed or retractable. The propeller may have either a mechanical or an electrical transmission. The mechanical transmissions for the swiveling system are of two types, L-drive and Z-drive. The L-drive is a single bevel gear right-angle drive that converts a vertical axis rotation to a horizontal one. The Z-drive uses two right-angle drives, so a vertical axis rotation is converted to a horizontal, which in turn is converted to vertical (Aniker 2021, Chapter 1).

The tunnel thruster is a maneuvering propeller mounted inside a tunnel running transversely through the hull. If mounted in line with the ship's center of gravity, it produces a "fixed direction transverse force," rather than a change in heading (El-Hawary 2000, 3–7).

The Bureau of Naval Personnel (1970, 633) makes reference to bow and stern thrusters that are essentially tunnel thrusters whose tunnels run "through the bow or the stern. These propellers can be reversed to provide thrust in either athwartship direction." These thrusters turn the ship by virtue of their position (their line of thrust does not pass close to the ship's center of gravity) rather than by swiveling.

Waterjets

The term "propulsor" is used in this section's title because thrust can be developed by means of a waterjet rather than a propeller, and a waterjet can likewise be mounted so it can be used for steering. The earliest waterjet propulsion patent was in 1661, to Togood (King 1878, 110).

Ruthven's "hydraulic propeller" system, the first to use a centrifugal rather than a piston pump (pumps are further discussed in Chapter 12), was tested on the *Albert* (1853), the *Nautilus* (1865), and the *Waterwitch* (1866). This used a "turbine wheel" (centrifugal pump) to generate the jet. On the *Waterwitch*, the speed achieved was just 6.5 knots (King 1878, 111–112). It was estimated that only "one-fourth" of the engine power "was actually utilized in propelling the ship" (Franklin 1866, 392).

The wheel of Ruthven's centrifugal pump was about 14 feet in diameter. The water leaving the pump was "led with a spiral curve to the jets at the side of the vessel." Here there was a two-way cock so the jet could be directed either forward or backward (The Franklin Institute 1866, 397; drawing in Elliot 1868, 593). If the jet on one side were directed forward and the other backward, the vessel would be turned, but the turning moment created would be limited by the ship's beam (32 feet).

Pagenstecher's 1864 patent likewise placed the jets at the side of the vessel, but instead of having a two-way cock, they could be independently rotated on the lateral axis to face forward or backward. Pagenstecher explicitly stated that the ship could be rotated "by turning one of the nozzles toward the stern and the other toward the bow."

Brewer's 1879 patent (Brewer and Ward 1879), finally, contemplates "providing the nozzle or pipe through which the water is to be ejected with a turning joint so arranged that the direction of the water or nozzle can be changed or turned while the water is being ejected." More particularly, this turning joint "allows the nozzle to be swung either to the left or right," rather than forward or backward as in Pagenstecher's design. Brewer's Fig. 2 shows the water jet emplaced at the stern.

Waterjets first became popular in the second half of the twentieth century for "small high-speed pleasure craft and work boat situations where high maneuverability is required with perhaps a draft limitation." While some modern waterjets have steerable nozzles, in others, "deflector plates are used to control the direction of the flow" (Carlton 2011, 357).

Paddle-Wheels

Steering by means of a paddle-wheel has also been proposed, even as late as 1900. Gensichen and Ehrke proposed placing a single paddle wheel at the bow, and two at the stern, "one on each side of the stern frame." These "rotate about horizontal axes, their lower halves being under water." Their patent figure 3 shows that these axes are parallel to the centerline. Thus, "to turn the vessel quickly the forward wheel is rotated in the opposite direction to the after ones, and to move the vessel bodily to port or starboard all three are rotated in the same direction" (Great Britain Patent Office 1903, 1900 Patent 7545). (Such paddle wheels, which turn on a horizontal axis and have fixed blades, should not be confused with the cycloidal propellers discussed below.)

Flettner Rotors

A rotating cylinder, placed in an air flow with its axis perpendicular to the flow direction, develops a lift force (the Magnus force) that is perpendicular to both. Thus, with vertical cylinders, one can use it to propel a ship. If, by way of example, the cylinders are rotating counterclockwise (as viewed from above) and the wind is from the east, the lift force will move the ship northward (Borg 1986, I:18). The wind may be natural or an apparent wind generated by also propelling the ship in a more conventional manner. In the latter case the purpose of the Flettner rotor is to reduce fuel consumption.

In the 1920s, Flettner conducted experiments and "concluded that a rotor could produce 8 to 10 times the driving force of conventional sails having the same projected area" (Flettner 1926, 6). He retrofitted the barkentine *Buckau* (renamed the *Baden-Baden*) as a "rotorship," with rotatable cylinders fore and aft. It crossed the Atlantic in both directions. Not only could the rotorship "be sailed much closer to the wind than was possible with the conventional sail plan," it "was also possible to steer the vessel by changing the rotational speed of the rotor sails and even to sail in reverse" (Flettner 1926, I:8).

Borg (1986) does not explain how this steering was performed, but consider a rotorship relying on the apparent wind generated by a conventional propeller pushing it northward. If the rotor rotated counterclockwise (as viewed from above), the lift force would push the rotor to the left, and if clockwise, to the right. If the rotor is near the bow (i.e., away from the center of gravity), this creates a turning moment, counterclockwise or clockwise, respectively. (The turn would be in the opposite direction if the rotor were spun the same way but was near the stern.) If the rotor ship lacked a conventional propeller, and must rely on the natural wind, steering by rotor would require that the natural wind be other than a pure crosswind, and a headwind or tailwind would be best.

The Magnus effect may also be generated by the flow of the water, rather than the air, around rotating cylinders. Roos (1928) proposed use of a single rotating cylinder at the stern as a rudder; the turning action would depend on the direction of rotation of the cylinder. If rotating cylinders were placed at both the bow and stern, not only could the turning speed be increased (if they were rotated in opposite directions), but the ship could be moved "broadside" by rotating them in the same direction. Roos's "rollers" lacked "the necessary end plates for high coefficients of lift" (Borg 1986, II:34); this oversight was remedied by Gasparini (1928) (Borg 1986, II:39).

Cycloidal Propellers

These combine propulsion and steering into a single unit. A cycloidal propeller (cyclorotor) looks something like a paddlewheel with the blades at the end of spokes (the rotor), but as the blades revolve around the center, the blades also oscillate around their own axes, thus changing their angle of attack. Depending on the relative motions of the blade and the rotor, it can create thrust in any direction perpendicular to the axis of rotation of the rotor. Thus, if the axis of rotation is vertical, the thrust can be in any horizontal direction. In modern cycloidal propellers, the blades have a symmetrical airfoil shape.

An individual blade will generate a lift force if it is at a nonzero angle of attack (the angle its chord makes to the direction of fluid flow around it). The direction of fluid flow is opposite the direction of movement of the blade in space. Initially, the direction of fluid flow encountered by each blade is tangent to the blade circle. Once the ship is moving, the blades move on a cycloidal path (hence the name) and thus the direction of fluid flow is the tangent to that path (Jürgens 2011, 5).

The oscillation of the blades has the same period as the rotation of the rotor; the amplitude determines the magnitude of the thrust generated, and the phase of the oscillation determines the direction of the thrust generated.

If the blades are oscillating, then at any given moment, each blade will be at a different angle relative to the blade circle, and thus the lift force on each blade will be different. However, since the blades are arranged symmetrically around the rotor, for a given phase setting, there will be a favored direction of thrust; the transverse

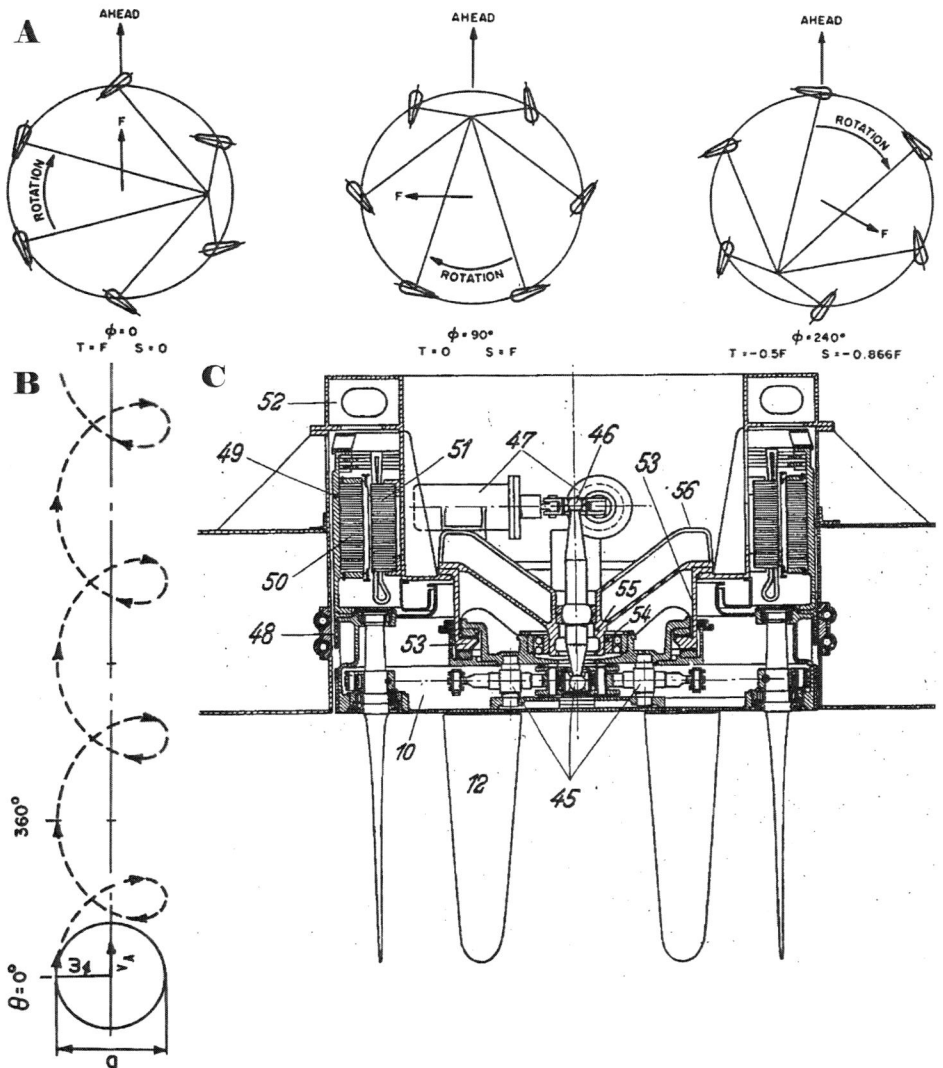

A. Schematic showing how the displacement of the control center of a cycloidal propeller changes the blade orientations and the direction of the resultant force (shown by arrow labeled "F") relative to the "ahead" (bow) direction (Ficken and Dickerson 1969, 17). B. Schematic showing the cycloidal path of one of the blades of such a propeller (Ficken and Dickerson 1969, 15). C. The Voith-Schneider propeller, vertical cross section. Servomotors 47 act on lever 46 (top end of the control rod), the lower end of which acts on the "blade shifting mechanism" 45. The blades 12 are mounted on the blade wheel 10 (Ehrhart 1935, Fig. 7).

force components on the symmetrically opposite blades will cancel each other out (Jürgens 2011, 5–6; see generally Fork and Jürgens 2002).

The "Fowler Wheel," patented in 1870 (King 1878, 111), was an early form of cycloidal propulsor, "a four-blade horizontal feathering paddle wheel ... with its blades controlled by links to a camshaft." Unlike a conventional paddle wheel, it turned on a vertical axis, that is, in the horizontal plane. "To move forward, all the paddles would be feathered—pointed in the direction of travel—except as they approached the starboard beam position. There the paddles were rotated perpendicularly to the ship, providing forward thrust." The catch was that the "wheel required about three times the power needed by an ordinary screw to achieve the same speed" (Caiella 2019).

On the Voith–Schneider propeller (VSP), the "amplitude and phase of the blades' motion," and thus the magnitude and direction of the net thrust, are controlled by the position of a mechanical "steering center." When this is at the "center of the rotor casing," the blades remain tangent to the blade circle for the entire revolution. There is thus a zero angle of attack, zero lift force, and zero thrust. When it is moved away from that center, to an "eccentric" position, the blades oscillate, creating lift and thus thrust. The degree of displacement determines the magnitude of thrust and the direction of displacement, the direction of thrust (Jürgens 2011, 3, 7).

In the original Voith patent (1933), this steering center was ring 42 and "rotated synchronously with the propeller rotor." It was connected to rings 43 and 44, and these could be shifted in any horizontal direction by the combined action of two operating levers. Ring 42 would then also be displaced.

In the Ehrhart 1935 patent, assigned to Voith, lever 17 pivoted at its center 18 in the socket 19. This caused the movement of ball 16 at the lower end of the lever, and thus of the steering center plate 15. In one embodiment, two orthogonally positioned servomotors acted on the upper end of the lever. That is essentially, the modern VSP system (Wärtsilä 2024a).

However, both the servomotors and the steering center plate are limited to horizontal movement. How, then, do we reconcile this with the pivoting of the lever?

I posed this question to Dirk Jürgens, Vice President of Research and Development, Group Division Turbo Marine, at Voith, and he responded, "You are right that the top end of the control rod descends and the bottom end ascends. But this is not a problem because we have a special mechanism to equalize both changes in height." Unfortunately, the nature of that mechanism is considered proprietary by Voith.

The VSP control rod has been compared to the early modern whipstaff (mvsmith 2007) . That is a fair comparison in that both control a steering mechanism by pivoting. However, with the whipstaff, the "business end" was kept in the plane of rotation of the tiller by sliding the whipstaff downward as the pivot angle increased, whereas Voith obviously takes a different (albeit cryptic) approach to the problem.

7

Going Fishing

(Fishing Vessels and Fishing Gear)

Fishing is not as old as civilization itself; it is much older! Salmonid fish remains and self-barbed points have been found at Upper Paleolithic habitation sites in Spain (Knecht 2013). By the Neolithic, primitive boats (dugout canoes, reed boats, and possibly hide boats) were available. The most obvious uses for such boats would have been for transportation and fishing. Modern reconstructions show that these boats were capable of at least island-hopping in the Mediterranean (Bachhuber 2011), and thus, presumably, of fishing for pelagic (open ocean) fish.

De Moor asserts that "archaeological finds demonstrate that all essential fishing techniques were known in the Near East before the invention of the art of writing around the beginning of the third millennium BC" (1998, 85).

Some techniques were actually quite a bit older than that. The "Antrea" fishing net, found in Finland, had a radiocarbon date of 9140 BP. The net was made of willow bast, and both bark floats and stone sinkers associated with the net were preserved (Miettinen et al. 2008). In the Maedun Cave in South Korea, researchers found grooved limestone sinkers, dated to 29,000 BP (Upper Paleolithic), and fish bones (AFP 2018). There are, of course, many different ways that nets can be used in fishing.

Fishing Gear

The International Convention for the Prevention of Pollution from Ships, Annex V, defines fishing gear as "any physical device ... or combination ... that may be placed on or in the water or on the seabed with the intended purpose of capturing or controlling for subsequent capture or harvesting marine or freshwater organisms" (He et al. 2021, 4).

The Food and Agriculture Organization has a multitiered classification of fishing gear. Even looking just at the first tier, it is diverse: surrounding nets, seine nets, trawls, dredges, lift nets, falling gear, gillnets and entangling nets, traps, hooks and lines, and miscellaneous gear (which includes harpoons, spears, pumps, electric shock, diving, etc.) (He et al. 2021, 2–3). There's also "auxiliary gear," which "support fish capture but do not directly catch fish"; this includes devices for finding and attracting fish, bycatch reduction, and fish transfer (4–5).

152　　　　　　　　　　　　　Work at Sea

Our concern here is with work at sea, and hence with fishing gear used by fishing vessels, not shore-based fishers.

Fishing Nets and Lines, Generally

Fishing nets take a variety of forms, but they all consist of a mesh created by interwoven ropes. The size and shape of the mesh openings determine which fish are captured by the netting, and consequently they should be uniform. With rough thin

Color woodblock print of Japanese fishing boat netting flatfish in Wakasa Province (Wakasa, gyosen karei ami), from the series "Famous Places in the Sixty-odd Provinces (Rokujuyoshu meisho zue)," 1853. Artist: Utagawa Hiroshige. Bruce Goff Archive, gift of Shin'enkan, Inc. (Art Institute of Chicago, reference 1990.607.135).

yarns, twisting is sufficient to obtain a uniform mesh, but with smooth thin yarns, knotting is necessary (Gabriel et al. 2005, 184–9).

Until the nineteenth century, nets were made by hand, typically by the fisher or the fisher's family. This was a time-consuming process, and if a net had to be replaced with a new one, a fisher might have to "relinquish fishing for a whole season" (Jamieseon 1829, 595). In 1796, J.W. Boswell received a 50-guinea reward for the invention of a fishing net machine (Felkin 1867, 141), and Robert Brown received a patent for another such machine in 1802 (156). "The fishing net machines combine the general features of the power-loom and the lace-machine … One woman can do the work of upwards of 100 hand-netters. The meshes are made rectangular, in the direction of the length of the net, and not diagonally, as in hand-made nets" (Whitworth and Wallis 1854, 25).

Until the twentieth century, the ropes were made from natural fibers; the choice of fiber varied from place to place and time to time.

The fibers could be treated to improve their qualities. In mid-eighteenth-century Norway, the flounder nets were made of hemp dyed brown with boiled oak and alder bark. Linnaeus reported that this was said to have two advantages: "the nets thereby become much more resistant to rotting; secondly, the fish do not shun these dyed nets as much as if they were white and un-dyed" (Hansen 2022).

In the mid-nineteenth century, Scottish fishers discovered that cotton nets were more durable than those of hemp, and could also "be made much finer and lighter" (Bremner 1869, 313).

Nylon (polyamide 66) was "the first of the man-made fibres to be widely applied in fishing nets" (Kristjonsson 1959, xxv). In Japan, in 1955, nylon, vinylon (polyvinyl formal), and polyvinylidene chloride were the three synthetics used in the manufacture of fishing nets, and synthetic fiber fishing nets had about a one-quarter market share. By 1961, polyvinyl chloride, polyester, and polyethylene were also in use, and synthetics accounted for 85 percent of the netting market, although "about 90% of the ropes used [were] still of natural fibers" (Schaefers and Alverson 1964, 2). Also, at the 1963 "Fishing Gear Congress," reports were made of experiments with polypropylene as a net material.

One of the factors that propelled the adoption of nylon was catch efficiency. Nylon gillnets were two to three times as effective as cotton or linen ones in several studies (Radhalekshmy and Gopalan Nayar 1973, 154). There were several theories as to why this was so. One was that "the elasticity of nylon allows the fish to force the twines apart and become caught." Another was that nylon nets are less visible to the fish because they are translucent, and nylon's strength permits use of finer twines (156).

In general, synthetic fibers do not rot (Radhalekshmy and Gopalan Nayar 1973, 146). This is an advantage as long as the nets remain in the hands of the fishers. However, once they are lost or abandoned, they become "ghost nets," continuing to catch fish without any human benefit. Biodegradable fishing nets have been developed. Unfortunately, they generally are inferior in catch efficiency to nylon—for example, 25 percent less efficient in catching the Northeast Atlantic cod (*Gadus morhua*) (Cerbule et al. 2022).

Floats and Sinkers

If the design required that part of the net be held at the surface, a float could be attached. Premodern floats were typically made of light woods or cork. Hollow glass floats were first used by Norwegian fishermen in 1840 (Pritchett and Seliger 2010, 128). They were reportedly invented by Christopher Faye, of Bergen, Norway (Baird 1879, 711), and were especially favored in the Lofoten Islands (Simmonds 1879, 35). In 1910, Japanese fishermen adopted them (Pritchett and Seliger 2010). In the twentieth century, plastic and aluminum floats became available.

If part of the net had to be sunk, the traditional option was to use a stone weight. At least by the nineteenth century, lead sinkers were also used. In 1869, Scott declared that "all sinkers should be of lead, as one of the most ponderous metals" (1869, 88).

Purse Seine

This is a type of surrounding net. "A purse seine is a wall of netting designed to encircle a school of pelagic fish near the surface and use a purse line to close the bottom of the net" (He et al. 2021, 7). The fishing boat sails around the school, deploying the net; the top of the net is held up by floats. Nowadays, "this gear accounts for about a third of total marine landings" (8).

"According to Scofield, closing the net bottom by pursing is said to have originated in 1826 in the Rhode Island menhaden fishery" (Iitaka 1971). According to DeBlois, this purse-seine was "284 meshes deep and 65 fathoms long … The first time the seine was set, there were fourteen men to help; they set around what they called a 500-barrel school of menhaden, and, while they were pursuing, the fish rushed against the twine so hard, that they twisted and snarled the twine around the purse and line and weight." After six hours of futile attempts to "gather the seine up, or get her into the boat again," they towed it ashore and unsnarled it after the tide went out. "It was a number of days before they could muster the courage to set her again" (1881, 47).

However, Gabriel et al. (2005, 455) refer to a report that in Duhamel du Monceau's *History of Fisheries and Fishes*,* Volume 2, Section 3, Chapter 11 (1772), the Basques are said to have used a sardine gear that opened and closed "comme une bourse (like a purse)." Monceau is said to have specifically described a purse line running through rings of horn mounted on the lead line.

In the nineteenth century, to harvest the fish, the net would have to have been hauled in and brought on deck, so the fish could be scooped out. In the twentieth century, it became possible to bring the net alongside and pump the fish out of the net and into a receiving tank. Even then, the empty net would ultimately be brought on board so the fishing boat could sail on.

---

* I presume Gabriel is referring to "Traité général des pesches : et histoire des poissons qu'elles fournissent tant pour la subsistance des hommes que pour plusieurs autres usages qui ont rapport aux arts et au commerce."

Also, until the twentieth century, the net was hauled in by hand, and its weight was not trivial. The schooner *Grampus* carried a 150-fathom, 700-mesh deep purse seine, for which the weight, including the web (cotton twine), tarring, corks, leads, ropes, and purse-blocks, was 1,185 pounds. Once "put into the water and wet through, [it] will weigh 400 to 600 pounds more" (Collins 1890, 471).

Perhaps the greatest advance in purse seining was the Puretic power block, which greatly eased the hauling in of the seine. In 1953, Puretic "conceived of the basic idea of passing the entire purse seine net through an elevated free-swinging, self-powered V-sheave, so constructed that gravity would wedge the net into the sheave, giving it the necessary traction to pull the net out of the water" (Schmidt 1959, 400–1).

If the sheave is viewed perpendicular to the sheave axis, it looks like two truncated cones, placed nose-to-nose. Also, the surfaces of the sheave are ribbed, but parallel to the sheave axis (Puretic 1956).

Originally, this was driven by an endless rope drive, in turn operated by the purse winch (the winch used to pull shut the purse-line). Later, a hydraulic drive was introduced. On the 28-inch power block used to haul in 300-fathom purse seines on the Pacific Northwest Coast, the hydraulic drive could exert a pull of "up to 1½ tons" (Schmidt 1959, 403).

Boat Seine

The Food and Agriculture Organization (FAO) defines this as "a cone-shaped net with elongated wings, seine ropes and a codend [closed end], and operated by one or two boats, capturing fish by encircling and herding" (He et al. 2021, 15). The ropes are attached to the ends of the wings. The net is laid on a smooth seabed and the ropes and the wings help guide the fish to the mouth of the net. The net is towed and gradually closed by hauling in the ropes, thus changing the shape of the net.

The earliest form is probably Danish seining, in which one of the seine ropes is attached to an anchor buoy and then the rest of the gear is deployed in a circle. It was invented by Jens Vaever in 1848 (Noack 2017).

Trawls

In 1376, fishermen complained to Parliament that certain competitors were using a "close net ... fixed that even the smallest fish cannot escape therefrom." Moreover, "the net touches the ground under the water." Clearly, this was a bottom trawl. Not only was this "wondyrchaun" harvesting so much fish that the surplus could only be sold as "pig food," but there was environmental damage: The net "touched the ground so evenly and heavily that spawn of fish and the flowers of the sea and other things wherewith the greater fish might live and be nourished are destroyed" (Anderson et al. 2012a, 109).

The modern trawl "is a cone-shaped body of netting, usually with one codend, towed behind one or two boats to catch fish through herding and sieving." Trawls are designed to be towed across the seabed (bottom trawls) or in midwater (midwater trawls) (He et al. 2021, 19). To maintain the opening, bottom trawls may have a rigid beam, on sliding supports, that fits directly across the mouth. A fourteenth-century description of the "wondyrchaun" referred to it as having a "great and long iron" (Aquatic Community 2024); that is, it was a "beam trawl."

Later, the otter trawl was developed. This used a pair of vertical "otter boards" a little ahead of the junction of the net and the drag line. If these were positioned obliquely to the current, they created a sidewise force (hydrodynamic lift), moving the boards away from each other and thus opening the net (Gabriel et al. 2005, 98).

Otter boards are known to have been used in line fishing by 1855 (Gabriel et al. 2005, 98). Their first successful use in trawling was in 1895, by James Scott, who received an 1894 patent (395). The otter boards were originally flat, but practical cambered boards (airplane wing-like) were introduced several decades later. These produced a strong "shearing force" with low drag, but were more cumbersome to store on deck than flat boards (397).

According to Payne (2022), the "otter trawl ... proved far more efficient than the beam trawler but was unworkable with sailing vessels."

While it obviously was possible for a sailing vessel to pull a beam trawl, it was limited in when it could work (there had to be a strong enough wind, and the vessel had to set in the same direction as the tide or the trawl would be lifted off), and where (it "could not work areas of rough ground," presumably because the available wind power wouldn't overcome the frictional resistance) (Roberts 2009).

Steam trawlers did not have these limitations. Moreover, their steam power could be used to drive winches, and winches allowed the use of larger nets, and of chain or steel cables rather than hemp rope (Roberts 2009).

The first purpose-built steam trawler was David Allan's *Pioneer* (1877) (Crane 2010, 47). By 1900, there were more than a thousand "first-class" steam trawlers on the Register of British Sea Fishing Boats (Engelhard 2005).

In the 1950s, stern trawlers demonstrated some advantages over side trawlers. They were "equipped with a chute in the stern" and the "shooting and hauling of the net" could be "achieved mainly by winches alone." Also, the net could be "shot in any weather conditions" (Gabriel et al. 2005, 406).

Dredges and Tongs

Dredges and tongs have the same purpose—harvesting shellfish—but dredges are much larger, more elaborate devices.

"A dredge is a cage-like structure often equipped with a scraper blade or teeth on its lower part, either pulled or towed to dig animals out of substrate and lift them into the cage or bag." Dredges delve even deeper than trawls, targeting "mussels, oysters, scallops and clams" (He et al. 2021, 31).

In 1791, Parliament enacted a law declaring that "if any person shall unlawfully and wilfully use any dredge ... for the purpose of taking oysters or oyster-brood, ... or shall ... drag upon the ground or soil of any such fishery; every such person shall be deemed guilty of a misdemeanor," punishable with a fine of up to 20 pounds and/or imprisonment up to three months (Eyton 1858, 7). Likewise, in 1811, Virginia banned use of "any drag, scoop or rake, or other instrument, except tongs," to harvest oysters in Virginia waters. In 1820, Maryland adopted a similar ban, except it allowed use of rakes (Gordon 2020).

Nonetheless, the use of dredges to harvest oysters became commonplace. The size of the dredge depended on the size of the vessel and whether it was powered by sail or steam. Windlasses could be used to haul up the filled dredge, and on steam vessels, they were driven by steam (Kellogg 1910, 133–4).

Another efficiency improvement was the hydraulic dredge. A "high-pressure hydraulic jet" was used to "fluidize the substrate and wash out animals from the sediment and into the cage." There were also alternative harvesting methods. Rather than haul up the dredge, the shellfish could be sucked up by a pump (He et al. 2021, 34).

"Tongs consist of a pair of rakes or rake-like baskets attached to two long handles joined together like scissors" (He et al. 2021, 75). Tongs can be used from small boats, but it is a slow process, and the conventional ones were limited to shallow water (30 feet maximum, but less than 15 was the norm) (Kellogg 1910, 130). The tong may seem a simple device, but deep-water tongs were patented in 1887 (Kellogg 1910, 131; Marsh 1887).

The Marsh tongs were of the scissor type, and similar to those used to handle ice blocks. However, they were equipped with three lines and weights. The center line (C) was attached to the upper end of a small crosspiece at the pivot point of the scissor, and a weight was attached to the opposite end. The other two lines (B) were tied to eyes at the ends of the operating arms (A), and there were tubular weights (E) on these arms. These could be fixed by a set-screw in any desired position on the arm.

"The tongs are lowered by means of the center line, the weights E serving to automatically open the arms" (Marsh 1887). Once the tongs were at the level of the oyster bed, the lines B were pulled, causing the rakes to come together.

Marsh also disclosed a locking mechanism that would be sprung open "as soon as the tongs strike the bottom …, thus allowing the arms to close together."

Falling Nets

Falling nets are one of the oldest fishing technologies; traditional forms include cast nets and lantern nets. These, if used from boats, are likely to be small boats in shallow waters.

A cast net "has a disc-like shape when cast." It "is constructed from a series of tailored netting sections joined together to produce a cone-shaped net with weights and a drawstring attached to the perimeter and cast by a fisher to catch fish." The weighted perimeter sinks quickly, preventing horizontal escape, and the "drawstring closes the bottom" (He et al. 2021, 39). The lantern net has "a wooden frame covered with netting" (Gabriel et al. 2005, 317).

A newer form is the "boat-operated falling net." It is used in deeper waters, and is "usually operated at night with the use of lights to attract target species" (He et al. 2021, 41).

Gillnets and Entangling Nets

"Gillnets and entangling nets are long rectangular walls of netting that catch fish by gilling, wedging, snagging, entangling or entrapping them in pockets" (He et al. 2021, 43). Typically, there are floats on top and weights at the bottom. They may be anchored to the seabed, or allowed to drift.

With anchored gillnets, the role of the fishing boat is to set the net and retrieve it when full. A drifting gillnet may drift with the boat, or it may be left out with a marker buoy.

The gillnet may be deployed as a straight wall, or in an open circle around a school of fish (with the boat closing the gap). In the latter case, the boat will attempt to scare the fish so they swim into the net.

Limiting the visibility of a gillnet is important. Fishers learned to dye the nets, typically blue-green or brown, to better match the environment. "The efficiency of these nets has been increased sometimes by several hundred per cent when natural fibers were replaced by less visible synthetic fibers (PA [polyamide] multifilament), especially by transparent monofilaments (PA) or monotwine (twines made of monofilaments of PA and also PE [polyethylene])" (Gabriel et al. 2005, 277).

Hook-and-Line

Before there were hooks, there were "gorges": essentially a double-pointed stick hidden in the bait and with a line tied around the middle. Initially, the stick is parallel to the line. The movement of the line or the fish causes it to turn to the transverse position and be held in the fish's throat or belly. "In Europe, gorges have been used since the Paleolithic period" (Gabriel et al. 2005, 86–87).

Next came the bent hook. The desired shape could be found naturally in the thorny parts of certain plants. However, it was more common for hooks to be fabricated by combining pieces of shells, bones, or wood, held to the desired angle by some kind of cordage (Gabriel et al. 2005, 88).

The invention of bronze, around 3300 BCE, made possible the fabrication of the hook as a single piece of metal. (Later, iron was used, but it had to be protected from rusting.) Metal hooks could be given a more effective shape, with a semicircular bend between the shank and the point (Gabriel et al. 2005, 89–90). They could also be barbed; "older Egyptian and Roman hooks made of metal were without barbs" (91).

"Today, most hooks are made from high-carbon steel, steel alloyed with vanadium, or stainless steel ... Many hooks are covered with some form of corrosion-resistant surface coating" (He et al. 2021, 61).

Lines were made originally from natural plant (or animal) fibers. Nowadays, polyamide or polyethylene is preferred (Gabriel et al. 2005, 95).

There are several variations on hook-and-line gear. The simplest was the handline; the line could be cast or trolled behind the boat.

"The oldest known painting of a fishing rod comes from an Egyptian tomb built around 2000 B.C." While the rod allowed the hook to be cast further away, in the painted scene, range wasn't an issue; the painting showed a seated nobleman fishing at a garden pond (Woods and Woods 2000, 44).

The fish, having been hooked, needed to be brought in. The reel was first described by Thomas Barker (*The Art of Angling* 1651), and it is pictured in his 1657 edition (Radcliffe 1921, 8). It was used, and perhaps invented, by an expert "trowler" for pike. His 12-foot hazel rod had a "Ring of Wyre" in the top, "for his Line to runne thorow; within two foot of the bottome of the Rod there was a hole made, for to put in a winde, to turne with a barrell, to gather up his Line, and loose at his pleasure" (Barker 1651, 9).

The first European description of longlining I have found was from 1865: "The hooks used are very large and powerful, and are fastened to the main line, at intervals of six feet apart, by means of smaller lines, each about four feet in length, and called 'snoods.' The main line itself is somewhat thicker than a pencil, and a number of such fastened together form an entire 'fleet' of lines. The extreme ends of the whole set of lines are securely fastened to the ground by means of anchors." The main lines themselves extended for "very many miles" ("A.B." 1865, 517).

However, several sources indicate that the technique was previously developed by the Japanese. Watson and Kerstetter (2006, 6) say that there are local claims of its use in the Izu region around 1850, and in Wakiyama prefecture a century earlier. Kalland pushes it back even earlier: "Sea bream longlines have … been used since 'ancient times' in Kanezaki, Nishinoura and Nogita, where a fleet of longline boats perished in 1695" (Kalland 1995, 121). However, these longlines were small by nineteenth-century British standards; the Japanese traditional sea bream longline "consisted of ten sections, each [1456 meters] long, and had in total 2,300 baited hooks" (102).

Complicating the picture further, Gabriel asserts that "in Norway, longlines were known at least since the middle of the sixteenth century" (Gabriel et al. 2005, 114).

Longlining was a labor-intensive process. The hooks had to be baited; the branch lines (snoods) attached to the main line; the line set; the line hauled in; the branch lines removed; the fish removed and gutted; and the hooks cleaned. Efforts were made first to automate portions of this process, and later to automate it all (Gabriel et al. 2005, 129–133).

Harpoon

A harpoon is a spearlike weapon with a barbed head or other means of anchoring the head in the prey's flesh. Some anthropologists leave it at that (Kipfer 2008, 142). If so, the oldest harpoon head found to date is the "Semliki" or "Katanda" harpoon of central Africa, 80,000 to 90,000 years old. It has four barbs, all on the same side of the head (Smithsonian 2022).

However, a more specific definition requires that the harpoon heads are detachable, and have some means of attaching a line (Kipfer 2008).

Harpoons were used to hunt not just whales, but also other marine mammals and large fish. While many indigenous peoples used harpoons, I focus here on the

Inuit. I compare their harpoons to those used by American commercial whalers. The improvements (or at least variations) in the harpoon have related to the head, the line, and the method of propulsion.

The pre–Contact Inuit harpoon heads were primarily made of antler, ivory, and whale bone, although native copper and meteoritic iron were used "in limited amounts ... especially for harpoon end blades" (Park and Stenton 1998, 14). The head could have "one or more backward-pointing barbs cut into their lateral margins": either on one side (unilateral) or both sides (bilateral) (17–18). Additionally, it could have spurs, "backward-pointing, tapered projections from the proximal (base) end" (22).

A common feature of Inuit harpoon heads was the ability to "toggle": to rotate 90°.* "In the toggled position it is far too wide to come back out the narrow entrance wound and thus forms a very secure anchor within the harpooned animal's flesh." This toggling was achieved as a result of the relative placement of the spur and the line hole. As the animal struggled, the head detached from the staff, the line was tensed, the spur caught, and the head pivoted around the line (Park and Stenton 1998, 22–24).

The harpoon shaft came in two parts, the foreshaft and the main shaft. Non-toggling heads had tangs, "held in the harpoon socket by friction." Toggling heads "were held in place at the end of the foreshaft by tension on the harpoon line, which was stretched back to the shaft and secured by an eyelet that slipped over a peg set into the shaft." For open water hunting, the foreshaft was likely to be loose, to facilitate the detachment of the head. If available, the shaft was made of wood. "Otherwise, the shafts were fashioned from narwhal tusk or from several pieces of bone or antler spliced together" (Arnold 1989, 80–1). The lines were made from hide (skin) (Bennett and Rowley 2004, 268), in particular walrus rawhide (Gabriel et al. 2005, 64).

In open water hunting, Inuit hunters did not fasten the line to their boats. Rather "a float was attached to the line to tire the animal out." These floats could be as large as one "made from the entire skin of a seal" (Arnold 1989, 80).

American harpoon heads were made of wrought iron. The shafts were hardwood, and fit into a socket at the base of the harpoon head. It was not secured with a screw or pin because it was "meant to pull out after the whale was struck." Early nineteenth-century American whaling lines were made of a tarred hemp, two inches in circumference and laid in three strands, and they were double hitched to the socket of the head (Tyler 2022b).

Early American whalers also tied the lines to floats, but these were logs rather than inflated bladders. But between 1761 and 1782 it became customary to fasten the line to a windlass in the whaleboat, and let the whale tire itself out by towing the latter (Tyler 2022c).

The word "harpoon" is from Basque *arpoi*, and "the seal of the Basque town of Biarritz in 1351 shows a whaling scene in which a harpoon clearly has a two flue

---

* There is some scholarly debate as to whether this was developed originally by the Inuit, or by another northern people (Cunliffe et al. 2002, 94).

head." That is, there is a single rear-facing barb on either side of the head (Tyler 2020c). That traditional shape was adopted by American whalers.

Unfortunately (for the whaler), these heads could be pulled loose. "When the whale ran and force was applied to the whale line, the harpoon shaft bent, and in doing so the two flue head could be bent and repositioned such that only one flue caught in the flesh to hold fast. The opposite flue would then be positioned with the sharp edge presented to uncut flesh. The force on the whale line and shaft, plus the motions imparted by the fleeing whale, caused the sharp flue to cut its way out" (Tyler 2022d).

In the mid-eighteenth century, some two-flue heads were given "stop withers," secondary barbs that extended inward from the primary barbs. "Their purpose was to make it more difficult for the whale to shake out the harpoon." Unfortunately, this was not a complete solution (Tyler 2022c).

Counterintuitively, a single flue iron held better than one with two flues. When the head was turned inside the animal, only the flat side of the head faced outward, and this would not cut its way out. "The first recorded use in the American fishery occurred in 1824" (Tyler 2022d).

The first European toggling harpoon heads were grommet harpoons. "The earliest record of a grommet harpoon … is a sketch made by Francis Thompson circa 1772." Like the Inuit head, this had a backward-facing spur at its base, but instead of the head pivoting around a line hole, it pivoted around the grommet, a pivot pin, at the end of the forestaff (Tyler 2022e).

However, this toggling could occur prematurely, reducing the depth of penetration of the harpoon. In 1848, Lewis Temple, a formerly enslaved blacksmith, devised a pivoting head with two barbs on the same side. A tie covering the blade of the rear barb held the head parallel to the shaft. The motion of the whale would cause the head to strain against the tie and cut it, allowing it to toggle. The earliest form of this head used a grommet, but later it and the tie were replaced with a wood shear pin holding the head in the penetration orientation. Once this sheared off, the head toggled (Tyler 2022e).

Originally, harpoons ("irons") were simply thrust or thrown. The former was the norm; a harpoon could be "darted a maximum of three fathoms" into a whale with a reasonable hope that it would "go far enough into the blubber to hold fast" (Tyler 2022b).

Some thought to use the power of a cannon to harpoon a whale at a greater, hence safer, distance. In 1731, the South Sea Company tried but failed to persuade its harpooners to use a harpoon gun (cannon). The first true attempt to use a cannon to fire a harpoon was apparently in 1733; "two whales were successfully shot" (Tønnessen and Johnsen 1982, 17). Nonetheless, there was much skepticism among harpooners about the new technology.

A 20-guinea reward was given in 1772 to Abraham Stagholt for an improved whale gun. Normally, the rope was attached to "a ring fastened near the head or halfway along the shaft, but as the harpoon left the barrel the rope tended to drag it down." Stagholt came up with the idea of "a harpoon with a slotted shaft along which the ring could run freely to its base" (Blackmore 2000, 303).

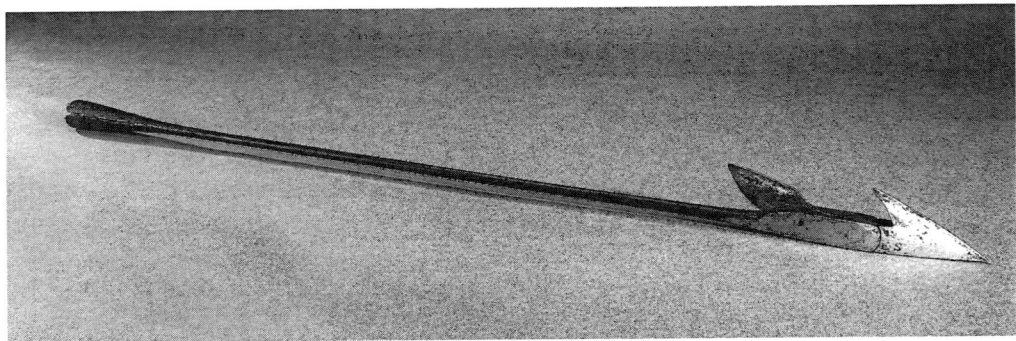

The Temple toggle iron. Note the hole for the insertion of the sheer pin. This version made in 1882 (National Museum of American History, Smithsonian Institution. Credit Jonathan Bourne through J.T. Brown. Catalog 056244. Media is CC0. Image cropped by author.)

Development of whale guns continued, and in 1820 they had an effective range of "near forty yards" (Tønnessen and Johnsen1982, 18).

Harpoon guns could be shoulder-held, or mounted on a swivel in the whaleboat. (Breechloading swivel guns were traditionally used by sailing warships as antipersonnel weapons, firing half-pound to two-pound shot.) There was a trade-off: The swivel-mounted weapon could fire a heavier harpoon, but it stressed the whaleboat (Tyler 2022). In 1882, Mason and Cunningham patented a shock absorption system. They "mounted their swivel gun on heavy rubber cushions," and they also placed rubber "on both sides of the trunnion mounting, and around the mounting post" (Tyler 2022a; Mason and Cunningham 1882).

A general problem with early harpoon guns was that with flintlock ignition, "there was always some doubt as to whether the priming would get damp and cause a misfire" (Blackmore 2000, 304). Wallis's 1815 swivel gun had "a flintlock with two locks," and "the flintlock was protected from spray by covering it with a hinged door" (Tyler 2022a). The Greener gun (1837) used a percussion lock, which improved reliability. Indeed, at least some Greener guns were "made with a percussion hammer that struck two nipples, both of which fired the gun" (Tyler 2022a). By "the middle of the nineteenth century the majority of harpooners were using guns" (Blackmore 2000).

The problem of recoil, damaging either to one's shoulder or to the whaleboat, could be circumvented by the use of rocket propulsion. Thomas Roys received several patents on rocket-propelled harpoons and he used them successfully in 1865–66 (Roys 1866, 19–20). His 1861 patent was for a "shoulder gun": A percussion lock ignited a primer, which in turn ignited the rocket fuel. Behind the rocket head there was a harpoon stock, and this carried a spring-loaded bar that would pivot into a crossed position after firing "to prevent the harpoon from being forced clear through the whale." At the rear end of the stock was a hook, which, as the rocket emerged, would catch a link to which a line was spliced. Finally, there was a flange around the barrel, to protect the shooter from backfire, with sight holes closed by valves upon firing. The rocket harpoon was nonetheless eclipsed by the harpoon cannon.

Once the whale was harpooned, it had to be killed (without allowing it to sink and take the whaleboat with it). European whalers initially used a lance, but there was experimentation with explosive devices. In 1821, Congreve rockets were used to kill whales that had already been harpooned by conventional means (Blackmore 2000, 306). However, the "bomb-lances" evolved were essentially explosive shells fired from shoulder or pivot-mounted guns. Of these, Svend Foyn's is especially worthy of note. "This fired a harpoon with long hinged barbs which opened like an umbrella inside the whale and held it like an anchor. It was tipped with a cast-iron shell full of gunpowder. The action of the barbs opening crushed a glass phial of sulphuric acid which ignited a fuse and exploded the charge … It thus discharged the functions of harpoon and lance in one" (Blackmore 2000, 306).

That description omits a few important features: One was that a "hinged joint linked the shaft to the head … the hinge prevented the blast from knocking free the line" (Hirsh 2013, 276). Also, Foyn used a "special gunpowder," and between the powder bag and the harpoon Foyn rammed down "tow, and then rubber wadding, and … a certain amount of wool" (Tønnessen and Johnsen 1982, 25).

Electrified harpoons were the subject of experiments in the nineteenth century (Gabriel et al. 2005, 68) but their first successful use in whaling was in 1929 (Webb 2011, 265).

Stupefying Fish

A traditional method of catching fish is to stupefy them by means of a plant-derived fish poison. It is difficult to say how long this has been practiced in pre-Columbian America, but it certainly predated contact with Europeans. Asian fishers likewise have a long history of use of fish poisons. In the ancient Mediterranean, saponin-containing mullein blossoms were used by the Phoenicians and later the Greeks. Because a fish poison needs to be in a minimum concentration to be effective, it was generally used only in small bodies of water (Gabriel et al. 2005, 44–47).

Fish may also be stunned by mechanical shock, such as a blow to the head with a club or a near miss with a bullet. "Explosives are thought to have been used for fishing as early as c1600!" Dynamite fishing is extremely wasteful, killing far more fish than are gathered from the surface (Gabriel et al. 2005, 43).

Electrical fishing began after World War I. Direct current is more effective than alternating current, and the lower the conductivity of the water, the more current will pass through the fish. Hence, electrical fishing is mostly used in fresh water (Gabriel et al. 2005, 48–9). However, in saltwater fishing, there has been use of electrified hooks (51).

Fishing Vessels

The switch from wood to steel construction, and from wind power to combustible fuels, made it possible for fishing vessels to venture out to sea more frequently and to sail further out.

In the Far North, steam made it possible for whalers to work through the drifting pack ice at the entrance to Melville Bay within a few days, whereas it was a much more "laborious and lengthy job for a sailing vessel." In 1861, the sailing whaler *Truelove* only penetrated a mile of ice "after toiling all day"; it then was towed by the steamer *Narwhal* into clear water (Jenkins 1921, 258).

The developments in hoisting machines discussed in earlier chapters also affected the fishing industry. For example, in the eighteenth century, capstans were used for hauling in nets in the "large-scale herring driftnet fishery" (Gabriel et al. 2005, 9). Later, "power-driven winches reduced the number of crew while increasing profit and safety at the same time" (9).

For the remainder of this section, I wish to focus on a few specific adaptations of vessels for fishing purposes, that is, some auxiliary gear they may carry.

Finding Fish

Sonar

In the 1920s, commercial fathometers became available. These measured the depth of the ocean floor below a ship by timing the echo of an emitted sound. Thus, they were a kind of sonar device, although that term wasn't coined until 1943.

Fish are detectable by sonar because their swim bladders contain air, and sound is reflected at the air–water interface. "In 1927, Rallier du Baty was the first to attribute false echoes on echogram recordings from the Grand Banks in the western North Atlantic as being indications of the presence of cod, *Gadus morhua*." In 1928, "the first tank experiments establishing the basics for detection of fish were performed." And "in 1929, Kimura published the results of experiments in which he demonstrated that fish could be detected with the relatively simple acoustic equipment of his time" (Fornshell and Tesei 2013).

In the 1930s, there was only modest use of sonar by the fishing industry (Fornshell and Tesei 2013). However, after World War II, surplus military sonar apparatus became available and this encouraged experimentation.

The original fish-finding sonar pointed straight down, emitting ultrasonic waves in a conical beam. The cone angle varied from model to model, and some later models had the ability to vary the cone angle. Lower sound frequencies were able to penetrate more water, but higher frequencies offered better resolution.

There are two basic "horizontal" sonar approaches. In "searchlight" sonar, the sonar beam is pointed sideways and a little downward, and the transducer emits the sound and receives the echo. It then is rotated around the vertical axis to point in a slightly different direction, and the process is repeated. All of these scans are assembled by software into a complete 360° representation of what's in the water on all sides of the ship. In "scanner" sonar, rather than physically move the transducer, there is a 360° array of transducers that can be sequentially turned on and off ("phased array"). This is much faster than physical rotation (Furuno Electric Co., Ltd. 2014).

Underwater Viewing

"Underwater television was successfully used for the first time in connection with the Bikini atom bomb tests of 1947." In the 1950s, it was used "for investigating the biology and ecology of fish and other animal populations" and "the operation of fishing gear." The housing had to be able to withstand water pressure and have low hydrodynamic resistance, and the camera had to have high light sensitivity (to minimize the required external illumination, which might alter fish behavior) (Clark et al. 1959).

For studies of fish behavior when caught in a trawl, or approaching a lure, the camera can be pointed in a fixed direction, since you know where the trawl (or lure) is. But for fish finding, you need to be able to see in any underwater direction.

One approach was a panoramic camera. McNeil received a 1964 patent for a panoramic camera "adapted for underwater use and capable of photographing a full 360 degree field."

In 1967, Hellenkamp received a patent on an "underwater scanning device," a closed-circuit television mounted so it could scan underwater. However, the mounting was of a peculiar type. The camera was arranged to view a mirror. The camera and mirror rotated (presumably horizontally) as a unit, but the mirror was independently tiltable, thus providing the ability to look down.

The camera need not be fixed to the ship's hull. The Navy's Project Fisheye featured a tethered but self-propelled camera, powered by a 2,000-foot-long cable. While the Navy's interest was in surveying underwater wreckage or guiding the rescue of a disabled submarine, a crewman used the camera to guide a fishing hook to a fish he wanted to catch (Johansen 1957).

Nowadays, a variety of digital underwater video cameras are available for fishers. They appear to be mostly used for recreational fishing, most likely because the commercial fishers have more sophisticated sonar equipment and the range of visibility in the ocean is rather limited.

Even if the camera is equipped with a strong light source, there is the problem that the image is fogged by light scattering.

AERIAL SPOTTING

The original form of aerial spotting was watching the behavior of seabirds; they were likely to congregate if they spotted a school of fish. While radar will not spot fish, they will spot flocks of birds that might be following a school of fish (Hayes 2020).

Spotter aircraft have been used to assist fishing fleets in the detection of schools of fish. Some tuna boats actually carry helicopters (Hayes 2020). Aerial drones have also been used for scouting the water (Maguire 2023) and even for actual fishing (DeFrangesco and DeFrangesco 2022).

Attracting Fish

FLOATING DEVICES

It is well known to fishers that many pelagic fish are attracted to floating objects, such as driftwood, mats of seaweed, and coconuts. This phenomenon has been known and exploited for centuries. Around 200 AD, Oppian wrote that

Mediterranean fishers "gather reeds and tie them together in bundles which they let down into the waves and underneath they tie a heavy stone by way of ballast. All this they let sway gently in the water, and straight away the shade loving tribes of the *Hippurus* (mahi mahi) gather in shoals and linger about delightedly rubbing their backs against the reeds. Then the fishers row to them to find a ready prey" (Taquet 2013).

While Oppian assumed that the attraction was that the floating object provided shade, others have pointed to the possibility that it harbors prey. Wilcocks (1975, 243) observed that "if a log of timber is found floating at sea and covered with barnacles, it is often surrounded with fish attracted by the various small Crabs, &c., which also make it their home."

A "fish aggregating device" (FAD) is an artificial or modified natural object set floating in order to attract fish. It may be anchored, as in the case of Oppian's reed bundles, or it may be allowed to drift. Drifting FADs are equipped with radio buoys. The FAD may float at the surface, or it may be weighted or tethered so it is "midwater."

There are also "artificial reefs," which remain on the sea bottom. Those using natural materials (rocks and plant materials) date back thousands of years, possibly even to the Neolithic (Pinto et al. 2017, 129). In the United States, the first purpose-built artificial reef "was off South Carolina in the 1830s using log huts." Purpose-built artificial reefs have included deliberately sunken ships, aircraft, automobiles, and railroad cars. Perhaps the best material for artificial reefs is concrete. Oil and gas drilling platforms, while not built for the purpose, also provide a reef-like habitat, and companies have been encouraged to leave them in place rather than towing them to shore and cutting them up for scrap when operations cease (Lukens and Selberg 2004; New Haven Reef Conservation Program 2016).

Lights at Night

Nocturnal fish (e.g., trout) are best fished for at night (Ekirch 2005, 171). Some fish, moreover, are attracted by light; these include herrings, sardines, sardinellas, and anchovies. Squid are also drawn to light (Ben-Yami 1988, 5–7).

The use of light as a lure is most effective in clear water on moonless nights (Ben-Yami 1988, 4). The principal historical change in night fishing has been the replacement of torches with kerosene, gas, or electric lamps (21–31). Electric lamps, suitably insulated, can be placed underwater (21).

Keeping Caught Fish Edible

"The spoilage process starts immediately after the death of fish," and fish are highly perishable because of their "high moisture content and availability of nutrients for the growth of microorganisms" (Singh 2021).

Hence, in premodern fishing, the goal was to fill the hold quickly and rush back to port, where the fish would either be sold fresh or preserved. Fish were traditionally preserved by drying, salting, smoking, or pickling.

In the fifteenth century, the Dutch built fishing vessels (*busses, buizen*) "which were large enough to allow fish to be gutted, salted and barreled on board." But they still needed to return to port, unload their cargo, and sail back to the fishing grounds. Some of this down time was eliminated in the mid-sixteenth century by *ventjagers* ("sale-hunters"): fast ships that would meet the *busses* and pay for their cargo (Wilson and Rich 1967, 171–2).

LIVE-WELLS

However, there was an alternative to on-board preservation of dead fish: keeping them alive on board or beside the ship. On board, one can use some sort of deck tub, or a built-in structure. Or the fish may be towed behind the ship in a basket (creel), or on a "string."

Archimedes has been credited with the construction in the third century BC of a ship (*Syrakousia*) with a large vivarium: "a huge reservoir of 21,000 gallons, her water-tight well, made of planks lined with lead, and filled with sea-water, in which a great number of fish were always kept" (Radcliffe 1921, 225–6). However, this was not a fishing vessel, but rather the Hellenistic equivalent of a royal superyacht.

According to Pliny (*Natural History*, IX, 29), in the first century AD, Roman naval vessels under the command of Optatus were used to transport "live parrot wrasses (*scari*) from … the southern part of the Aegean" to the "shores of Italy" (Boetto 2010, 247). Presumably, they were in some sort of tank on deck.

More germanely, "a boat with a live tank (*navis vivaria*) was unearthed near what was the entrance to the Claudian harbor of Rome" (Evelpidou et al. 2018, 291). This *Fiumicino 5* was dated to the late second century AD. It was 5.6 m long, 1.50 m wide. The fish-well was centrally located, and could hold about 300 liters of water. Nineteen holes assured the continuous flow of water into the "fish-well"; these could be plugged as needed (Boetto 2007, 244–6).

Ships were also equipped with "live wells" during the Renaissance. Seven sixteenth-century well-ships were found in Lyon, and two sixteenth-century *waterschepen* in the Zuiderzee (Boetto 2007. 248).

"Well-smacks were introduced to England in 1712, being first used at Harwich" (Stevenson 1899, 341; cf. Sahrhage 2012, 88). We next look more closely at the use of "well-smacks" in nineteenth-century Barking, once "the most important fishing port in England" (Short Blue Fleet 2023).

Well-smacks were small fishing vessels used for lining and trawling. The *Saucy Jack* (built in 1836) was 60 ft overall and cutter-rigged (one square sail). "The midships portion of the ship, ten feet or thereabouts, was entirely separated from the bow and stern by watertight bulkheads. Augur holes were bored through the hull below the water-line—well down so as to prevent air from entering at large roll angles. These holes enabled the water to enter and circulate and therefore remain fresh within the well … Access to the well was via a hatch on deck, and in front and on either side of it was the welldeck, which narrowed the mouth of the well into a funnel, thus keeping the level of the water within certain limits when the smack was rolling or pressed down under sail." The well was usually divided into two or three compartments. "Wellheads, decks and funnel were made of the best wood, exceeding in strength any other part of the smack as the design meant that they were subjected to great strain" (Short Blue Fleet 2023).

At Barking, and perhaps elsewhere, a "fleeting system" developed "whereby a smack would take fish from other vessels when leaving for home." This was good for customers, but not necessarily for the sailors, as it meant that they were at sea longer, as long as six months. This led to a strike in 1844, and thereafter the cruises were limited to six to eight weeks (Short Blue Fleet 2023).

Carlton wrote that "fish die in transportation for three reasons": (1) "lack of oxygen"; (2) "poisoning by the [carbon dioxide] they themselves evolve from their gills"; and (3) "mechanical injuries," from "jarring, dashing against the bottoms and sides, wounds from other fishes, etc." (*American Angler* 1899, 293).

While the natural circulation of water on the traditional well-smacks provided for some level of oxygen and carbon dioxide exchange, it could be improved upon.

The earliest proposal of improved circulation appears to have been by Sir Richard Steele and Joseph Gilmore in 1718. Their proposed "Fish-Pool" was a ship with not only a grating and air pipes above the fish well in the hold, but also "grates 'fore and aft'" (Steele 1790, 194). Neither the text nor the accompanying figures are particularly clear but it seems that these were a little above and below water level rather than on the deck, so that the ship's motion created a current in the hold in the opposite direction to the movement of the ship (199–200). They declared, "in this invention the air and water flow together, come into the ship horizontally, and pass through it in a constant succession, yielding fresh air and fresh water, to the relief, sustenance and delight of the fish" (199–200). They found investors and several Fish-Pool vessels were launched in 1720 (Aitken 1889, 254). They were, however, a failure, as the fish "battered themselves against the sides of the vessel" (Steele 1790, 188).

In 1830, the naturalist de Moulins dealt simultaneously with the problem of providing oxygen and removing carbon dioxide in an aquarium by "keeping plants in the water" (Taylor 1876, 11). However, I have not found any instance in which a well-boat carried plants to recondition the water.

Instead, at some point in the nineteenth century, some British well-smacks were equipped with "artificial circulation," which presumably was capable of more frequent water changes than the old augur hole system. A steam-powered pump brought untainted water into the bottom of the well, creating a vertical circulation that both aerated the well water it and caused "all the scum and refuse to rise to the top." The latter escaped through overflow pipes (Stevenson 1899, 342).

The possibility of pumping air (as well as water) into a fish-well using a steam engine is mentioned in Leach's 1873 British patent (No. 358). (When live saltwater fish were transported inland, on railroad fish cars, air pumps were a necessity, as it was impracticable to change the water [Stevenson 1899, 349].)

Water can be intensively aerated "by spraying the water onto a metal plate in the air and returning the water back to the fish, or by forcing the water through a raceway where a lot of turbulence and foam is created. The increased aeration displaces the carbon dioxide as well as adding oxygen" (Branson 2008, 229). The fish may also be chilled to slow their metabolism, thus reducing oxygen demand and carbon dioxide output (*American Angler* 1899). (Cold water also holds more oxygen [Stevenson 1899].)

Modern well-boats may have wells as large as "3000 m^3, holding up to 300 t of live fish" (Lekang 2019). They may operate as open (sea water flows in and out) or closed systems.

A closed system would still need an oxygen source (e.g., a compressed air or oxygen tank) and a carbon dioxide scrubber.

Fish parasites and pathogens are also a concern. The incoming water may be filtered to keep out parasites and the fish given antiparasite treatments. The water may also be disinfected using ultraviolet light (Lekang 2019).

Well-boats traversing large distances may experience significant changes in local sea temperature and salinity, and if they are open systems, these changes may stress the fish (Lekang 2019, 262). Ships have also been used to tow fish in cages (Fernö et al. 2020, 344).

Chilling, Superchilling, and Freezing

Chilling is the reduction of the temperature of fish to a point near but above the temperature at which its fluids begin to freeze, typically about -1°C (Alam 2007). Superchilling has been defined by FAO as "reducing the fish temperature to about -2.2°C (28°F), at which point half the water is frozen, and keeping it there." Freezing usually refers to reducing the temperature of the fish to -18°C or less (Koutsoumanis et al. 2021). Chilling, superchilling, and freezing result in progressively longer preservation times, but there can be issues with food quality (e.g., "freezer burn").

Freshwater ice can only be used to bring fish just down to its own freezing temperature, 0°C. Colder temperature may be achieved by exposure to seawater ice or an artificial brine. Still colder temperatures require the use of "dry ice" (frozen carbon dioxide), ammonia, liquid nitrogen, or other refrigerants.

The speed of freezing is an important factor in determining frozen fish quality, and this depends not just on the temperature of the refrigerant but also on the efficiency of the heat transfer between the fish and the refrigerant. The fish may be immersed in or sprayed with a liquid refrigerant (water or brine typically, but liquid nitrogen is possible). Alternatively, they may be cooled by a cold air blast or contact with a cooled metal surface (plate freezer) (North and Lovett 2005).

In the Far North, the Inuit were well aware that food could be frozen for later use. In winter, "fish froze whole within a few seconds of being removed from the water." If they were caught around freezeup (September–December, with freshwater freezing sooner), "whole fish were often placed on a gravel bar to freeze overnight, and then thaw again the following day." (Burch 2006, 146). John Bell visited Astrakhan in 1716 and said that its fish "caught in autumn are carried to Moscow frozen and sold there and in the places adjacent" (Bell 1763, 39–40).

Natural ice could be used on fishing vessels to freeze and thereby preserve fish. In addition, it could be used on other vessels to transport fish to a city where fish were in high demand. For example, in the late eighteenth century, a Scottish employee of the British East India Company observed that the Chinese deliberately flooded paddy fields in the winter, collected the resulting ice in storehouses, and then used it in summer "when transporting the fish long distances inland by boat on the river systems" (Short Blue Fleet 2023a).

He told fellow Scotsman George Dempster about this practice, and Dempster arranged in 1785 for his salmon supplier to build ice houses and transport salmon from Scotland to London by sea, packed in ice. The average passage time was six days and there was "very little" deterioration (Short Blue Fleet 2023a).

Note that Dempster's salmon fishers weren't carrying ice out when they went fishing. However, ultimately British fishing vessels started carrying ice with them to sea, so they could remain out longer and still return with fresh fish, which sold at a premium over salted fish (David and Norman 1995, 238). Reportedly, the first use of this stratagem in the American fishing industry was by a Gloucester smack in 1838, to preserve halibut that had died inadvertently (Stevenson 1899, 359).

The ice could be of local origin or imported from a colder clime, such as Norway for European customers or New England for American ones. The ice trade was not, of course, limited to the supply of ice for fish preservation; the ice was used with a variety of perishable foods and to keep drinks cold.

There was, even in the 1930s, still a prejudice against frozen fish because of various undesirable sensory characteristics. In 1936, George Reay set forth the three requirements for overcoming these problems: "newly caught fish, quick freezing to a low temperature, and maintenance at that temperature": -10 to -20°F (Waterman 1987, 44). However, that required artificial refrigeration.

There are two basic approaches to artificial refrigeration: vapor compression and absorption. In both cases, heat is extracted by evaporation of a low-boiling-point refrigerant, but they differ in how the refrigerant is returned to the liquid state.

In 1756, William Cullen created artificial ice by using a "pump to create a partial vacuum over a container of dimethyl ether. As he reduced the pressure, the ether boiled, absorbing heat from its surroundings." In 1834, Jacob Perkins patented an ice machine that operated on the vapor compression principle. The first commercial refrigeration unit, developed by Linde in the 1870s, used ammonia (boiling point -28.01°F) as the refrigerant (American Society of Mechanical Engineers 2020).

When artificial refrigeration was developed to the point of being competitive with natural ice, it likewise could be exploited either by fishing vessels to preserve their catch until they returned to port, or by cargo vessels to carry frozen fish caught by others to a customer. The latter usage came first.

Beginning in 1892, in Sandusky, Ohio, fish were slow frozen inside a cold store by exposure to air cooled by pumping cold brine through pipes. Beginning in 1894, Pacific salmon frozen in this manner were shipped from British Columbia to the United Kingdom (Waterman 1987, 15).

The steamer *Karmay* was "probably the first fishing vessel ever to be equipped to freeze her catch at sea." In 1915 it was fitted "with a carbon dioxide compressor and an Ottesen brine tank installation, claimed to be capable of handling 10 tons of fish a day" (Waterman 1987, 17). In 1926. the SS *Holder* was similarly equipped, with the intent that it serve as a mother ship to "freeze line-caught halibut from dories working off Greenland" (25–6).

However, one general problem with the "mother ship" concept was that the bigger the ship, the larger was the initial and operating cost, and the larger was the fishing fleet needed to fill its freezer capacity and thus render it profitable. There were often delays in supplying the fish to the mother ship, and that meant that there was deterioration if the fishing vessels didn't have adequate means of preservation on board.

Nonetheless, the Russians built a "factory trawler," the *Pushkin*, in 1955; it could freeze 10–20 tons of fish a day. The USSR, East Germany, and Poland all expanded their factory trawler fleets, and "by 1975 almost 90 per cent of all Russian frozen fish was being produced at sea" (Waterman 1987, 63).

The alternative to the "mother ship" would have been to equip each fishing vessel with its own freezing unit, but that required systems that were compact yet economical. In 1929, the Societé Anonyme pour la Conservation Industrielle du Poisson (SACIP) fitted a steam trawler "with a carbon dioxide compressor, a brine freezer that could handle about 0.4 tons an hour at -20°C, and a cold store to hold 40 tonnes at -15°C." The fish were loaded into the top bucket of a "bucket wheel" drum, which slowly turned to immerse the fish in the brine and thereby freeze them. The freezing process took 0.5–2.5 hours (Waterman 1987, 29). Unfortunately, some thought that the "first cost" of the SACIP system was "too high for the average trawler" (41).

In 1955, a compact Torry-Hall vertical plate freezer was installed on the Grimsby steam trawler *Northern Wave*, and it conducted a series of experimental voyages in which "some 30 tons of the earliest-caught white fish were frozen and stored at -30°C until landing." Freezing the early catch meant that the trawler could remain at sea longer (Waterman 1987, 63).

An alternative to installing icemaking equipment on board was to load the fishing vessel with an artificial ice made on shore. Better yet, one could load it with "dry ice." This was attempted (as a supplement to normal ice) "on a Grimsby trailer in 1934" (Waterman 1987, 41), but the temperature distribution was unsatisfactory.

The *Junella* (1962) was a British stern trawler. The primary refrigerant was R22; the secondary refrigerant was trichlorethylene rather than brine. Whole fish were frozen in blocks in vertical plate freezers (Waterman 1987, 70). Later British trawlers followed the *Junella*'s pattern (72). By 1970, there were 36 British freezer trawlers (72). West Germany had about 60, Norway and the Netherlands each about a dozen, and the USSR more than 500 (73).

After freezing, the fish must be kept that way, and the quality of insulation of the hold becomes significant. In the old ice trade, sawdust (R value 2.44/inch) was commonly used as insulation to inhibit the melting of the ice. Cork (R value 3.33/inch) was widely used in the refrigeration industry. However, there are modern materials that will better insulate a cold store, such as polyurethane foam (R value 6.25–7/inch) (Shawyer and Medina Pizzali 2003, Chapter 5).

8

Flags and Blinkers

(Visual Communication at Sea)

Sailors need to communicate, from one part of ship to another, and sometimes ship to ship or ship to shore. Communication may be during the day or night, and by sound, light, or radio.

A communication system must be capable of expressing all of the messages the sender might need to convey, and of doing so both accurately and rapidly. There is a tension between comprehensiveness (which requires a more complex signal code, as discussed later) and the other two goals (as a complex code is harder to learn). Moreover, the standard of what is adequate rapidity for naval use became more stringent as the speed of ships increased (Wolters 2013).

The development of a more comprehensive set of signals also increased the ability of the commander of a fleet or squadron to respond flexibly to changes in circumstances, rather than being forced to rely on the prescience of a set of fighting and sailing instructions (and the ship captains' willingness to adhere to them) (Corbett 1908, 19).

Communication and Coding

The word "encoding" in the communications field has several different but overlapping meanings. One form of coding is the conversion of message text into a form compatible with the communication channel. For example, if you are communicating electronically on the Internet, you type characters at the keyboard, and your computer converts each character into a sequence of bits (zeroes and ones) in the computer's random access memory, which ultimately is transmitted electronically. The conversion is according to an internationally recognized code, such as the American Standard Code for Information Interchange (ASCII) or the UTF-8 implementation of the Unicode Standard.

In the cryptographic field, a code is a cryptographic system in which, where possible, the plaintext is converted to a cryptic form either word by word or phrase by phrase. Letter-by-letter encoding occurs only as a last resort. Codes—in the sense of a shorthand for a word or phrase—were sometimes used simply for brevity (to minimize telegraphy charges) rather than for secrecy.

Just to complicate matters further, a coded message, in the cryptographic sense, may be further encrypted (superenciphered) for greater secrecy.

8. Flags and Blinkers

With a visual communication system, unless you can display the actual message as text characters on physical or electronic signs large enough to be visible to the intended recipient, the text must be encoded by the sender into a combination or sequence of transmissible and distinguishable forms, and then decoded by the recipient.

Claude Shannon, in his "Mathematical Theory of Communication" (1948), spoke of a transmission of information over a communication channel as being a "sequence of choices from a finite set of elementary symbols."

This was not a new realization. In 1872, Myer wrote, "A sign or signal is any thing, or sound, or act, or indication by which to excite attention or convey a meaning" (15). And if you have a set of what Myer called single, primary, or elementary signals—each distinguishable from the others—they may be joined together to make a combination signal (16). This is similar to Shannon's sequence, but the elementary signals are displayed simultaneously rather than sequentially.

Each elementary signal must have a characteristic, or set of characteristics, that distinguishes it from its peers. If you have red, white, and blue flags, you have three elementary symbols distinguished by color. If their meaning is altered by where they are hoisted, then the number of elementary symbols is the number of color-hoist position combinations.

The reason for confronting the question of coding up front in this chapter is that the same coding system can be used with different communication channels and devices (flags, lights, electrical pulses, etc.)

The simplest coding system ("alphabet") uses two symbols (binary). In the computer field, a single signal that can be either of two values is referred to as a bit (binary digit). The international teleprinter (Baudot–Murray) code used a series of five bits per letter. A binary code of five "bits" that can be in either of two states offers a total of $(2 \times 2 \times 2 \times 2 \times 2) = 32$ possible sequences. It thus is sufficient to represent all 26 Roman alphabet letters.

The concept, however, is much older, as it was the basis for the "Baconian cipher" (steganographic system) devised by Francis Bacon in 1605. (Steganography is the art of hiding a secret message in a "cover" message or a physical object so if the latter are intercepted, the very existence of the secret message is not obvious. For example, there might be an understanding that there are two subtly different ways of forming each letter in the cover message, and these are read as a binary code, with each group of five denoting a particular letter in the secret message.)

The number of possible combination signals offered by a set of elementary symbols depends on (1) the number of elementary symbols, (2) the maximum number of elements in a combination, and (3) whether the code allows signals that are subcombinations. While a fixed five-element code with two possible elementary symbols per element offers 32 possible combinations, if the code allows a signal to have anywhere from one to five elements, there are $2 + 4 + 8 + 16 + 32$, or 62, possibilities. And that brings us to Morse Code.

Bacon used a five-bit binary code for every letter, regardless of how rare or common it was. American Morse Code, developed in the 1840s by Samuel Morse and Alfred Vail, is also (arguably) binary, but its character representation is of variable length, with the shorter sequences assigned to the more common letters. "E" and "T" are one bit, while rare letters like "Q" need four bits. This had the advantage of shortening the average time to transmit a character. This evolved into International Morse Code (1865), which also encodes numbers, and these require four or five bits.

Morse Code's binary implementation is not "on" and "off" but rather short "on" (dits or dots) and long "on" (dahs or dashes). In the latest international standard, a dah is three times as long as a dit. After each "on" there is an "off" (space) that is as long as a dit, so dits and dahs may be distinguished.

Because the character representations are variable length, it also had to have a way of signifying the end of a letter sequence. Between one character and the next there is an "off" (space) as long as three dits, and for the sake of readability, between words there is an even longer signal "off." People argue whether Morse code should be characterized as binary (dot, dash), ternary (counting the "letter space"), or even quinary (also counting the other two spaces). Morse code has also been called "pseudo-binary" (Desurvire 2009, 132, 134), which is the term I prefer.

The "General Service Code" of the late-nineteenth-century U.S. Signal Corps was also pseudo-binary (Myer 1872, 68), but with different assignments than in Morse. ("A" was "22" rather than "dot-dash," and "E" was "12" rather than "dot"; "1" and "2" designated particular motions in wigwag, and were reversed in some versions. Motion 3 was used to show the end of a word, sentence, or message.) With a variable allotment of one to four elements, it had 30 possibilities and thus could encode all 26 letters. The decimal digits were encoded by five-element sequences.

With a three-symbol (ternary) code we only need three "trits" to designate a letter of the English alphabet. A ternary-encoded Latin alphabet for steganographic use was described in *Steganographia* (written circa 1499, published 1606) by Johannes Trithemius. Myer proposed a ternary-coded alphabet in 1872 (Myer 1872, 96), but with different assignments than Trithemius's.

With a five-symbol system, we need just two symbols to encode a letter of the Roman alphabet if we use one pair to represent both I and J or both P and Q. This was the basis for the "Polybius square," an encryption device invented by Cleoxenus and Democleitus and improved by Polybius in the second century BCE. It was used for encrypting letters of the 24-letter Greek alphabet based on the row and column coordinates of their positions in a 5 × 5 scrambled alphabet square.

With a six-symbol system, we can enlarge the square to 6 × 6 and thereby represent the 10 decimal digits as well as letters A–Z. The ADFGVX cipher used by the German Army in 1918 was of that type (the letters ADFGVX served as the row and column labels, and were further encoded by Morse code).

Of course, with a 10-symbol system, only one symbol is needed for each decimal digit, rather than two. And four such elements can provide 10,000 different signals—more if the code gives meaning to one-, two-, or three-element signals, too.

If a combination signal can specify a particular letter of the alphabet, it follows that a sequence of combination signals can spell out any message, however long. But it can be tedious to spell out every possible observation or command, letter by letter, hence the need for relatively short signals with meanings particular to naval or merchant marine practice. For example, it is better to be able to hoist a single flag to warn "enemy in sight" than to have to spell out all twelve letters of that message. And that in turn has led to the evolution of some rather complex naval codes. Ideally, the more urgent the message, the smaller the combination assigned to it.

Communication Protocols

Holzmann (1994) points that the sender and recipient need to be able to distinguish between control information and message data. Control information includes the sender indicating a readiness to send a message and the intended recipient, a readiness to receive it. As the message is transmitted, the recipient may need to request that a portion be repeated, or that the message transmission rate be reduced. (Modern U.S. Navy semaphore also includes "move signs" to direct the sender to move up, down, left, or right.) The sender may wish confirmation that the message was successfully received, and this could take the form of a simple acknowledgment, noting the number of characters received, or of repeating the message back to the sender. If the recipient is supposed to relay the information to others, the control information would indicate who are the next correspondents, and the urgency of re-transmitting it.

Limitations of Visual Communication

With certain exceptions, visual communication is limited to line of sight. The length of that line is dependent on the height of the "transmitter" and "receiver." The distance to the "geometric" horizon is the square root of twice the product of the radius of the earth and the height of the viewer (Young 2021). If we assume a spherical earth, then the radius of the earth is constant. Of course, the transmitter is not on the horizon but at some distance above it. Conveniently, the line-of-sight distance is the sum of the distances to the horizon for the transmitter and the receiver.

Refraction (the bending of light, here by the atmosphere) can cause transmission to be possible beyond the geometric line of sight. On the other hand, the atmosphere, by scattering light, can limit visibility (Young 2021).

There has been experimentation with tethered balloons and kites, which of course can attain heights much greater than any ship's mast, to improve signaling. Niblack reported that "at Heligoland, in 1891, a German squadron used, with great success, a captive balloon for distant and squadron night signaling" with incandescent lights (1892, 479). In 1904–1907, Colonel Samuel Franklin Cody's manned kites were tested as carriers of signal flags (Kuntz 2012, 140). In 1914, a brief *Scientific American* article reported that the French Mediterranean squadron had tested the use of box kites for signaling. It noted that "the box kites, being light, can be hoisted even when there is but a slight breeze; and their size makes them easily discernible from great distances" (*Scientific American* 1914, 81).

All signals may be differentiated on the basis of number and timing. Visual signals may additionally be differentiated on the basis of size, shape, colors, pattern, location, and, for combinations, spatial arrangement.

Visual communication is dependent on the contrast between the signal and the background. Standard human vision ("20/20") is able to resolve a spatial pattern separated by a visual angle of one minute of arc (Evans 2006, 2). If, for example, alternating black and white stripes on a flag were narrower than that, they would merge into a single gray mass. At 1,000 yards, that angle would correspond to 0.29 yards. Black and white of course offer maximum contrast. As contrast decreases, so does visual acuity. Visual acuity also decreases with decreasing luminance (Johnson and Casson 1995). In daytime, that would be related to both the amount of sunlight falling on the signal device and its reflectivity.

The telescope and the binocular were key enabling technologies for visual communication, as they greatly extended the range at which visual signals could be distinguished. With both instruments there is a trade-off between magnifying power and field of view. Additionally, the light-gathering power increases with the diameter of the objective, but so also do size and weight. In 1872, the U.S. Signal Corps telescope was 30 power (Myer 1872, 231).

Binoculars typically provided less magnification than the telescope, but of course a wider field of view, which helped in keeping the signal in view despite the roll of the ship. Indeed, they were called "marine glasses." Myer considered them better than the telescope for distances up to five miles, and usable up to 10 miles (1872, 232). In 1889, the Chief Signal Officer reported that a person on foot could use binoculars of up to ten power, hand-held, but five power was the norm for reconnaissance (Chief Signal Officer 1890, 50–1).

"If no atmosphere were present, range would be directly proportional to magnification. The effect of the atmosphere is to decrease the apparent contrast of the target as distance increases." Thus, there is a trade-off between apparent size and apparent contrast. A study of nighttime binocular detection of a target (rather than discrimination of a signal) indicated that "with constant pupil size, nighttime detection ranges increase in direct proportion to magnification, up to a limit of 10x or so. If the binocular is mounted rather than hand-held, ranges continue up to at least 20x" (Gordon 1957, 24–25).

Visual communications are degraded by weather conditions, notably fog, rain, and snow.

Sail Signals

In 1618, Sir Walter Raleigh instructed his squadron on how to communicate sightings of other ships. If it were a great ship found alone, "you shall strike your main-top-sail, and hoist it again so often as you judge it to be an hundred tons of burden ... and so answerable to her greatness." If they saw a small ship alone, "you shall do the like with your fore-top-sail" (presumably only once). But if they encountered "many great ships," the spotter was told, "you shall not only strike your main-top-sail often, but put your ensign in the main top" (Cayley 1806, 406).

Even in the nineteenth century, there may have been some limited use of sail signals. Burney (1815) lists seven sail or sail-gun signals (474). For example, the signal to unmoor was the main-top-sail loose in the top, and to moor, the mizzen-top-sail hoisted and clewed up. The combination of the fore-top-sail loose, and one gun fired, signaled to prepare for sailing.

Flaghoist Signals

Flags have been a popular daytime signaling method for centuries. They have the advantages of being relatively inexpensive and able to be seen by several intended recipients simultaneously. They do have the problem that they can't be read at night, in a calm, or if the wind is blowing the flag directly toward or away from the intended recipient. Flag colors could also be difficult to read around sunrise and sunset (when the light is red-shifted and less intense), and when the sun was behind the signaling ship, thus silhouetting it (Raper 1828, 95).

Flag signal systems are, unfortunately, relatively easy to tinker with, and this has led to difficulties. In 1864, Admiral Farragut of the Union Navy complained, "You will have to do something to simplify the signals, they have changed so frequently that we can scarcely learn the flags before they are altered" (Howeth 1963, 9).

Flags are primarily distinguished by color and pattern. Niblack states that the most useful flag colors are red, yellow, black, white, and blue. However, their distinctiveness and contrast with the background may be degraded as the flags fade and soil (the latter as a result of "smoke, powder gases, sunlight and rain"). He warns that "no more than two colors should be used in any one flag, as the chances are increased of mistakes where the flag droops and only a patch of color is concerned" (1892, 457). However, historically, there was some use of tricolor flags.

With regard to the patterning of multicolored flags, in 1780 Captain Young (Rodney's flag captain) wrote, "chequered flags should be abolished. Quartered, halved, three-striped, striped corner ways, half up and down, and pierced, are the only ones that are properly distinguished at a distance" (Raeside 2020a).

Flags have also been differentiated by shape, typically rectangular (or square), swallow-tailed, triangular, and a long trapezoid. Niblack notes that flag shape can be used to indicate "the character of the signal" (1892, 456) and says that "the shapes of flags can be made out at five miles or more," while the "limit of visibility as to color" is reached at "between four and five miles" (464).

I have my doubts about there being a visibility advantage of shape over color, unless there is also a size distinction. Sir Samuel Hood complained in 1814 that triangular flags "are in general difficult to discern" (Bob Raeside 2020a), and in 1828 Raper noted (92) that drooping can cause a rectangular flag to look triangular.

Early signaling was haphazard, with a limited number of distinct signals specified on an ad hoc basis before a particular campaign or even a particular engagement. The most common ad hoc signals were to report sighting the enemy, to order the squadron to engage the enemy or to withdraw, and to call a council of captains.

In 1515, Antoine de Conflans prepared detailed signal instructions for the

French navy. It used "flags, banners, firing of cannon, the use of lanterns at night, and even the set of sails" (Miller 1944, 62). De Tourville's 1689 signal book made use of "twelve flags and three pennants" (Miller 1944, 62).

The first proper signal code in the British navy was promulgated in 1653 by the Commonwealth Admirals Blake, Deane, and Monck as part of their Fighting Instructions (BCW 2010), as shown in the table.

Signaled action	Signal	Instruction number
frigates to approach and scout enemy fleet	"wefted" (knotted) general's ensign	1
engage the enemy	shoot off two guns and red flag on fore topmast-head	3
squadron requests reinforcement	pennant on fore or main topmast-head	4
ship requests reinforcement	wefted ensign	5
ship needs to bear away from enemy	pennant on mizzen yard-arm or ensign staff	6
form line to windward behind admiral	blue flag at mizzen yard or topmast	7
tack to gain wind on enemy	red flag at spritsail, topmast shrouds, forestay, or main topmast stay	8
all flagships come up	red flag on mizzen shrouds or mizzen yard arm	9
at night, retreat without anchoring	fire two guns, wait three minutes, fire two more	14

As can be seen, there were six signal elements (a wefted ensign, a pennant, red, white, and blue flags, and a signal gun), and the meanings of the flag could be altered depending on where on the ship it was flown. The only combination signals were gun–flag combinations and, with the exception of instructions 5 and 6, the signals were from the admiral to the fleet, not the other way around. Moreover, there was no provision for the admiral to send a signal to just one component of the fleet.

The Duke of York's "Instructions for the better Ordering of His Majesties Fleet in Sayling" (1673) was rather similar, but we see some fine-tuning. Instruction IX allowed the admiral to use a red flag on the mizzen topmast-head to call the vice-admiral to bring the ships of the starboard quarter to come to their starboard tack, and a blue flag in the same location to command the rear-admiral and the ships of the larboard (port) quarter to come to their larboard (port) tack. Instruction X said that the union flag on the fore topmast-head told the van to tack first, while if it were on the mizzen topmast-head, it ordered the rear to tack first (Corbett 1905, 157–8). Wrixon (1998, 392) states that 15 flags were used, while Dempsey lists just 10.

Corbett notes that a later Admiralty manuscript discussing these instructions made the observation, "There have happened several misfortunes and disputes for want of a sufficient number of signals to explain the general's pleasure" (1905, 157 n1).

Naturally, the number of signals was increased to keep the admirals happy. Vice Admiral Lord Howe's 1776 code used 21 different flags (18). There were solid colors (red, yellow, blue, and white), bicolor horizontal stripes (red–white, red–blue, blue–white, red–yellow, blue–yellow, and their inversions), a few tricolor horizontal stripes (red–blue–white, white–red–blue, and their inversions, plus blue–white–red) , a red cross on a white field, and the Union Jack. The meaning of these flags depended on where they were hoisted.

Some earlier eighteenth-century systems had used additional patterns (diagonal stripes, diagonal cross, checkered, square-in-square, six horizontal stripes, etc.) and some retained and even elaborated upon the pennant, making it a long isosceles triangle instead of a rectangular flag.

There were both physical and psychological limits on this proliferation. The physical limit was that the different flags had to be distinguishable from a distance despite the vagaries of wind direction and speed, and perhaps intervening sea spray, fog, rain, or smoke. That limited the choice of both colors and patterns. The psychological limit was that both the sending and receiving signalmen had to be able to memorize the signals; you didn't want to be paging through a signal book while a battle was raging. That surely limited the total number of signals.

Nonetheless, "by 1780 there were 50 flags each hoisted on an average in seven different positions providing for about 330 signals" (Dempsey 2018).

There was still the problem of individual admirals devising their own signal books. In the British Navy, this came to an end with the *Signal Book for the Ships of War* (1793), although, strictly speaking, it was not made binding on the whole service until 1798 (Corbett 1908, 78).

This was "used successfully at the Battle of 1st June in 1794 and the Battles of Cape St. Vincent in 1797 and The Nile in 1798" (Craddock 2021, 13).

It appears that the naval architect Joseph Furtenbach was the first to propose use of signal flags that represented numbers rather than specific orders. His *Architectura Navalis* (1629) contains depictions of flags and pendants labeled with the numbers one to eight and, somewhat more cryptically, the alchemical symbols for male, female, sun, and moon (66–67; cf. Miller 1944, 62).

During the eighteenth century, there was more serious flirtation with the idea, first in France and later in England. Some of these proposals were never implemented, and others treated the numerical flags as supplemental to the older system (and thus merely added an additional complication).

In 1781, Kempenfelt introduced a signal book using 16 numeric flags in two flag hoists, with the upper flag signifying the page and the lower one the entry on that page. "The mast or locations in which a flag was hoisted no longer formed a part of the code" (Earle 1912, 1040).

In Howe's 1799 system, there were nine triangular flags used to identify portions of the fleet, eight pendants, and 25 rectangular flags with multiple meanings. The latter could be used singly, but 10 of them could be used in combination as numeral flags. For example, the red flag could mean "enemy in sight" or the numeral 1. The number code "15" meant "prepare for battle"; "18," "it is noon"; "21," "alter course to starboard in succession"; and "101," "close around the Admiral" (Edles 1799; Raeside 2020).

The advantage of a numerical system was that with a "symbol set" of 10 different flags, one could transmit all the old orders by assigning them a numerical code, which could be more than one digit if the corresponding flags were displayed in either a sequence, or as a simultaneous combination in which the flags were to be read in a particular spatial order. Thus, in both the Howe (1799) and the Popham

(1803) system—historians disagree (Naval Marine Archive 2022) as to which was used by Nelson at Trafalgar (1805)—the word "England" was represented by the flags 2-5-3.

I will discuss the Popham (1803) system in more detail. The preparatory (telegraph) flag, a red-and-white diagonal, was hoisted first to indicate that a numerical codebook message was being sent. Then the message would be sent, and finally the finish flag would be hoisted (or the telegraph flag lowered). If a number was to be sent qua number, rather than as an element in a numeric code, it would be preceded by a numeral pennant (Popham 1803, 6, 8).

Flag codes 1–25 designated letters of the alphabet, with 9 standing for I and J, and V preceding U (Dempsey 2018). The two-digit codes, of course, were represented by two flags. (McMillan [2001] states that this was the first naval signal system with an explicit design to spell out letters.)

The hoist location determined the order in which flags were read. The hoist locations, in order, were the main mast, fore, mizzen, gaff, and ensign-staff (Popham 1803, 8). Starboard was before port and upper yard before lower. On a single halyard, the flag combination (hoist) was read from top to bottom (Raeside 2011).

However, signalmen were not obligated to use every mast for a long message. They had to take visibility into account. While some depictions of Nelson's message to the fleet at Trafalgar show three masts in use (agefotostock 2024; Gregg 2023a), the Admiralty Library of the Ministry of Defence points out that "HMS Victory was at the head of the fleet, the wind was aft, and the repeating ships were in their station, so the mizzen masthead was the most generally visible location as well as the most usual. The flagship's signals would have been bent onto the mizzen topgallant signal halyard" (Raeside 2018).

So, based on either the 1799 or 1803 system (the numeric codes were the same but which flag denoted which number changed), when HMS *Victory* signaled the fleet at the Battle of Trafalgar, the flags were read as follows:

Hoist	Code	Meaning
1	253	England
2	269	expects
3	863	that
4	261	every
5	471	man
6	958	will
7	220	do
8	370	his
9	4	D
10	21	U
11	19	T
12	24	Y

The Popham system included a single substitute flag. If hoisted underneath another flag, it indicated a repeat of the number signified by the flag above it

(Popham 1803, 7). So, in hoist 7 above, the middle flag could have been the substitute flag.

The one-, two-, or three-digit codes in the Popham system were denoted entirely by flags. However, his code extended up into the two thousands. "The thousands are denoted by pendants or balls, as may appear most likely to be seen … superior, 1000; inferior 2000" (Popham 1803, 7).

The recipient signaled "message understood" by hoisting the affirmative flag or repeating back the signal. Message not understood was signified by hoisting it with a white pendant underneath. If a particular word in the message was not understood, the recipient could hoist the numeric pennant followed by a number flag indicating the ordinal position in the message (Popham 1803, 6).

The first U.S. Navy signal book was created by then-captain Thomas Truxton, commander of the USS *Constellation*, in 1797. It used three-digit numerical codes, represented by pennants (Robison 1932). While the Truxton system was not formally adopted by the government, it served as the basis for the subsequent Barron signal book (1802) (Howeth 1963, 513). I am not going to discuss the later nineteenth-century U.S. Navy signal books, because they didn't feature any major innovations.

For merchant vessels, the first popular signal system was Frederick Marryat's *A Code of Signals for the Merchant Service* (1817). This used 10 numerical flags (with three shapes: rectangular, triangular, and swallow-tailed) and seven control flags. The code was divided into six lists, and some of the control flags designated which list the transmitted numerical code was referring to (Gregg 2023).

In 1857, the British Board of Trade published the *Commercial Code of Signals*. This had 19 flags, representing 18 of the consonant letters of the alphabet, with the nineteenth being the code/answering pennant, which was not itself an element of any of the codes. (The omission of the vowels was so the flags could not be used to spell out any taboo words!) Three of the alphabet flags, used singly, denoted "affirmative," "negative," or "quarantine." Two-, three- or four-letter codes had various nautical meanings, such as "JD": "You are standing into danger" (Gregg 1999).

The *International Code of Signals* evolved from an 1889 international conference. It came into use in 1901 and superseded the 1857 code in 1902. It was substantially revised in 1932 and 1965. The ICS is a full alphanumeric code, that is, with individual flags for the 26 letters of the alphabet and the 10 decimal digits. The "A" and "B" flags are swallow-tailed (proportioned 2:3), and the remaining alphabet flags are square. The numeric pennants are trapezoidal (as is the code/answer pennant), proportioned 5:9. There also three triangular "repeater" flags (proportioned 7:11) to indicate the repetition of the value of an earlier flag in the same hoist. This means that only one set of flags is needed to send any four-element hoist. Thus, there are a total of 40 ICS flags (Raeside 2020b).

Prior to World War II, each navy had its own set of signal flags, which didn't necessarily correspond to the ICS (or each other) (Kent 2001, 188). After the formation of NATO, the NATO country navies abandoned their individual naval flag sets and adopted the same set. NATO uses a set of 68 flags. These include the 40 ICS flags, but they have additionally 10 numeral flags, 17 special flags and pennants (in addition to the ICS code pennant), and a fourth repeater (substitute) (Raeside 2024).

For communication between naval vessels, the ICS flags have classified meanings, and the ICS numerical pennants aren't used. If a NATO ship has to communicate with a civilian ship, it will hoist the code pennant and then use ICS.

Flag Hoisting

The signal halyard is a rope that passes through a pulley on the masthead or spar. As described by Myer (199) in 1872, it has a loop at one end and a toggle on the other. The individual signal flags likewise have a toggle at the top and a loop at the bottom (Myer doesn't say how they are fitted onto the flag), but Gower (1808, 202) refers to each flag having a line about a fathom long, so the flags when connected are spaced at "equal and distinct distances."

The toggle of the top flag is fitted into the loop of the halyard. If a second flag is part of the signal, its toggle fits into the loop of the flag above. Finally, the bottom flag's loop is placed over the halyard's toggle. The string of flags could then be hoisted.

Signal flags flown "from the halyards of an unidentified escort vessel at the Charleston Navy Yard, South Carolina," April 19, 1945. The colors used are black, white, red, and in one instance, yellow. The patterns are horizontal, vertical, and diagonal stripes, checkerboard, diamond, and "X" (National Archives. Naval History and Heritage Command, Catalog 80-G-K-4136. Converted to monochrome and cropped by author.

The major innovation in flag hoisting was by Edward Fitzmaurice Inglefield (not to be confused with his father, Edward Augustus Inglefield, who invented the Inglefield anchor). While flag lieutenant to Rear Admiral D'Arcy-Irvine in the *Agincourt*, he had observed several problems with the toggle-and-eye system. In cold weather, the eyes were "frozen hard, and the signalmen had to open them with their teeth" in order to send up a new signal. Also, if the eye were not sufficiently taut over the toggle, it might slip off and a man had to be sent aloft to retrieve the flags.

In 1889, while first lieutenant of the *Melita*, he improvised a C-shaped wooden clip, with a "slight opening," that would go on either end of the short rope length holding the signal flag. The clip of one flag could be engaged with the clip of another if they were to be hoisted together. It worked and he then had a pair made out of Muntz metal. He sent this to Mark Kerr, the flag lieutenant for Admiral Sir Anthony Hoskings, who ordered a complete set to be made. The signalmen on Hoskings's flagship (the *Victoria*) complained about them, and Hoskings sent the clips to Rear Admiral Lord Walter Kerr. The latter's flagship, the *Trafalgar*, used the clips and beat the other ships, including the *Victoria*, in a signaling competition—whereupon Hoskings overrode the protests of his signalmen and adopted the clips for the *Victoria*. Thereafter, they were "only supplied to ships which specially asked for them" until 1894 or 1895, when they were mandated for all ships (Inglefield and David 1940).

Inglefield and Arthur Davis received a British patent on the clip in 1894. In one embodiment, "each clip has a portion which is cut away and formed with pointed ends, the two clips being connected and disconnected by bringing the cut away portions opposite each other, and in a position at right angles to each other, so that one clip or link will slip over the other." The purpose, of course, was to prevent accidental disengagement. In addition, the drawings showed that it was not (or no longer) a simple C-shape, but either an approximate "figure-8" with a chamfered split in one of the rings, or an approximate C-shape with an eyelet screwed into the base.

The Inglefield clip is still in use by sailors, but it is sometimes called a sister clip or a Brummel hook. In the late-twentieth-century U.S. Navy, the signal flags were secured to the halyard with a snap-and-ring system (Naval Education and Training Support Command, United States [NAVEDTRA] 1996, 5–6).

Signal flags hoisted by the sender are usually "closed up," that is, hoisted so the top touches the point of hoist. Answer or repeater hoists are normally initially hoisted "at the dip," which means about "three-quarters of the way up." After they are acknowledged, they are closed up (NAVEDTRA 1996, 5–4).

After the Age of Sail, the number and height of masts were limited. In 1926, the Naval War College assumed that a large vessel was limited to five simultaneous flag hoists, an intermediate vessel to four, a small vessel to two, and a surfaced submarine to one (Kidd 1924, 65). The maximum size of a flag hoist was three flags (57).

Flag Storage

In the British Navy, flags came to be stored in a flag or signal locker, with a separate pigeonhole for each flag type. Gower (1808, 203) says that the flags, with the attached lines, "are to be placed in a range of cells, fitted up at some convenient part of the poop or quarter deck, each cell being marked with the number of the flag or pennant it contains. Such an arrangement will greatly facilitate the operation of making signals." (This suggestion did not appear in his 1796 treatise.) A signal locker was mentioned in a American court martial proceeding, also from 1808 (Navy Department 1822, 205), but without any explanation of the layout of the locker. However, cabinets with pigeonholes date back at least to 1688 (*Oxford English Dictionary* online, "pigeonhole," 6a).

After the invention of the Inglefield clip, "flags were stowed in the locker with the clips outward" (Kent 188). Twentieth-century navies specified which flags were to go in which pigeonholes, rather than leaving it up to the individual captain or signalman (Proc 2022).

Another storage system was the "flag bag," nowadays an all-metal bin but originally a metal frame covered with canvas (NAVEDTRA 1996, 5-4). Since they were "large metal racks with notches" (Raines 2019), they are perhaps best compared to a "hanging file" system in an office.

Flag Size

The larger the flag, the greater is the distance from which it can be seen, but the greater is the wind force needed to stretch it out, and the more cumbersome it is to store and handle it. With a strong wind, the larger the flag, the greater the strain on the halyard and spar.

In the early eighteenth century, a ship of the line might fly a flag that was 15 feet high and 27 feet long. When the Admiralty adopted three-digit numerical codes, so three flags had to be displayed from a single spar, the flag size was reduced to 12 feet by 14 feet (Nicolls 1991, 220). Marryat recommended still smaller flags (6 by 8 feet rectangular flags, and 4 by 18 feet pennants) (Marryat 1841, Notice). Since modern ships have short masts, flag hoist signal flags nowadays are no larger than 3 feet 4 inches by 5 feet (Nicolls 224).

Flag Material

Myer taught that flags for hoist signals were made of bunting, "that it may fly with a light breeze." In contrast, flags used for signals that are based on the motion of the flag, rather than its color, pattern, or shape, were to be made of cotton or linen, so they "glide easily through the air" (Myer 1872, 199–200).

"Distant" (Shape) Hoist Signals

By the late eighteenth century, there was a perception that as range increased, it became difficult to distinguish the colors of the various flags, and it was thought that an abbreviated "symbol set" based on shape might be more effective at long range.

While hoisting of signal balls probably predated it, a comprehensive three-dimensional shape system was proposed by the naval architect Richard Hall Gower in 1796. This made use of "cones and cylinders" (Gower 1796, 158). (It appears from examining the code table that there was a third primitive shape, a double cone [bases touching], which would look like a diamond in profile, and there may also have been two different sizes of cones.) Since these, like spheres, are "solids of revolution," they have the advantage of looking the same regardless of the horizontal direction from which they are viewed.

Gower's preference was that the shapes be "made of light wicker, and painted black," with a diameter of about three feet. But if these forms were too bulky, "they may be formed of canvas, set out with hoops; thus made, they will collapse into a small compass when they are out of use." He recommended that they be hoisted to the yardarms or stays, rather than to the masthead.

Gower's code (based on his Plate III) is set forth here.

Code	Shape
0	Cylinder over inverted cone
1	Upright cone over cylinder
2	Inverted cone over cylinder
3	Cylinder over upright cone
4	Inverted cone
5	Upright cone
6	Double cone over cylinder
7	Cylinder over double cone
8	Double cone over upright cone
9	Inverted cone over double cone
Substitute	Upright cone over upright cone

Several of the later systems seem to be something of a step backward from Gower's proposal, as they still made use of flags (wind issues) or colors.

In the system that Rear Admiral Raper (1828, 19) proposed to the British admiralty in 1825, the numerals 1–9 were represented by rectangular flags of different proportions (numerals 1–4), a black ball (5), or the flags in combination with a black ball (6–9), and zero by a weft. Raper warned that "spreaders must always be employed when there is not wind enough to display them clearly." Raper proposed that his signal ball be made by fastening together "two light ash hoops with iron swivels, and throwing over them a loose bag of old bunting, of any colour, left open at the bottom, which the wind will spread to its full extent" (96).

In 1835, Philipps (146) presented a table of "distant signals" that combine the use of the upper sails (royal or top-gallant) of the main or fore mast with one or two balls on the main or fore mast. The sail could be lowered on the cap, or hoisted and clewed up. The system allowed sending 30 of the more important flag hoist code groups. For example, the main upper sail hoisted and clewed up and a ball at the main denoted code 4, "an enemy is in sight."

Richardson's 1864 edition of the Marryat code included "geometrical signals." It did not entirely dispense with color indications, as the red signals represented 1–5 and the blue (or black) ones indicated 6–0. The shapes, in order, were an upright cone, an hourglass (cones with apexes touching), a double cone (bases touching), a cylinder, and an inverted cone. There were also red, black, and yellow "hexagons" (in profile) used in place of distinguishing pennants (Richardson 1864, Plate VI).

Richardson claimed that the signals had "galvanized iron expanders, and are so constructed that they can be distinguished from all points of the compass, yet they occupy trifling space when not in use" (vii). Flag code 7468 delivered the message "your flags cannot be distinguished, use the geometrical signals."

The "distant signals" of the International Code of Signals use three symbols: "a ball, a square or rectangular flag, and a triangular pennant." Every distant signal hoist included at least one ball. One ball alone was the preparatory and answering signal, and two balls the annulling signal. The other signals used combinations of three shapes for the 18 consonants of the ICS. The ICS provided that "in case of need, and in the absence of the proper signals," one could substitute handkerchiefs for the flags, horizontal oars or spars for the pennants, and pails or buckets for the balls (Great Britain Board of Trade 1894, 3–4).

Lieutenant House reported (Niblack 1892, 492) that based on an 1889 proposal, experiments were conducted by the U.S. Navy with a cone and a cylinder, "each about two feet high and made with barrel hoops and bunting." These were collapsible shapes, rigged so that when held aloft, "a downward motion on the lines exposed the shape, and an upward motion collapsed it." One man could work "one shape with each hand," with a rapidity "about equal to that of wigwag" (see later discussion).

Nonetheless, as of 1892, "the use of shapes was still only a matter of experiment" in world navies. Niblack considered the most practical collapsible shapes, from a mechanical standpoint, to be cylinders and cones, and commented that if only one shape is displayed at a time, that the transmission rate is too slow for squadron purposes. Niblack proposed a hoist construction that would simultaneously display four shapes (independently, a cylinder or a shape with a hexagonal profile, like a Japanese folding paper lantern) on a vertical mast and use with it Myer four-element pseudo binary code (Niblack 1892, 461).

One important design issue was how large to make the shapes. The British found that collapsible drums that were 4 feet tall and 3.5 feet in largest diameter were "much too small" (Niblack 1892, 463). But "shapes larger than 6 feet are too difficult to work and collapse in any kind of breeze" (467).

Niblack was also concerned that at long range, refraction might distort the appearance of the shapes (Niblack 1892, 468). Lieutenant Bower thought that collapsible shapes were "excellent on board ships when not engaging an enemy; but in battle, presenting as they do a large target, they would become worse than useless," being damaged so that either they could not be collapsed when open or expanded when closed (497).

The modern incarnation of shape signals are the "day shapes" specified in the *International Regulations for Preventing Collisions at Sea*, Part C and Annex I (1972). Four basic three-dimensional shapes are used: ball, cylinder, cone, and diamond (a double cone). The standard size is a diameter of at least 0.6 meters. The cone has a height equal to the diameter and the cylinder a height twice its diameter. They are black fabric (or vinyl) stretched over a collapsible frame. Up to three shapes are used in combination, usually in a vertical line. Up to three such lines may be displayed simultaneously.

Intraship Flag Signals

Until the late nineteenth century, the boatswain's instructions for squaring the yards (laying them at a right angle to the keel) were bellowed out to the boatswain's mate. The Royal Navy then adopted a system that "saves the lungs of the boatswain, allows him to go to any distance from the ship, is not liable to be misunderstood, is far more expeditious, and in a close harbour, saves all the intolerable bawling that follows the close of any exercise with the fleet." "Three small flags …, bent to short staves," were used: red for the main mast, white for the foremast, and blue for the mizzen mast. The boatswain (who was being rowed about the ship in a boat) "faces the ship" and indicated which yard to "top" (topping a yard is using a tackle to lift it into position). "For lower yards, the flag is held horizontal; top sail yards, flag-staff perpendicular; top-gallant yards, he stands up in the boats, and holds the staff at an angle of 45°; royal yards, flag-staff at the same angle, but the arm lowered." The boatswain's mate, on the ship, raised his arm to acknowledge the signal, and the boatswain waved the flag "when the yard is square" (Alston 1860, 399–400). Alston does not identify any flag signals to indicate that the yard should be turned further inward or further outward in order to come square, so that was apparently left to the judgment of the boatswain's mate, subject to the boatswain's wave of approval.

Myer Wigwag

Various signaling systems have been devised that used gestures of the hands or arms, or with a handheld object (rods, flags on poles, handkerchiefs, disks, hats, swords, torches, or lanterns), to convey a message.

Here, I confine my attention to systems known to have been used on ships. The most popular was definitely Myer "flag telegraphy," better known as the "wigwag." In its original form, approved for Union Army use in 1860, it had three elemental symbols and a rest position. In the rest position, the single flag was held straight up. Motion 1 was to bring the flag down low to the signaler's left and then return it to rest. Motion 2 was the symmetrically opposite action. Motion 3 was lowering the flag directly in front of the signaler, and then returning it to rest (Raines 1996, 7). Later, motions 1 and 2 were reversed, as documented in Myer's *Manual of Signals* (1864). It appears that this change was made because the rightward motion would be in the left field of view of the receiver, and we expect that in a series of numbers, the smaller are on the left (Myer 1872, 95). Nonetheless, perhaps in deference to those who learned the older system, he said that the parties could agree to do it the old way.

In the 1864 system, motion 3 appears to have been used to denote the end of a word (Myer 1864, sec. 34). Myer thus had a pseudo-binary (two-element) coding system, like that of the telegraph's Morse code (sec. 6). Also, like Morse code, it had to represent letters of the alphabet using variable-length groups, for example, "one" for the letter "I" but "two-two-two" for "d" (Wolters 2013). (To represent all letters of the alphabet, a sequence of one to four binary digits would be needed.)

In his 1872 signal manual, Myer proposed a three-element coding system (Myer 1872, 96), presumably using the "front down" motion for "3" and a longer rest to denote the end of a word. This is a ternary (three elemental symbol) coding system, and Myer recognized that a combination of three symbols was sufficient to encode all 26 letters of the alphabet (Wrixon 1998, 406), as there were $3 \times 3 \times 3 = 27$ possible three-digit combinations.

Niblack (1892, 440) complains that in the wigwag, motion 3 will be taken as motion 1 or 2 unless the sender is facing the receiver squarely, and that this is problematic if the transmission is to or from a rapidly moving or turning vessel, or if the transmission is a simultaneous one to two vessels that are not more or less in the same line of sight.

The first use at sea (presumably of the two-element system) was by Myer on June 15, 1861, when Myer, on a tugboat, gave guidance to a shore battery bombarding the Confederate works at Sewell's Point, Virginia (Signal Corps Association 2024). It was used again for land–sea coordination (using Army signalmen) during the joint-force attack on Port Royal Ferry on January 1, 1862 (Signal Corps Association 2024).

This success prompted the Union Navy to adopt the system in 1862 (Raines 1996, 13), and "when Rear Adm. Farragut ran his fleet past the defenses of Port Hudson, Louisiana, in March [1863], signal officers high in the mastheads kept the vessels in contact with each other" (27).

The Confederacy had a signal officer who had been trained by Myer before hostilities began, and an Army signalman was stationed on the *Merrimac* during its 1862 engagement with the *Monitor* to receive intelligence from the shore. "Subsequently, signal officers accompanied almost all blockade-runners of the Confederate navy" (Signal Corps Association 2024).

The 1872 wigwag flag was usually four feet square (Myer 1872, 190) and attached to a 12-foot jointed staff (202). The signal kit included white, red, and black flags. Myer recommended use of a red flag at sea for best contrast, as the background would be the woodwork or sails of the vessel, the sky, or the water (264).

For night signaling (Myer 1872, 81–83, 205), the flag was replaced with a "flying torch" attached to the signal staff. This torch was a cylinder of copper closed at one end and filled with a combustible fluid, preferably turpentine. The wick was cotton and the motion of the staff helped supply oxygen to the flame. Myer was of the opinion that torches were brighter than lanterns and could be read from 15 miles away. Since the signalman's body couldn't be used as a point of reference, a "foot torch" (for shipboard use, often a lantern) was placed at the feet and in front of the signalman. Myer warned that care must be taken to make sure that the foot torch is always visible to the recipient, elevating it if need be. For example, it could be placed on the rail (82).

Mechanical and Manual Semaphore

Mechanical semaphore systems use one or more rigid signaling arms mounted on a vertical post. Popham's land semaphore (1816) had two arms on a single post but with different pivot points, whereas for shipboard use, he put one arm on each of two posts. The latter were to be about 12 feet tall, with arms 6 feet 4 inches long and 10 inches wide (Knight 1861, 8:68). Pasley (1823, 26), another semaphore inventor, criticized Popham's sea semaphore on the grounds that it was not always possible to see both posts, or to keep track of which was which. Others (6) complained that the semaphore was sometimes inadvertently read in reverse (i.e., from behind rather than in front). Popham's land semaphore was certainly successful—the sea one, not so much.

In any event, in the 1820s, Popham's semaphore was largely replaced by Pasley's, which used two moveable arms with a common pivot point on a single post (it also had a fixed "indicator" arm, mounted low on the post, to prevent reading in reverse) (Craddock 2021, 66).

The extent to which the Pasley semaphore was used on ships is unclear. The anonymous author of the 1842 *Handbook of Communication by Telegraph* complained (45) that the mechanical semaphores were too "cumbrous" for use aboard ship. Nonetheless, he admitted that they had an advantage over flags in that in calm weather, the flags did not fly out unless "stretchers" were attached to them.

Fast forward to 1881. Vice-Admiral Dowell admitted that it was "often difficult to see a flag, but there is a greater difficulty in finding a substitute. We have heard of a masthead semaphore … [T]he enemy have Gatling guns, and they would more easily knock your masthead semaphore to pieces than they could damage your flags" (Dowell 1882, 373).

Nonetheless, in 1886, the *Electrical Review* carried an article referring to the torpedo ram *Polyphemus* (1881) practicing signaling with "her new masthead semaphore." (The *Polyphemus* had low-voltage electric lighting and it is probable that the semaphore was equipped with lights for nighttime use.)

By 1889, it was reported that "the use of semaphores, consisting of two arms, … is much favored in the British Navy. Many ships now have a semaphore on each quarter and on each bow, so that signals may be made clear of the ship's masts; and masthead semaphores, having longer and larger arms, are now fitted to mastless ships. They are operated by a crank and endless chain, the position of the crank corresponding exactly with that of the semaphore arm" (Wainwright 1889, 66). (I confess to being perplexed why several authorities credit Admiral Wilson with inventing the masthead ["truck"] semaphore in 1893.)

Manual semaphoring mimics the signal positions of the mechanical semaphore, but using two flags, one in each hand. Note also that this differs from wigwag not merely in terms of the number of flags used, but also in that the signal is formed by the position of the flags, rather than by their direction of motion.

According to the U.S. Navy's 1908 *Boat-Book*, the arms of the mechanical semaphore were painted so the middle third was white and the remainder black. For manual semaphoring, the hand flags used were 12 to 15 inches square, and affixed to a 2-foot-long wooden staff (115). The flags were later enlarged to 15 to 18 inches square, and made fluorescent (Bureau of Naval Personnel 1961, 27). Hand semaphore has been adapted to night use by means of "wands fitted with red lenses" (NAVEDTRA 1996, AI-2).

There are eight possible positions for each flag, but only 28 of the 64 possible two-flag combinations are used. In 1996, you needed to transmit and receive plain language semaphore message at 10 words a minute to qualify for Signalman 3, and 15 to advance to Signalman 2 (NAVEDTRA 1996, AII-5).

Heliograph

The heliograph communicates by means of reflected sunlight. Its signal mirror was held in a two-axis mount on top of a tripod. A horizontal sight arm held a vertical vane with a sighting mark in front of the mirror. The mirror had an unsilvered spot in the center. These were used to aim the mirror so that the reflected light would be toward the receiving station. When the sun-heliograph-receiving station angle was greater than 90 degrees, a second mirror was brought into action. The sunlight would be reflected off the "signaling" mirror and onto the "duplex" mirror, and from there to the recipient.

Flashes were produced by a "keying" the signaling mirror (slightly deflecting the light bean toward or away from the recipient); the heliograph mirror was aligned so the light was visible when the key was depressed. "The normal rate of signaling may be taken as 8 words per minute for morse" (Great Britain Naval Staff 1918, 43).

In 1905, the heliograph in use by the Royal Navy was the Mance Mark V. This had a 5-inch mirror (Royal Signals 1905). A rule of thumb was that 10 miles of range were obtainable for every inch of diameter of the mirror (Great Britain Naval Staff 1918, 23) However, that range was probably achieved only in mountain peak to mountain peak communication. The actual range is limited by the state of the atmosphere and by the curvature of the earth. Ignoring refraction and assuming a spherical Earth, for an observer on the bridge of a battleship whose eye is 40 feet above the waterline, the distance to horizon is 7.7 miles, and this is doubled for that observer to see the flash emanating from a heliograph mirror 40 feet above sea level.

The rays of light from the mirror form a cone. And the ratio of the width of that cone to its length is the same as the ratio of the diameter of the sun to its distance from the earth, which is about 1:107 (Great Britain Naval Staff 1918, 23). The permissible lateral divergence—the permissible aiming error measured at the receiving station—is thus the range to the receiving station divided by 214. While that seems rather narrow—and the heliograph was equipped with aiming aids—that also means that an enemy was unlikely to intercept a heliograph communication, whereas he could see hoisted flags.

In 1892, Albert Niblack criticized naval interest in heliography: "Heliography is not possible from one ship to another, nor from a ship to the shore, nor from the shore to a ship. Absolute immobility is required in the receiving and transmitting stations. This we do not get in either a ship at anchor or underway" (437–8). If he was correct, then its naval use would be limited to marines on shore.

In 1904, Thursfield admitted that its use afloat had been "restricted," but saw signs that this might change. During practice maneuvers in 1903, with Admiral Wilson commanding the B Fleet and Admiral Domville the X Fleet, "the leading B1 cruisers had got beyond ordinary signal distance, but by heliograph the *Hawke* was just able to send back information that *Europa* was chasing the X cruiser *Diana*" (Thursfield 1904, 68). Moreover, when the two wings (B1 and B2) of the B fleet first met, "the first signals received by the *Majestic* from the *Revenge* were transmitted by heliograph" (69). Thursfield comments that Admiral Wilson had modified the heliograph apparatus to overcome the difficulties presented by the "unsteady platform of a ship at sea," but "the nature of the apparatus employed has not been made public" (69).

The details were, however, revealed in Bradford's 1923 biography of Wilson. The signaling mirror and its sighting vane were clamped on a long, tripod-mounted telescope, with the vane axis parallel to the line of sight of the telescope, and the telescope eyepiece was equipped with crosshairs. One man kept the crosshairs, and thus the signaling mirror, aligned on the receiving station, "in spite of the motion of the ship," while a second keyed the mirror (Bradford 1923, 182). (Strictly speaking, this arrangement would have a parallax difference between the telescope and the sighting mirror, but I suppose the operators would learn how to compensate for this, perhaps adjusting the sighting vane slightly off parallel based on the estimated range.)

Tracing the subsequent history of the heliograph at sea is difficult because the term "heliograph" was sometimes applied to signal lamps with blinkers (as is apparent from the caption to a photograph in the 1966 *Naval Reservist*).

The last vestige of the heliograph in the maritime world is the emergency signaling mirror provided with survival kits.

Visual Communication—Night

The illuminance provided by "direct sun" is about 100,000 lux, dropping to about 10,000 on an overcast day. In contrast, that provided by a full moon is about 0.1 lux, and by a quarter-moon, 0.01 lux (Engineering Toolbox 2004). Distinguishing the colors and shapes of flags, based on the moonlight they reflected, would have been possible only at very close ranges. While I know of no scientific study of that task, the Army found that men dressed in khaki could be seen against the sky at 300 yards under full moon, 150 under half-moon, and 100 under starlight (Gordon 1957, 5). All that would have been seen would be silhouettes, so this is relevant only to the perception of shape, not color.

Until radio was developed, self-luminous light signals were the only practicable means of nighttime communication at a distance. Bear in mind that light communications may be more difficult to detect and intercept than radio ones, and thus remained useful after ships were equipped with radios.

During the daytime, our eyes are most sensitive to light with a wavelength of 555 nm (green). When our eyes are dark-adapted, only the rods of the retina are active, and the peak sensitivity is 507 nm (blue-green) (*Wikipedia*/Luminous Efficiency Function).

A signal may be conveyed by the presence of absence of the light (implying a binary signal code), by the number and relative position of the lights, or by the color of the lights. It is worth noting that it takes "5 to 20 times as much light to distinguish the color of a light than to simply distinguish" its presence or absence (Lewis 1986, 34).

The ability to distinguish color is also dependent on the intensity of the light reaching the eye. "When a light is just visible it is nearly colorless ... As the intensity of the light is increased, or the observer approaches closer to it, red lights may be distinguishable from the lights of other colors, but these continue to look alike. At still higher levels of illuminance at the eye of the observer, green lights become recognizable" (and even higher illuminances are needed to distinguish yellow and blue) (Breckenridge 1967, 13). In 1841, railroad engineers concluded that "the visibility of a

"Night signaling was done by a number of lanterns of various colors and displays in various combinations and positions in the ship, either on deck, in the shrouds, in the tops, or at the peak. The signal of a wish to communicate at night was given by rockets or Bengal lights" (Naval History and Heritage Command, Catalog NH 71127. Scanned from *Iconographic Encyclopedia* (1851), volume III).

red light was but one-third that of a white light of the same intensity; that of a green light one-fifth; and that of a blue light one-seventh" (17). However, a 2003 paper reported that the perception time for blue LEDs was shorter than for red or yellow lights of equal luminosity (Lin 2003).

A contributing factor, especially at longer ranges, is the effect of the scattering of light by air molecules. This is called Rayleigh scattering, and it is why the sky looks blue and the rising or setting sun (from which the rays pass through the greatest thickness of atmosphere) looks red. The intensity of the scattering is inversely proportional to the fourth power of the wavelength, so it is 9.4 times greater at 400 nanometers (blue light) than at 700 nanometers (red light) for equal initial intensities. That means that with focused beams, blue and green light will be visible at shorter ranges than red light, being scattered out of the line of sight of the recipient, and blue-white or green-white light will look whiter (but dimmer) as the range at which it is viewed increases.

Pyrotechnics

A two-color pyrotechnic signal system was conceived by Benjamin Franklin Coston in the 1840s, and a three-color one was developed and patented in his name by his widow, Martha Coston (Coston 1859; Pilato 2024). According to this patent, the numerals 0 to 9 were represented by red, white, and blue flares, either individually (for "1" to "3") or by a sequence of two (e.g., white then red for "4") or three different colors (white then red then blue for "9"). The signals were fired from a signal gun. There were three different sizes of paper boxes that could be set off (either by hand percussion of by the percussion cap of a signal pistol); the larger sizes contained two or three different pyrotechnic compositions that would be burned through in succession, corresponding to the key for that numeral. This was intended, of course, for use with a signal code book in which words or phrases were represented by numerical codes.

It was not possible to achieve a bright blue, and in the American Civil War implementation, green was used instead. The code used differed from that in the 1859 patent. Communication began with the preparatory signal (white–red–white) and its acknowledgment (red–white–red). The single-color flares represented the numbers one, four, and seven, rather than one, two, and three (Coston 2024).

Sometimes the Coston flares were combined with rockets. For example, the force blockading Charleston, South Carolina, in 1864 used a rocket followed by Coston No. 0 for "blockade runner going out."

Niblack (1892, 470) reports that in an 1892 trial of an "improved" Coston system, 42 percent failed to be correctly read at distances of three to eight miles. The principal problem was confusion of white and green. Niblack also complains that because of their brightness, they destroy the crew's night vision. Madame Coston had warned that "a signalist should be careful not to look at the brilliant flame of the signal burning near him, as thereby the eye is not fitted to discern accurately the colors of distant lights" (Coston 2024).

In 1878, the U.S. Navy began using the Very code, which used a pattern of four bursts, each of which could be red or green, to encode numbers. Despite being a binary code, it did not correspond to Morse code in its original implementation (Wrixon 1998, 430). Niblack reports (1892, 479–80) that the altitude achieved was limited to about a hundred feet because the "stars" couldn't withstand more powerful charges. Nonetheless, this yielded a range of visibility of 13.7 nautical miles, as opposed to 8 miles for the Coston flares displayed at deck height of 32.5 feet. (The height of the observer's eyes was not stated.)

Nowadays, pyrotechnics are mostly used for emergency situations. The 1996 Signalman's training manual has some interesting comments (Naval Education and Training Support Command 1996, 4–12) on the limitations of pyrotechnics. First, with regard to the color of the signal, it says: "Experiments have proved that the standard colors red, white (or yellow), and green are the only satisfactory colors under varying conditions of visibility. Under certain atmospheric conditions, white signals may appear yellow. Likewise, a white signal may be mistaken for a green signal under certain humid conditions. It is easy for tracer signals to be confused with red ones." Second, the "range of visibility for a pyrotechnic signal is variable and unreliable because it depends largely on weather conditions." Finally, it is difficult to determine who fired the pyrotechnic (the enemy?) and from where, and whether the signal was observed by the intended recipient.

However, pyrotechnics do have the advantage that pyrotechnic rockets may be fired to a great height (Myer 1872, 211), with a corresponding increase in the distance to the horizon. Early twentieth-century models attained an altitude of 800 to 1,000 feet. In addition, if the payload is equipped with a parachute, its period of visibility is increased (Faber 1919, 10).

Signal Fires and Lamps

There are two major considerations with this class of signaling device: how the light is generated, and how it is arranged or manipulated to produce a signal.

Light source. There was a general progression toward brighter and brighter light sources. All else being equal, the brighter the light source, and the more it can be concentrated toward the recipient, the greater is the effective range. It also helps if the light source is positioned high above the deck, to increase the distance to the horizon. Over-the-horizon communication is possible if there are cloud bases that can be illuminated with a beam bright and tight enough that the lit ellipse is perceived.

In 1617, Raleigh used a fire signal aboard his flagship to send commands to the other ships in his squadron (Wrixon 1998, 417). Greenwood (1715) described 14 British Navy night signals that made use of lanterns, usually in conjunction with signal guns. For example, the signal to moor was a lantern at the top of each of the three masts, while unmooring was ordered by three lights on the main mast and two gun reports.

In 1862–1863, Philip Howard Colomb of the Royal Navy used an oxyhydrogen limelight as the light source for a signal apparatus (Sterling 2008, 209; Colomb and Bolton 1870, 25). In 1870 he and his colleague Bolton asserted a signal range of 25 (24) or 27 (5) miles.

Colomb also proposed, alternatively, an oxycalcium light (Colomb and Bolton 1870, 22). This was an older form of limelight in which the oxygen raised the temperature of the flame directed against the lime, creating a highly luminous hot spot. It was less dangerous than the oxyhydrogen system (Noye 1998).

A kerosene lamp with a focusing lens (Begbie lamp) came into use in the 1880s and was used until World War I ("QSVC" 2012). Subsequently, signal lamps were of the handheld incandescent ("Aldis") or pedestal-mounted carbon arc type. (Signal lamps would require less power than searchlights of the same effective range.)

In 1894, a battleship might carry a 40,000-candlepower projector with a 90-centimeter (36-inch) mirror (Heinz 1894, 765), and the maximum range of the searchlight for night signaling was said to be 25 nautical miles (Heinz 1894, 781). A 12-inch searchlight, under favorable conditions, may be seen as far as 9 miles by day and more than 16 miles at night (Howeth 1963, 11).

Some signaling searchlights have been equipped with mercury or mercury–xenon arc lamps, rather than carbon arcs.

The cutting-edge technology nowadays is the laser. The laser offers much higher data transmission rates than radio because the carrier waves are light waves (ultraviolet, visible, or infrared), which are of much higher frequency and thus provide a greater bandwidth (Ronny 2012, 5). The narrow beam angle of the laser is something of a double-edged sword. On the one hand, it increases the range and reduces the opportunity for an enemy to detect and intercept the transmission. On the other hand, it requires very accurate aiming of the beam to hit the receiver (7), and ship motion makes that more complicated.

In 2006, lasers were used to "successfully send large data, movie and audio files, as well as enable live ship-to-ship video conferencing between the USS *Denver* and the USS *Bonhomme* [*Richard*] at ranges from 2.5 to 11 nm" (Raible 2011, 9).

Flashing signal methods. The simplest mode of signaling is by making the light visible for a short or a long interval, that is, a pseudo-binary coding system. This has the virtue that only a single light is needed, but there must be some way of controlling whether the light is visible. There were three basic options: turning off and on the light; deflecting the light toward or slightly away from the recipient (as with keying of a heliograph); and covering and uncovering the light.

The deflection method was used in the British Aldis signal gun. This used an incandescent light in conjunction with a parabolic mirror, and the trigger tilted the mirror (Wrixon 1998, 438).

The covering method was used by Colomb. In his 1870 monograph (Colomb and Bolton 1870), he describes several different approaches. First, he presents a signal apparatus with a semi-cylindrical shade. It fell by its own weight, and was raised by a lever (12). Second, he proposes one with a "collapsing drum" or "collapsing cones," intended for daytime use (34–5). Finally, he devised a "Venetian blind-like contraption": "a series of shutters, each working on a pivot, and all connected together in such a manner as to move simultaneously by the motion of the handle" (36). Several decades later, Admiral Percy Scott coupled the Venetian blind-type shutter with a carbon-arc searchlight, and this became known as the "Scott shutter" (Wrixon 1998, 438).

The shutter concept is a simple one, but the exact mechanical design affects the maximum transmission speed. Feldman alludes to a model in which "each shutter need move only 1.5 inches to reach the maximum 'open' or 'closed' position." It could be operated at 360 cycles per minute, equivalent to a signal speed of 18 words per minute. The shutter was driven by a rotary solenoid (Feldman and Monahan 1962, 13).

The on–off method may seem the simplest, but there were problems. With high-powered carbon arcs, the light does not "flash or die down quickly enough" (Niblack 1892, 469). Others had a service life limited by the number of times they were switched on and off.

One expedient used with carbon arc lights was to engage an electrical shunt in parallel with the arc (Luckiesh 1920, 198), thus dimming the light, but the shutter system remained dominant. A dimming approach was tested with the xenon–mercury arc lamp, but it was not able to achieve the target "bright–dim" ratio of 20:1, needed for differentiation by human eyesight (Feldman and Monahan 1962). Feldman suggests that with physical detectors capable of discriminating smaller brightness differences, the dimming method would be workable.

More recently, the U.S. Navy tested the use of light-emitting diodes (LEDs), as they can be switched on and off at very high frequency (LaGrone 2017).

Myer (1872, 88) pointed out that at sea, where both ships are in motion, it is

"USS *Yorktown* (CV-10). Signalman uses a blinker lamp to send a message to the carrier's escorting destroyer, at night, circa late 1943." The shutter mechanism is visible as faint horizontal lines (National Archives. Naval History and Heritage Command, Catalog 80-G-419960. Image cropped by author.) https://www.history.navy.mil/content/history/nhhc/our-collections/photography/numerical-list-of-images/nhhc-series/nh-series/80-G-419000/80-G-419960.html.

possible that the recipient will lose sight of the sender between flashes and that the subsequent flash will be missed. "It is well to have another lantern standing close to it as a marker, to enable to the telescope at the other station to be kept on the signal station at night."

Other Light Signaling Methods

In 1803, Commodore Preble of the United State Navy relied on the spatial arrangement of three or four lanterns to encode numbers and a few special signals (Wrixon 1998, 419).

Colored light systems were proposed, too. In the 1850s, Ward proposed signals using combinations of red, white, and no light (Wrixon 1998, 422). The Berg system used red, white, and green.

In 1891, the U.S. Navy adopted the Ardois system. At the time, this used a cluster of five double lamps read top to bottom or sender's right to left. Within the pair, the upper light was red (Morse dot) and the lower light white (Morse dash), and only one light was on at a time. Thus, the Morse code was implemented with dots and dashes represented by color and position rather than by duration. The light sources were 32-candlepower incandescents (Niblack 1892, 473).

In 1892, Niblack urged adoption of a four-light system instead, arguing that it was cheaper, more suitable for ships with short masts, and easier to learn. The French, Austrian, and Italian navies were already using four double-light designs, whereas the Germans used three lights (Niblack 1892, 475–8). (The French experimental system, uniquely, used a four symbol code: fixed red, pulsating red, fixed white, and pulsating white [Niblack 1892, 475].)

It appears that Niblack prevailed soon thereafter, as the four double-light system was installed on the battleship USS *Indiana* (commissioned 1895) (*Electrical Engineer* 1896, 264).

An 1897 report from the commander of the naval militia of Maryland called attention to Lieutenant L.F. Smith's modified Ardois apparatus. This used four white oil lights instead of pairs of electric lights. These could be made to appear red or off by drawing down a red filter or an opaque screen. The apparatus was recommended "for use on small vessels not fitted with an electric plant" (Gibbons 1897, 109).

In general, a colored light was achieved by placing a suitably tinted glass filter in front of a "white" light source (an incandescent light bulb has a color temperature of 2,700–3,300°K, and carbon arc and acetylene of 3,700, as compared to 1,800 for a candle flame and 5,500–6,000 for zenith daylight [Choudhury 2014, section 1.7]). Wrixon (1998, 422) notes that if tinted glass "was too thick it dulled the light and if it was too thin it appeared as white light to observers." The filtration may be increased by increasing the concentration of the colorant in the glass rather than by increasing glass thickness (Breckenridge 1967, 19).

A truly colored light source may be achieved by use of a gas (argon, neon, krypton, xenon, etc.) discharge lamp, but that does not appear to have been used for naval signaling, even though "neon" signs are ubiquitous on land.

Transmission speed and accuracy. The most straightforward but slowest system was to have the sender mentally encode the message as a series of signal elements (e.g., long and short flashes of Morse code) and manually operate the lamp to send each element.

Colomb actually developed a mechanical system for sending a four-number code group, in which the signalman set keys and then the apparatus played out the corresponding short and long flashes. His signal box contained a cylinder turned by a handle, and the surface had four series of pins and bars, with the series being set at slight intervals. Between the end of the fourth series and the beginning of the first, there was an interval corresponding to a quarter of the circumference—this would ensure a break between code groups. The operator placed the keys in the slots corresponding to the numbers to be sent and then turned the crank (Colomb and Bolton 1870, 13ff). Regrettably, Colomb doesn't explain how the position of the key moves the pins and bars or how they act upon the shutter. The handle was to be cranked at a "uniform rate of 100 to 150 turns in a minute."

Numbers one to five were represented by the corresponding number of short flashes, six by a long flash, and the remaining digits by the combination of a long flash with one or two short flashes, with the long flash at the beginning or the end of the sequence. Colomb says that "the lengths of the flashes and intervals are regulated by the machinery" (Colomb and Bolton 1870, 9), but doesn't explain how that machinery handles number signals of different duration.

There were actually five keys, with the fifth operating on a set of pins and bars at the right extreme of the barrel. It was used to transmit the auxiliary signs and otherwise rested in a blank slot.

The Ardois lights were operated by a keyboard. The five-light system had 62 possibilities (with one to five lights lit, and lit lights red or light), and apparently there were a corresponding number of selections (Niblack 1892, 101). With a four-light system having one to four lights lit, there are 30 possibilities.

An 1898 illustration of the Ardois keyboard shows that the control keyboard is not a keyboard with individual keys, but rather there is a dial selector and, presumably, a single key (Countiss 1898, Fig. 11).

Countiss (1898) also provides a photograph (Fig. 14) of Claudius V. Boughton's "Telephotos" keyboard, which was used with a similar four red–white lantern system. This had 31 individual keys, in two rows. It also had a bulb display grid above the actual keyboard. A different typewriter-like keyboard is the subject of Boughton's 1898 patent (application filed 1895), but it specifies that lamps of one color represent dots and lamps of the other represent dashes (Boughton). A similar Telephotos keyboard is depicted in an 1895 article in the *Electrical Engineer* (1895, 489).

As of 1906, some warships were equipped with Ardois night-signal sets and others with the Telephotos (Cowles 1907, 381–4).

Recently, the U.S. Navy tested a system in which the sender and receiver both have a device that runs a software flashing light-to-text converter. For the sender, this controls the shutter of the signal lamp. For the recipient, the flashes are captured by a GoPro camera and then interpreted by the software (Szondy 2017).

Security. Light signals may be omnidirectional (a masthead or yardarm light) or unidirectional (focused beam, as with a searchlight). The latter, of course, was more secure. In addition, some lights could be equipped with filters so only infrared light was transmitted. In the U.S. Navy, the radio code words "Nancy Hanks" indicated that the ship was switching to infrared transmission, and the recipient needed to use an infrared-to-visible converter in order to read the signal (Proc 2022a).

Range and Speed of Signaling

Figures quoted for the range of visibility of the various signaling systems vary. This isn't surprising, as there are variations in equipment (especially over time), signalman expertise, and atmospheric conditions. In addition, the figures quoted usually aren't based on rigorous experimentation.

In the 1901 Naval War College wargaming rules, searchlight signaling was limited to 30 miles. The Very signals could be read up to six miles, and the Ardois signals, up to three to four miles. "Day signals"—I am not sure whether this referred to flag hoists or to manual flags—were assigned a range of three miles (Naval War College 1901, 26).

In 1924, Kidd (65), referring to "War College Standards," opined that for large combat vessels, the limit of transmission, in thousands of yards, was 20 for flash, 10 for flag, 5 for wigwag, and 2 for semaphore. For small vessels, he reduced flash to 8 and flag to 5. ("Flash" here refers to a shuttered light and not to heliograph.)

With regard to transmission speed, this depended on whether the message was in plain language or encrypted, the range (signaling would slow down if the range were greater than half the maximum range), and whether the receiving vessel was under fire or not. Under the best circumstances, the Naval War College assumed that to send 10 plaintext words required 6 minutes by flash and 12 minutes by semaphore or wigwag, At increased range, these increased to 9 and 24, respectively. For code, it figured 15 and 30 at short range, and 24 and 48 at long. A vessel under fire would take twice the time to receive the message (Kidd 1924, 63–4).

The total number (incoming plus outgoing) of simultaneous flash, wigwag, and semaphore communications that a ship could handle was deemed to be limited by the size of the ship (which presumably correlated with the number of trained signalmen on board): large, four; intermediate, three; small, two; and surfaced submarine, one (Kidd 1924, 65).

9

"Ahoy There!"
(Nonvisual Communication at Sea)

In 1803, at night off Gibraltar, there was a tense interchange between Commodore Preble of the USS *Constitution* and the captain of a British frigate. Each used a speaking trumpet to hail the other and demand, "What ship is that?" And each, displeased with the other's reply, threatened to open fire on the other (*Nautical Magazine* 1907, 366–7). The matter was ultimately resolved peacefully, but one might well imagine that if they had no way to communicate, the encounter would have led to an exchange of gunfire.

We first consider the physics underlying audio communication and then discuss the various audio communication methods used at sea. Afterward, we consider other nonvisual communication methods, notably courier vessels, carrier pigeons, and radio.

Physics of Audio Communication

Sound is a traveling wave phenomenon, a cyclic, in space and time, compression and rarefaction of the medium in which it travels. The frequency of a sound is the number of cycles per second (Hertz, abbreviated "Hz"). The wavelength is the speed of sound divided by the frequency. Sound intensity (sound power per unit area) is proportional to the square of the amplitude of the wave.

Sound has the advantage that it doesn't require line of sight; it can curve around walls. How far away a sound can be heard depends on its initial intensity (energy) and its frequency (pitch), as well as on the environment. There is a fall-off in intensity as a sound moves away from the source that is simply the result of the sound spreading out. In addition, there is a further attenuation with distance as a result of the viscosity of the medium in which the sound is traveling. According to Stokes's law of sound attenuation, the attenuation is proportional to the square of the sound's frequency.

The sound we hear from a particular source is the sum of the sound waves that reach us directly and those carried to us by reflection. If sound strikes an object, part of it is reflected, and part is absorbed, reducing the intensity of the reflected wave. Moreover, there will be a difference in time of arrival between sound reaching us directly and by reflection from the same source, creating an echo effect that may reduce intelligibility. Of course, our brains must also parse out sounds coming from different sources.

While sound intensity is a purely physical concept, loudness has physiological (human ear frequency response) and psychological dimensions. The perceived loudness of a sound is roughly proportional to the logarithm of the intensity, and hence intensity is often expressed on a logarithmic scale in which 0 decibels (dB) corresponds to a sound intensity of 10^{-12} watts per square meter, the threshold of hearing at a frequency of 1,000 Hz. A difference of 10 decibels corresponds to a 10-fold difference in sound intensity. Normal conversation is then 60 dB, and the eardrums burst at 160 dB.

In air, at 20°C, sound travels at a speed of 343 meters per second. However, the air is not a uniform medium in which the sound travels at equal speeds in all directions. The speed of sound is dependent on temperature, traveling faster in warm air than in cold. In the daytime, over land, the air is warmest near the ground, and as a result the sound waves turn upward (refract) and are lost more quickly. At night, one can have a temperature inversion, and waves that would otherwise be lost are refracted downward, enhancing reception (Hannah 2007, 25). The effect is particularly strong over water.

Wind is another issue. Winds increase with height. If you are shouting downwind, the sound waves are refracted toward the ground, and if upwind, the reverse is true (Hannah 2007, 22). On a ship, the relevant wind is the apparent wind, which is the vector difference between the ship velocity and the true wind velocity.

It is often believed that "shouting into the wind" (i.e., upwind) is ineffectual. In actuality, "sending a sound upwind, against the flow of air, makes the sound louder due to an acoustical effect called convective amplification." (The reverse is true for a sound sent downwind.) It only seems ineffectual because "your ears are downwind of your mouth," so "your own voice sounds quieter to you" (Conover 2023).

Direct Communication of Speech

Sometimes, for clarity, a word would be spelled out. But this had its own problems, as some consonants had little carrying power or were easily confused. "The natural solution to the problem ... was to name each letter ... When the system was first tried, each person was permitted to use names of his own selection for the letters." However, "there was a lack of uniformity which did not aid accuracy ... Some were briefer and consequently more efficient for naval purposes. To remedy this condition it was decided to adopt a standard list of names."

It was eventually realized that accuracy in "letter name" recognition would depend on "recognition of successive vowel sounds, following each other in a prescribed order ... the order of sounds being sharply different from that for any other name in the list" (Tawresey 1926).

Tawresey gives the revised list in use in 1926, and it is interesting to compare it to the current NATO phonetic alphabet.

	1926	*NATO*		*1926*	*NATO*
A	Affirmative	Alfa	N	Negative	November
B	Baker	Bravo	O	Optional	Oscar
C	Cast	Charlie	P	Preparatory	Papa
D	Dog	Delta	Q	Quack	Quebec
E	Easy	Echo	R	Roger	Romeo
F	Fox	Foxtrot	S	Sail	Sierra
G	George	Golf	T	Tare	Tango
H	Hypo	Hotel	U	Unit	Uniform
I	Interrogatory	India	V	Vice	Victor
J	Jig	Juliet	W	William	Whiskey
K	King	Kilo	X	Xray	Xray
L	Love	Lima	Y	Yoke	Yankee
M	Mike	Mike	Z	Zed	Zulu

Voiced numbers could also be confused, especially "five" and "nine." Hence, Tawresey reports, "five" was rendered as "fife," and "fifty" as "fife-ty." NATO has a complete phonetic alphabet for the decimal digits: WUN, TOO, TREE, FOW-ER, FIFE, SIX, SEV-EN, AIT, NINE-ER, ZE-RO (NAVEDTRA 1996, 1–3).

Now let's consider physical aids to audio communication.

Speaking (Hailing) Trumpet

A "speaking trumpet" was supposedly used by Alexander the Great to address his army (Barbieri 2004, 211). However, if the device was known to the classical world, it was forgotten in Europe during the Dark Ages. Claims for its invention, or reinvention, in 1670 were made on behalf of Samuel Morland (Barbieri 209). In his 1672 monograph, which certainly popularized the concept, Morland asserted that the mouthpiece end must be at least the diameter of the "orifice of the speaker's mouth," and that the tube "must be enlarged by degrees, and not too suddenly."

Of course, the narrow ("throat") end must be larger than the open human mouth when speaking, or some of the sound won't be captured by the horn in the first place.

A horn shape has several effects on sound. At the narrow ("throat") end, we have "proximity resonance." If the throat is narrow enough, then sound waves reflected off the wall are mostly in phase (sync) with newly generated sound waves of the same wavelength and thus reinforce the latter (sometimes called "loading the source"). This occurs if the diameter of the throat (the maximum round-trip distance to the wall) is not more than a quarter-wavelength (Heller 2013, 110).

This amplification occurs with a tube, too. To convey a sense of the degree of amplification possible, the power output of a monopole sound source (one that radiates sound equally well in all directions) "when put in [an infinite cylindrical] pipe of [cross-sectional] area S is $\lambda^2/\pi S$ times larger than when in free air, where λ is the wavelength of the sound" and the wavelength is much greater than the square root of "S" (Heller 2013, 117).

Admittedly, the human mouth is not a monopole sound source. However, the loading at the throat of a finite-length horn should be similar to that of a finite-length pipe, and the difference between them will be in terms of how energy is transmitted to the air at the other end, versus how much is reflected back to the source. That bring us to the issue of "impedance matching."

A horn offers better "impedance matching" than would a cylindrical tube. "Impedance" is the resistance of a "packet" of air to vibration. It is the force needed to give that air a particular velocity. The impedance of air inside a cylindrical tube is high because the tube walls tightly constrain air movement (high air pressure, low velocity). The impedance outside the tube is low (corresponding to low air pressure, high velocity). Whenever there is a sharp change in impedance—whether low to high or vice versa—sound is reflected. The wider the mouth of the tube, the smaller is the change in impedance as the sound wave leaves the tube (Heller 2013, 6ff).

So, we want a narrow throat and a wide mouth, and a smooth transition from one to the other. Horn shapes include conical, parabolic, exponential, hyperbolic, and tractix. They differ in terms of how quickly the horn flares out and how the flare rate changes with the position along the length of the horn.

With certain simplifying assumptions (Kolbrek 2008, 2), the optimum shape to minimize internal reflections is what is called an "exponential horn" (Heller 2013, 119), because it is the shape in which the cross-sectional area of the horn increases at a constant rate as one moves from the throat to the mouth. The acoustic resistance of the mouth of a finite-length exponential horn to sound of a particular wavelength will not significantly deviate from that of an exponential horn of infinite length if its mouth perimeter is more than four times the wavelength in question (Dinsdale 1974).

There is no definitive standard for width of the mouth of the horn. Authorities have variously recommended (for an exponential horn) that it have a diameter of one-quarter (Davis and Jones 1989, 217), one-third (NAVEDTRA 1973, 198), one-half (Crowhurst 2010), or equal to (Heller 2013, 118) the longest wavelength to be reproduced, or that it have a circumference equal to that wavelength (Lenard 2021; Borwick 2012, 34) (equivalent to it having a diameter that is 1/3.14 times that wavelength). In the last case noted, the acoustic resistance is 6 dB. For narrower mouths, the resistance increases nonlinearly (Dinsdale 1974).

There is a further reduction in required mouth area for a given acoustic resistance if the horn is not radiating in all directions (as would be the case if the speaker were standing in the masthead). Standing on the middle of the deck halves the solid angle presented to the mouth of the horn, and if the speaker's back is to the outside wall of a cabin, it is halved again (Dinsdale 1974).

At room temperature and atmospheric pressure, the sound waves audible to the human ear have wavelengths between 17 millimeters and 17 meters.

The acoustic analysis of the speaking trumpet is further complicated by the fact that the source of the sound is actually in the vocal folds (colloquially, "cords"), 15–17 centimeters away from the proximal end of the trumpet (Heller 2013, 368). In speech, the speaker subconsciously controls the force of the breath and the opening of the vocal fold, and changes the shape of the vocal tract by varying the "tongue position, jaw opening, and lip shape" (361). This in turn alters which sound wavelengths are reinforced by resonance, as well as which are suppressed. At the lips, there is an impedance mismatch as the sound waves escape the relative confinement of the vocal tract. Use of the speaking trumpet reduces the mismatch at the lips by preventing that escape, and therefore distorts speech (120, 368). Additional distortions may arise depending on the frequency response of the trumpet to the resonant frequencies of the speech. Since a real-life horn cannot be infinite in length, it necessarily will have both frequencies that it amplifies and frequencies that it suppresses as a result of reflection at the mouth of the horn (118).

The drive (fundamental) frequency for the vocal folds is about 100 to 125 Hz in males (Heller 2013, 357), corresponding to a wavelength of 11.25 to 9 feet. The resonance in the vocal tract will reinforce some harmonics (integer multiples of the fundamental frequency), creating the "formants" (vowel sounds). "The sound power in speech is carried by the vowels ... Intelligibility is imparted chiefly by the consonants" (Roy 2019). Shouting does not significantly increase the sound level of the consonants (DPA Microphones 2021).

The consonants are voiced at frequencies above 500 Hz, and most consonants are found in the 2,000–4,000 Hz range. Cutting out the sounds below 500 Hz reduces intelligibility by only 5 percent; at 800 Hz, by 10 percent, and at 1,250, about 20 percent. Background noise reduces intelligibility, especially if it is in the 1,000–4,000 Hz range (DPA Microphones 2021).

Morland's instrument was 5 feet, 6 inches long, 2 inches wide at one end and 21 inches at the other (Morland 1672). The normal maximum vertical human male mouth opening is about 50–60 millimeters (Christensen 2010), and the throat of Morland's trumpet is similar in size (51 millimeters) Loading by proximate resonance should occur for sound wavelengths greater than 8 inches (four times the throat diameter), and thus to a frequency of less than 1,688 Hz. That includes the most important sound waves for both sound power and intelligibility. For 1,688 Hz-sound waves in a 2-inch throat, if we treat the source as a monopole in a pipe and ignore all energy losses, the power amplification would be about 6.5-fold.

The horn was most likely a truncated cone. With an exponential horn, there is a clearcut lower cutoff frequency, but that is not the case for a conical horn.

If Morland's horn was an exponential horn with a mouth of 21 inches, then we would expect to see loss of sounds with frequency under 643 Hz (corresponding to a wavelength equal to the mouth diameter). However, that would not have impacted intelligibility.

Morland claimed that in his experiments at Deal Castle, his voice could be "plainly heard off at sea; within two and three miles ... when the wind blew from the shore." However, Radau (1872, 59) reports that a later British experiment with a 20-foot trumpet only achieved a 2-mile range. Morland demonstrated his speaking trumpet to Charles II, and his favorable reaction led to examples being "sent to sea on royal warships" (Davies 2017).

In 1708, Jacques Ozanam (407) declared that it was "of good use at sea, in a storm or a dark night, when one ship dare not come within reach of speaking nakedly to the other," or for an admiral "to convey orders to his whole fleet." The same year, the privateer Woodes Rogers, near the beginning of his epic circumnavigation of the globe, was given a speaking trumpet by another captain (Rogers 1712, 9). The first reference I could find to the actual use of the speaking trumpet at sea was in a 1745 court martial proceeding, the trial of the lieutenants of the *Dorsetshire* (Fearne et al. 1746, 11). Falconer (1769, "Speaking-TRUMPET") calls it "a trumpet of brass or tin used at sea, to propagate the voice to a great distance, or to convey the orders from one part of the ship to another, in tempestuous water, etc."

Ear Trumpet

All of us are familiar with the practice of cupping one's ear in order to hear better. One can improve on this by means of an "ear trumpet," and artifacts found close to Pompeii are believed to have been made for this purpose (Valleriani 2012, 2). The concept was rediscovered by Giambattista della Porta (1589), who sought to determine the "most efficient shape and design" for an "ear-spectacle" (3). Paolo Aproino tried to build such a device in 1613, but reported that resonances spoiled the intelligibility of the sound (Barbieri 2004, 207). A series of additional ear trumpets were made later in the seventeenth century.

The ear trumpet is essentially a speaking trumpet used in reverse. This exploits the reciprocity theorem of acoustics: As long as the intervening medium is not in motion, one may interchange source and receiver (Heller 2013, 128). However,

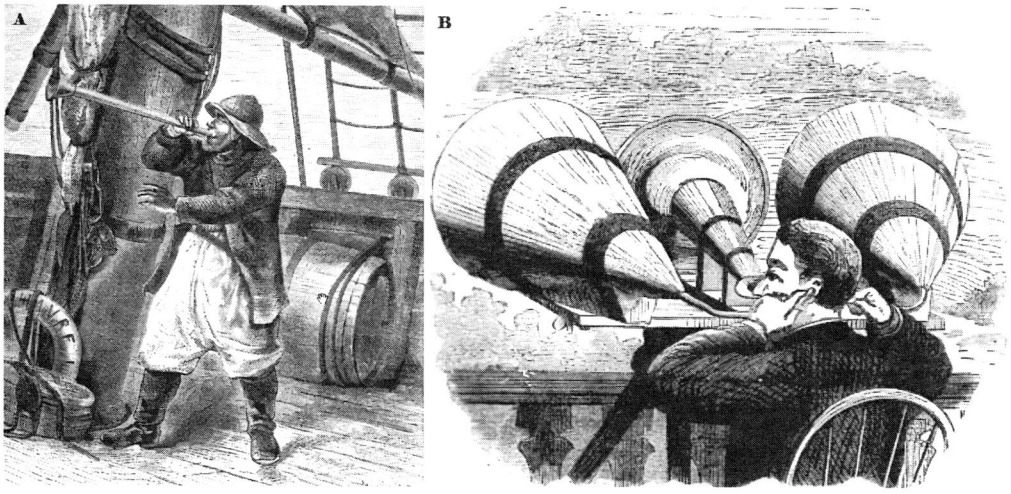

A. Artist's sketch of a merchant seaman using a speaking trumpet (Guillemin 1877, 111). There is a photograph of a nineteenth-century French speaking trumpet, made of brass, about 23 inches long, in the Crosby Brown Collection of Musical Instruments, 1889, of The Met, accession 894.1627. B. Edison's "megaphone": actually the combination of a speaking trumpet and two ear trumpets, the latter each 6 feet 8 inches long and 27.5 inches diameter at the larger end, and equipped with flexible ear tubes connecting the narrow end to the ear (*Nature* 1878). Similar devices (without the speaking trumpet) were used as aircraft locators before the development of radar (Self 2024). These were generally used on land, in conjunction with antiaircraft batteries. There was some exploration of the possibility of a shipboard sound locator for air defense in 1919-1920, but little came of it (Friedman 2014, 95).

while sound moves just as easily in either direction of the trumpet, if you used a trumpet with a throat sized for use as a speaking trumpet as a hearing trumpet, some sound would be lost, as the human mouth is much larger than the ear canal (130).

Voice Tubes (Speaking Tubes; Voice Pipes)

In *New Atlantis* (1626), Francis Bacon imagined that the inhabitants had "means to convey sounds in trunks and pipes, in strange lines and distances." Use a tube to connect two horns together, and you have a voice tube.

Voice tubes were used in stately homes, factories, mines, and ships. The earliest reference given in *Oxford English Dictionary* was from 1833 ("Orders being given by the waiter above through a speaking-tube"). However, Webster and Parker's *Encyclopedia of Domestic Economy* (1844) said that the speaking tubes first came into use about 35 years earlier, that is, around 1809.

Insofar as maritime usage is concerned, an illustration by Ackerley (1834) shows the mouthpiece of a voice tube mounted on a steering wheel, so the helmsman may communicate with men on the lower deck (in heavy weather, these manned tackles attached to the tiller head) (Appendix vii). Lardner (1851) wrote that "a speaking tube is sometimes used on shipboard, being carried from the captain's cabin to the topmast" (177).

On HMS *Black Prince* (commissioned 1861), a sail/steam hybrid warship, the

The flying bridge of the German cruiser *Prinz Eugen*, "soon after V-E Day." On the left is a bank of voice tubes. On the right (after cropping) is a "Seehund" infrared blinker (Naval History and Heritage Command, Catalog NH 96461). (Note also NH 103106 USS *Agamemnon*, 1918-19, seven voice tubes.)

communication between the bridge and the engine room was by voice tube (Dowell 1882, 373).

Voice tubes were also used to control ships' batteries, as noted by Noel (1874, 70).

Finally, there were voice tubes at lookout stations—for example, on the foretop of the Japanese battleship *Nagato* (commissioned 1920) (Tully 2003). On the battleship USS *Vermont* (launched 1905), there were 136 voice tubes, with a maximum length of 382 feet, and diameters of 1.5–2 inches (Snow 1909, 853).

Voice tubes are still found on modern warships. The 2003 self-study manual for an "interior communications electrician" in the United States Navy explains that "voice tubes are installed aboard ship in addition to the sound-powered telephone system to provide another way for transmitting information between designated stations." Actually, they were a backup to a backup, as electronic intraship communications were the norm (Self 2021).

The manual explains that on surface ships, the tubes run from the pilothouse (helmsman's station) to the exposed conning station on top, to the navigation bridge wings, and to the captain's sea cabin; from the navigation bridge wings to the chart room chart table; and from the flag plot chart table to the flag bridge wings. Since these are backup systems, they are equipped with "a call circuit to alert personnel to man the tube." The tubes are 3-inch diameter brass (Parker 2003, 5–32, 33).

It is apparent from a late-nineteenth-century account by Admiral Dowell that intelligibility could be a problem; during practice maneuvers, the captain thought he was told that the ship was making 13 revolutions when it was in fact making 33 (Dowell 1882, 373). However, the available alternatives were no better: "The ordinary mechanical engine-room telegraph … is constantly getting out of order … The telephone is out of question for service afloat" (Oldknow 1893, 108).

One potential disadvantage of the voice tube is that "communication between two water-tight compartments presents a clear risk of flooding, and requires shut-off valves to be fitted either side of the bulkhead" (Self 2021). Another was that voice tubes could be broken or disconnected. On the USS *Monitor*, the one from the pilothouse to the turret failed, for unspecified reasons, at the beginning of the Battle of Hampton Roads (Mindell 2000, 76).

Liversidge offered some guidance as to voice tube installation and use. The pipes were to be hung "on elastic brackets, so as to be insulated from the vibration of the structure." The user was to "speak slowly and distinctly," and "always repeat orders received through the voice pipe" (1899, 171).

We do know something about the constraints on the use of voice tubes. Oldknow (1893) asserts that the "maximum length for efficiency of a voice tube is 300 ft, and for every bend approaching the rectangular 15 ft must be deducted."

National Bureau of Standards (NBS) physicists proposed that "the loss of intensity in a section of tube is proportional to the intensity entering," and that the proportion is a constant for a given length of a given kind of tubing (Eckardt 1926, 164). "Suppose that for each 10-foot section the output intensity was, roughly, 85 per cent of the input intensity." This is equivalent to a prediction that the intensity follows a negative exponential decay curve with a decay constant of 0.016 per foot. For such a tube, the intensity after 200 feet would be 4 percent of the original value (167). From comparison with Eckardt's table 1 (170), this is a close match to a brass tube with a nominal diameter of 4 inches (the average decay constant was 0.015).

Only a few diameters were tested, but it appeared that the average decay constant was inversely proportional to the diameter. For four-inch brass, it was half the value of two-inch brass (Eckardt 1926), but the trend line drawn by Eckardt is curved (Fig. 3). Self (2021) argues that the attenuation is the result of "viscous friction between the vibrating air and the walls of the tube." Doubling the diameter doubles the wall area of the tube, but quadruples the mass of air in that section. "In fact the loss in dB is inversely proportional to the area, and hence to the diameter squared. The theoretical attenuation of a 5-cm (2-inch) diameter voicepipe is 30 dB per 100 metres" (Self 2021).

In any event, Eckhardt's table 1 only provides part of the picture, as it was determined using pure tones. The NBS also tested the intelligibility of syllables. It was found that 200 feet of one-inch tube gave the same intelligibility as 1,600 feet of 2.5-inch tube, and of nearly 2,400 feet of 4-inch tube (Eckardt et al. 1926, 181).

Various mouthpieces were also tested. With pure tones, short cones were found to be superior to long cones, and cones to an "exponential horn" shape (Eckardt et al. 1926, 177). However, mouthpieces improved intelligibility of syllables only when used with the two-inch brass pipe, and not with the wider pipes tested. With a 400-foot length of 4-inch pipe, 93 percent of syllables were heard correctly.

Transmission of sounds in the range of 300–3,300 Hertz (Hz) were tested. (The young adult human hearing range is 20 to 20,000 Hz, but human speech ranges from 64 Hz, music C2, to about 2,048 Hz, music C7 [Amplifon 2021]). Curiously, in speech, frequencies below 1,000 Hz contribute 80 percent of the sound energy, but those above most of the intelligibility (Eckhardt et al. 1926, 180). For 4-inch brass tubing, transmission over a 10-foot section was in the 80–90 percent band, with a low at about 2,300 Hz. For 1-inch brass, it was in the 60–75 percent range, with the low at 2,700 Hz. For one-inch iron, it ranged from 50 to 70 percent, bottoming out at 3,300 Hz (171).

Sound-Powered Telephones

In a conventional analog telephone system, an electric current is supplied by a battery or generator. The microphone in the sender's phone modulates (changes its magnitude up or down) that carrier current in response to sound waves from the person speaking. In the recipient's phone, that modulated current causes an element in the speaker to vibrate, creating sound waves that mimic the original ones.

In a sound-powered telephone, there is no carrier current. Rather, the current is generated by the microphone itself. A flexible diaphragm is attached to an electrically conductive coil, wrapped around a support (armature), and the coil is disposed within the magnetic field of a permanent magnet. Sound waves cause a slight but significant movement of the diaphragm, which causes the coil to move. When a conductor moves inside a magnetic field, a current is induced, and since the conductor is moving back and forth, the current is a pulsating one. Magnetic induction was discovered in 1831.

Bell's January 1877 patent filing listed, as an object of the invention, "the electrical transmission of musical tones, articulate speech, or sounds of every kind without the necessity of using a voltaic battery." The telephone of his Figs. 3 and 4 has a bar with a coil of insulated copper wire wrapped around one end (pole). He explained that it would operate without a battery if the bar depicted therein were made a permanent magnet. His Fig. 5 depicted another "form of telephone for use without a battery," and the permanent magnet therein has a horseshoe shape, so coils are wrapped around "pole-pieces" attached to both ends (poles). In either case, the coils are acted upon by a plate diaphragm.

The sound-powered phone was used in the Bell System for almost two years. The problem with the sound-powered phone was that "it could not transmit over the longer distances envisioned for Bell's 'grand system' of universal service." In December 1878, the Bell System switched to a battery-powered transmitter, but the "magneto" design survived for decades as a telephone receiver (Meyer 2020). The concept is also found in the "dynamic" or "moving coil" microphone.

Naval sound-powered telephones were implemented in two physical forms, the headset and the handset (Bakels 2024). The date of the first adoption of sound-powered systems for naval use is murky. A 1933 contract for the construction of light cruiser no. 43 called for the installation of "sound powered telephone equipment" upon it (Senate 1935, 4907), but it is doubtful that this was the first such installation.

Frank Massa of RCA-Victor was involved in the research and development project in the early 1930s. He recalled that the unit was "required to withstand several vertical drops from a 6 ft height directly onto the steel deck of a battleship. Additionally, the telephone could not change in sensitivity or frequency when exposed to several rounds of naval gunfire while mounted on the 7-psi-blast pressure contour" (Massa 1985, 1300).

The system received some publicity in 1944. *Popular Science* declared that it was "used on all Navy ships and most freighters," and claimed that it could withstand "concussion from the firing of navy ordnance." An illustration showed a "typical installation" on a battleship, with phones at the battery directors, engines and boilers, damage control, stowage, and the ship service telephone exchange (*Popular Science* 1944). *Life* carried a Western Electric ad reporting that on the battleship *Wisconsin*, there were "2200 instruments connecting all battle stations" (*Life* 1944).

The operation of the twentieth-century sound-powered telephones differed somewhat from what Bell described in 1876. The most important difference was that the coil was given a rocking motion, rather than a reciprocating one, by the diaphragm.

The use of a pivoted armature in the transmitter of a sound-powered phone wasn't new, having been proposed by Thomas Watson in his 1882 patent. None of my sources talk about why the reciprocating armature was replaced with a pivoted one, but my best guess is that the latter was more responsive, and thus better at transmitting the higher frequencies.

The instruction book for the United States Army's TP-3-T1 field telephone (1938) indicates that the sensitivity range of sound-powered telephones was -63 to -53 decibels (.000003–.00003 milliwatts), whereas the output range was -38 to -33 decibels (.001–003 milliwatts). The sensitivity and output ranges for battery-powered telephones were -53 to -43 and -8 to 0 decibels, respectively (Chief Signal Officer 1938, 8). Thus, the sound-powered telephones have greater sensitivity (they respond to weaker signals) but weaker output. The average speech power is 0.015 milliwatts (4). The talking ranges were up to seven miles for sound-powered versus twenty-two miles for battery-powered (Fieldphones 2022).

It provided a schematic of the transmitter unit. This shows a driving rod connecting the center of the diaphragm to one end of the armature. This rod passed through a hole in the "north side" of a large horseshoe magnet, and in the neutral position, the armature was on the axis of symmetry of the magnet. The armature is pivoted at the center, with restoring springs at the far end from the driving rod junction. Each pole had two protrusions ("pole pieces"), reaching toward each end of the armature. While not clearly shown in the TP-3-T1 manual, a coil surrounded the armature, as seen in the manual for the TP-3 (War Department 1944).

As of 1936, the sound-powered telephones approved by the Bureau of Marine Inspection and Navigation for merchant marine use were manufactured by the Automatic Electric Company, the RCA Manufacturing Co. or the Hose-McCann Corporation (Bureau of Maritime Inspection and Navigation 1936, 20–21). RCA's MI-2040-AS Sound Powered Telephone Handset, "designed for maritime applications," used an aluminum alloy diaphragm and an Alnico magnet (RCA 2016, 18). Hose-McCann Communications asserts that its founders "developed the first safe, reliable and rugged sound powered communication system for marine use" (Hose-McCann 2010) but provides no particulars.

"Telephone Talker on a shipboard gun station, using open circuit telephone, circa the 1930s. He is wearing a speaker-type gas mask, which enables him to use his microphone during a chemical attack" (Naval History and Heritage Command, Catalog NH 101675).

Electric Megaphone

The purely acoustic speaking trumpet was largely replaced in the second half of the twentieth century by electric megaphones, which couple a microphone, a battery-powered, transistorized amplifier, and a loudspeaker to an acoustic horn.

Nonverbal Audio Communication

Musical Instruments

The use of musical instruments for command and control is ancient. The Roman army had four trumpet-like instruments, the *tuba*, *bucina*, *cornu*, and *lituus* (Peddie 1996, 21). Little is known about the specific calls used.

The "boatswain's call," a non-diaphragm whistle, has been used on British ships since at least the thirteenth century (Admiralty 1951, 75). (In the United States Navy, it was usually called the "bosun's pipe.") There were two notes, low and high, and three "tones" (really, playing styles), plain, warble, and trill (76). Calls could comprise several note-tone combinations in sequence. The 1951 Admiralty manual lists 15 pipe calls (77).

The Admiralty manual adds that "the more important routine orders are passed by the bugle in ships which carry a bugler" (Admiralty 1951, 78). The 1953 edition of the United States Navy's *Manual for Buglers* presented 103 calls (Bureau of Naval Personnel 1953). The role of buglers today is much more limited, with only 15 calls in the 2003 edition (Navy Music Program Management Office 2003).

Ship's horn or whistle signals are still in use, primarily for collision avoidance. They may use short (1-second) or long (4- to 6-second) blasts. Depending whose rules you are following, they may signal what you intend to do or what you are actually doing.

A signal with duration or time interval as a distinguishing characteristic is referred to as "chronosemic" by Myer (1872, 130), and apparently there was a fog signal chronosemic system proposed by Benjamin Franklin Greene in 1864. Myer's version expected the recipient to distinguish among 10 different time intervals to distinguish the 10 decimal digits (Myer 1872, 169). Myer acknowledged that this required a timekeeping apparatus, such as "a chronometer, deck-clock, comparing watch, metronome, sand-glass [or] seconds' pendulum" (173). He also proposed a "signal chronoscope" with a wind-up clock and an interval unit setting. This could be started when the attention sign was heard and would sound a bell and advance a needle for each unit time interval until the termination sign was heard and a stop-detent released (174).

Using sound duration signals to send Morse code was also problematic: "the dot is a short toot, the dash is a blast of five or ten seconds, the long dash is a blast of ten or fifteen seconds … To make signal for a simple change of course requires some five minutes" (Niblack 1892, 440).

Signal Guns

Cannon were also used for signaling, especially at night. To save wear and tear on the "battle" cannon, by the eighteenth century, small signal cannon were provided for use in signaling and saluting. The information that could be so conveyed was rather limited, as the only practical signal elements were the number of shots and the intervals between them. The "Signals" essay in the 1842 *Encyclopædia Britannica* suggested that one could distinguish short (4 or 5 seconds), moderate (8 or 10 seconds), and slow (12 or 15 seconds) intervals. It also recommended first getting the intended recipient's attention by a double discharge with an interval of ½ to 1 second. But it also warned that at long distances, two gun reports with a short interval between them could "coalesce into one long-continued sound."

The bookseller Thomas Clio Rickman received English Patent 2288 (1799) for a "signal trumpet." As explained in Burney's 1815 revision of the Falconer dictionary, it is nothing more than a speaking trumpet with a pistol in place of a mouthpiece. Its advantage over the common speaking trumpet was its greater volume—Burney says that "the report thus produced equals that of a nine-pounder." But of course it provides "merely an alarm signal" (Burney 1815, 584).

Communication by Physical Delivery

Intraship Communications

If an order or report had to be delivered to someone out of shouting distance, it could be entrusted to a crewman, perhaps a "ship's boy." For example, on the *Pensacola*, a messenger boy was sent to tell the engineer in the engine room to "start fires under the port main boilers" (Bergman 1890, 22). This did not always work out as intended; after the *Ashuelot* ran aground off the coast of China, and "the order to abandon ship was given, a messenger boy was sent to notify the officer in charge of the engine room, but the boy was so frightened that he jumped overboard without delivering the message and as a result the engine-room force narrowly escaped being left on the sinking ship" (McKay 2013, 334).

Pneumatic tubes were used, beginning in the nineteenth century, to deliver mail and small parcels. These would be placed in a carrier cylinder that was either pushed by a burst of compressed air or pulled in by a partial vacuum (Self 2020).

Pneumatic tube messaging systems were installed on some ships. On the *Titanic*, if a passenger wanted to send a telegram, the message form would be turned in to the "Enquiry Office" (customer service desk), and then sent by pneumatic tube to the radio room three decks above (Brewster 2012).

They were also used on warships. On the battleship USS *Texas* (commissioned 1914), there was a pneumatic tube system for conveying classified messages from the decoding room to the bridge (Brewster 2012, 122). In 1922, Lieutenant Commander Tisdale wrote, "The pneumatic tubes are necessary for confirmations and for handling long messages. It is of course preferable to send all messages through the tube in writing if time permits" (Tisdale 1922, 46).

LONG-DISTANCE COMMUNICATIONS

The earliest form of direct, long-range (over-the-horizon) communication at sea was by dispatch vessel. In the late sixteenth century, the Spanish *aviso* ("advice") ships, equipped with oars as well as sail, could make the passage from Havana to Spain in 28 days (Corbett 1899, 337). The news of the British victory at Trafalgar was carried home by the schooner HMS *Pickle* (Pope 2013, 93), although communication was only one of its many roles during its career. Even in the late nineteenth century, some ships were designed and built for use as dispatch vessels, such as the steam-powered HMS *Iris* (1877) (*Wikipedia*).

Homing pigeons were used occasionally for (one-way) ship-to-shore communications. In 1893, homing pigeons on the USS *Constellation* were able to fly to their home loft at Annapolis, Maryland, from up to 150 miles away. The U.S. Naval Pigeon Messenger Service was established in 1896, with home lofts at Boston, Portsmouth, Newport, Brooklyn, Key West, and Mare Island. It was discontinued in 1899, when all Navy ships were equipped with radio (Arevalo 2021).

Nonetheless, in the 1901 strategic wargame rules used by the American Naval War College, homing pigeons were given a range of 50 miles over water or 300 over land, and a speed of 50 miles per hour, and 50 percent were assumed to reach their destination (Naval War College 1901, 26).

In World War I, pigeons were used by naval aviators on antisubmarine missions along the French coast (the birds weighed less and took up less space than a radio set). They were similarly used by airships on home water antisubmarine patrols during World War II (Naval War College 1901).

Pigeon 498 was released by Thomas Crisp, commanding the Q-Ship HM *Smack Nelson* in 1917, to report that he was under attack by a submarine. He was killed (and posthumously received the Victoria Cross), but the pigeon delivered his message, despite having one wing wounded by shrapnel. Crisp's crew was rescued (Imperial War Museums 2024; Smith 2011).

Communication by Radio

The first type of radio transmitter was the spark-gap type; Heinrich Hertz built one in 1887, and Guglielmo Marconi received a patent on his system of wireless telegraphy in 1896. Spark gap transmissions could not be used to carry audio, and they produced considerable radio frequency interference across a broad range of frequencies. Consequently, they were banned in 1934 (except for emergency use on ships).

The spark-gap transmitter faced competition from transmitters employing the Poulton arc converter (1904), the Alexanderson alternator (1904), and the vacuum tube (1912). In 1924, Kidd reported that new light cruisers carried high and lower power arc transmitters, a medium-power tube transmitter, and an auxiliary spark transmitter, and that when the high-power arc transmitter (range 1,200 miles) was in use, "no other transmission or reception can be carried on" (75–6).

The tube transmitter was the dominant technology in World War II, and it in turn was replaced by transistorized units.

The first radio exchanges at sea were carried out by the British navy in 1899 (Wrixon 1998, 460), and the American navy followed suit soon thereafter. In 1904, the battleships USS *New York* and USS *Massachusetts* and the torpedo boat USS *Porter* communicated "out of sight of each other and separated by 36 miles of ocean." The following year, a war game was conducted in which a fleet equipped with radio attacked, at night, a fleet limited to visual signaling, and destroyed it (NAVCOMM 2006, 3–4).

Ships could carry multiple transmitters, receivers, and antennae. With sufficient personnel, the battleship USS *Colorado* (1923) could simultaneously "(1) transmit four radio messages, (2) receive seven radio messages from vessels of the fleet, [and] (3) receive one message from a high-power shore station" (Kidd 1924, 75).

The 1901 Naval War College wargame rules assumed wireless (radio) signaling was limited to 30 miles (Naval War College 1901, 26). By 1924, the Naval War College expected that the main transmitters of battleships, cruisers and aircraft carriers had a range of 1,000 miles, and destroyers and fleet submarines 500 miles. The larger ships also carried secondary (150-mile range) and auxiliary (25-mile range) transmitters, giving them the opportunity for multiple simultaneous radio communications. All of these transmitters had variable power controls so that the range could be deliberately limited to minimize the chance of enemy interception. They also had a variety of frequencies they could transmit on (Kidd 1924, 68).

Radio communications initially used Morse code. The radiotelephone, permitting voice communication, was invented by Lee de Forest in 1907. It was first used on an American naval vessel in 1909 (NAVCOMM 2006, 5).

Radio is the mainstay of modern seaborne communications. Nonetheless, the 1998 Signalman Training Manual warns that visual signals (flag hoist, semaphore, and flashing light) are needed when "security considerations silence all electronic communications" (1–2).

The range of radio communications is dependent on the following factors: the power, frequency, and height of the transmitter, the height and sensitivity of the receiver, and the atmospheric conditions.

The geometric horizon is proportional to the sum of the square roots of the heights of the transmitter and the receiver. Over-the-horizon transmission is possible for low-frequency (below 3 MHz) radio waves, which can be diffracted around the curve of the earth—so-called "ground waves." Because sea water has a much higher conductivity than even moist earth, "ground waves" suffer much less attenuation when beamed over the ocean than when used on land (Radiocommunications Bureau 2014, 5).

Such long-range transmission is also possible for shortwave between 1 and 30 MHz, which can "bounce off" the ionosphere. Unlike ground waves, these "sky waves" skip over a part of the earth. Other radio waves may gain some range as a result of refraction.

In Operation Moon Bounce, in 1954, the Navy beamed radio waves at the moon from a transmitter in Maryland and picked up the reflected waves 2½ seconds later. In 1955, they used the moon to send a teletype message to a receiver in San Diego (American Physical Society 2012).

10

Underwater Activities

Some of the bounties of the sea are readily available only to those who venture beneath its surface. While some sponges may be found in the intertidal region, many inhabit greater depths, clinging to coral reefs or the sea floor. There are many references to sponges in Homer's *Iliad* and *Odyssey*. In the Aegean, divers sought not only sponges, but also pearl oysters, other molluscs containing mother-of-pearl, murex shells, and red coral (Marx 1990, 9; Frost 1968). The latter was gathered in the western Mediterranean (notably Genoa, Sardinia, Corsica, and North Africa), and Renaissance Genoa produced several notable free divers, including "Andrea the Cormorant" (McManamon 2018).

It was not unusual for the larger Spanish ships to carry a diver to make repairs below the waterline (Perez-Mallaina 1998, 209). Repair of bridges, docks, and breakwaters may also require divers.

Shipwrecks near port were not uncommon, and the Venetian Arsenal had divers on call to remove wrecks in the shipping lanes (McManamon 2016, 368).

Efforts might also have been made to salvage the wreck itself. Divers could be used to locate the wreck, and to attach lines to it so it could be brought up in its entirety. If the ship had broken up, they could be used to bring up individual items of value. "Portuguese divers contracted with the government to recover guns from ships that sank near Lisbon" (McManamon 2016, 368).

Salvage could be quite profitable. In 1687, William Phipps's divers recovered 30 tons of silver from the 1641 wreck of *Nuestra Señora de la Concepción* (Rothman 2013). (Phipps's success inspired a large number of investors to put money into new diving equipment inventions or salvage missions, but in general they lost their investments [Ratcliffe 2011].)

In antiquity, military divers were reportedly used to remove underwater harbor defenses, cut anchor ropes, and deliver messages or tow supplies to besieged towns (Frost 1968).

The Dangers of Diving

Ancient divers appeared to delight in frightening tourists with purely imaginary dangers of diving. For example, the first-century Isidore of Charax was told that if a diver thrust his hand into an open oyster, it might close up and cut off his fingers (Schoff 1914). Oppian believed that tens of thousands of wrasses would surround and bite divers; Pliny believed that the manta ray would hover over divers and prevent them from returning to the surface (Frost 1968). Of course, the divers also mentioned sharks, but one suspects that even the risk of shark bite was greatly exaggerated for effect.

But deep-sea diving is in fact dangerous. The most serious of those dangers are driven by the increase in water pressure with depth.

Nitrogen narcosis. The increased water pressure increases the pressure of the air within the lungs, causing more of the gases to enter the bloodstream. For reasons not well understood, this causes a change in consciousness, similar to drunkenness. There is mild impairment of reasoning and performance at 2–4 bar pressure (33–100 feet). The effects worsen as pressures increase, with hallucinations reported at 6–8 bar (165–230 feet) and stupefaction at 8–10 bar (230–300 feet) (*Wikipedia*/Nitrogen Narcosis). Fortunately, the effect is reversible if the diver returns to a shallower depth.

Nitrogen narcosis occurs in breath-hold divers but in them it tends to occur during ascent, whereas in scuba divers it usually occurs during descent (Tetzlaff et al. 2021). At first it was thought that only the nitrogen in the pressurized air was to blame for narcosis. However, it is now known that oxygen is also narcotic (Sawatsky 2008; Sawatsky 2009; Sawatsky 2018).

Decompression sickness. As the diver ascends, the water pressure decreases and dissolved nitrogen in the bloodstream and tissues that cannot be exhaled immediately forms bubbles. The bubbles can block blood vessels or injure tissue. Moreover, the accumulated nitrogen remains dissolved for at least 12 hours after each dive, so repeated dives in a single day increase the risk of decompression sickness (Moon 2023).

Decompression sickness is usually prevented by limiting the dive depth and by making "decompression" stops during the ascent. Modern scuba divers wear portable dive computers that calculate a decompression schedule by which the diver may return to the surface safely.

If decompression sickness occurs—through ignorance or carelessness, or because an emergency forced the diver to disregard the schedule—it is treated by administration of pure oxygen or by recompression (and gradual decompression) in a hyperbaric oxygen chamber. The effects of decompression sickness may warrant avoiding further diving for weeks or even months.

It was once thought that breath-hold divers were unlikely to suffer decompression sickness on ascent, as the gas uptake is restricted by the amount of air in the lung at depth. However, there is now some evidence of decompression sickness in repetitive and even single deep breath-hold dives (Tetzlaff et al. 2021).

Eye, middle ear, sinus, or lung barotrauma. The pressure of the water on the diver's body is transmitted throughout the blood and body tissues. Since they are liquid or solid, they are not significantly affected. However, the body includes gas spaces in the lungs, sinuses, and middle ears. An artificial gas space may also be created between a mask or goggles and the face. These gases expand during ascent as a result of the change in pressure, and this expansion may injure adjacent tissues. Lung and sinus barotrauma are avoided either by breathing suitably pressurized air throughout the dive or by freely exhaling, during ascent, any air inhaled at depth.

Avoiding middle ear barotrauma requires opening the eustachian tube, by yawning or swallowing (Feenker 1985, 10–18).

Hypothermia. The ocean is substantially colder than the diver's body, and the diver will lose heat by both conduction and convection. Water conducts heat about 25 times better than air. In terms of convection, the water is moving, as a result of either ocean currents or the diver's own deliberate movement. This greatly increases convective heat loss, and it can be a thousand times greater than in air. Heat is also lost by breathing. Our lungs warm the air and some heat leaves with each breath. In fact, this accounts for about a quarter of total heat loss from a diver (dipndive 2023).

A diver may acclimate to cold water by frequent immersion; "an improved tolerance … can be gained in as little as one week" (Ratcliffe 2011). That said, cold is dangerous. The normal body core temperature is 38.5°C. Mild hypothermia sets in at a body core temperature of 33–35°C. This stage is marked by impaired physical ability and possibly an altered mental state. In moderate hypothermia (29–32°C) there is little or no physical ability. With severe hypothermia (24–28°C) the victim may lose consciousness. Tables are available that show, for different water temperatures and without protective gear, the time to loss of dexterity, loss of ability to self-rescue, and unconsciousness (Downing 2024). Mathematical modeling of the heat transfer process indicates that in 15°C water, the time to reach hypothermia (body temperature 35°C) "increases from 1 h when no suit is used to about 4 h for a 3 mm thick suit" of neoprene (thermal conductivity 0.190 W/mK) (Agullella-Arzo et al. 2003).

Drowning. Obviously, if you inadvertently inhale when you are breath-hold diving underwater, water enters the lungs, and it doesn't take much to trigger a muscular spasm that closes down the airway.

Ascent blackout. As the diver ascends, the partial pressure of oxygen in the diver's lungs decreases. If it falls below the level required for consciousness, the diver blacks out—often near the surface. The problem is exacerbated if, prior to diving, the diver hyperventilated. This artificially reduces the carbon dioxide level in the lungs. It is the rise in carbon dioxide concentration that triggers the urge to breathe. With this urge suppressed, the blackout comes without warning (grayout). In any event, it's not the blackout itself that kills, it is water entering the lungs after the blackout (Bart and Lau 2023).

Diving Tables and Hyperbaric Therapy

In 1662, Nathaniel Henshaw created a hyperbaric chamber, believing that it could be used to treat digestive and respiratory ailments. This was a compressed air, rather than a compressed oxygen, chamber, and the pressure achieved was probably no better than 1.3 atmospheres (Stewart 2011, 37).

Oxygen was isolated by Joseph Priestley in 1775. However, there was fear that hyperbaric oxygen was toxic, and thus the hyperbaric chambers used in the nineteenth century also used compressed air. These were used in the hope that they would cure or ameliorate various diseases, especially pulmonary diseases (Krishnamurti 2019; Singh et al. 2014; Smolle et al. 2021).

In 1878, Paul Bert determined that "caisson disease" (decompression sickness) was caused by too rapid a decompression, and recommended that divers stop halfway back to the surface. In 1908, John Scott Haldane developed "practical dive tables" for "staged decompression" (Krishnamurti 2019).

Bernard and Heinrich Drager (1917) and Behnke and Shaw (1937) used hyperbaric oxygen to treat decompression sickness, and the U.S. Navy adopted hyperbaric oxygen therapy for divers in 1939 (Krishnamurti 2019; Singh et al. 2014; Smolle et al. 2021).

An "oxygen toxicity seizure" may occur during the course of hyperbaric oxygen treatment. However, it is a rare side effect, as the body's "antioxidant defenses are usually adequate during the hyperoxic exposure created by a typical clinical hyperbaric oxygen treatment" (Heyboer et al. 2017).

Free ("Breath-Hold"; "Apnea") Diving

Oppian, in the second century, described contemporary Greek sponge diving. The diver plugged his ears with oil-soaked sponges and carried a heavy stone to accelerate his descent. He used a sickle-like knife to remove the sponge from the rocks and then was hoisted by rope back to the workboat. Oppian comments, "No ordeal is more terrible than that of the sponge divers and no labor is more arduous for men" (Marx 1990, 8). Isidore of Charax said that Persian Gulf pearl divers dove to 120 feet (Frost 1968).

The Spanish discovered that certain indigenous peoples had superior diving ability. On Venezuela's Pearl Coast, they initially enslaved native Americans and had them dive for pearls. Later, they made use of West African slaves for this purpose. The water temperature off the Pearl Coast is 60–70°F, which is unusually low for the Caribbean, and this increased the hazard of pearl diving. The pearl divers were also used for salvage operations (Dawson 2006).

Divers from Gujurat, in northwest India, were used to recover treasure from the wreck (1629) of the *Batavia* (Western Australian Museum 2024; Bernstein 2009, 212). There are still pearl (and chank) divers working in the Gulf of Mannar, between Sri Lanka and India. Their methods were first described in the early thirteenth century. Hundreds of boats would be working simultaneously (Athiyaman 2018).

In Japan, there have been professional female divers (*Ama*) since 750, and they have been documented as reaching depths of 25 meters, with anecdotal reports of successful dives to 45 meters (Tetzlaff et al. 2021). Until the Meiji period they wore only a *fundoshi* (loincloth), a belt of rice straw, and possibly a *tenugui* (head scarf). The typical length of a single dive was 1 minute (Kalland 1995, 165), but 2 minutes is possible (Gakuran 2013). There were also women divers (*haenyeo*) in Korea, beginning sometime in the seventeenth century.

In southeast Asia, there are the Bajau, the sea nomads. They live on small houseboats and come "ashore only to trade for supplies or to shelter from storms. They collect their food by free diving to depths of more than 230 feet." The average dive is half a minute, but they accumulate more than five hours underwater daily (Yong 2018).

In northern Europe, breath-hold divers were relatively uncommon (probably

Color woodblock print of female abalone divers, ca. 1797/1798. Artist: Kitagawa Utamaro. (Clarence Buckingham Collection, Courtesy Art Institute of Chicago, reference 1942.221. Converted to monochrome by author.)

because of the coldness of the water), and John Jacob Janson ("Jacob the Diver") was able to charge three-quarters of the value of all guns recovered from the *Santa Cristina* (sunk off the southern coast of Ireland) in 1620 (Ratcliffe 2011).

Diving Accessories

Weights and buoyancy compensators. Greek sponge divers traditionally use a *skandalopetra* (a flat smooth stone with a hole in the middle, so a rope can be tied to it), as ballast to accelerate their descent (Pelios 2023).

Weight belts have the advantage of leaving the hands free. The belts may have fixed or detachable lead weights, and an emergency quick release. Weight belts date back at least to the nineteenth century; Ruschenberger (1838, 325) refers to a Ceylonese pearl diver having "from 4 to 8 lbs of stone in a waist belt."

With divers' suits made of a flexible material, one may vary the amount of air inside the suit, and thus the buoyancy. However, this is not possible with diver's "armor," a rigid suit.

Buoyancy compensators are much more complicated. In essence, to decrease buoyancy, water is allowed to enter a container, compressing the air within. In 1914, Stolle received a patent on a buoyancy control device using a compressed gas bottle connected to a container by a valved tube. The container also had a gas outlet valve and a tube by which water could enter the container. To sink, the container is allowed to fill with water. To rise, the diver opens the gas inlet valve and allows the gas to flow into the container. It expands and displaces the water. To sink once more, the diver opens the gas outlet valve, and the loss of gas permits the water to reenter the container.

Some later "buoyancy compensators" were derived from inflatable life vests. "In 1961, Maurice Fenzy designed the horse collar or Adjustable Buoyancy Life Jacket that was inflated by mouth. In 1968, Joe Schuch and Jack Schammel developed a buoyancy compensator vest that featured a smaller buoyancy ring behind the diver's head, and a midriff section with volume to lift the diver's head out of the water if one or both of its carbon dioxide (CO_2) cartridges were activated for an emergency ascent" (Raymond 2023).

The operation of one buoyancy compensator is described in the Braly (1971) patent. The diver would increase buoyancy by taking breaths, through the regulator mouthpiece, from the tank of compressed air and expelling the inhaled air into an inflatable bag, the "buoyancy pillow." The bag had a release valve so it didn't rupture as the diver ascends and the air inside attempts to expand further.

Protective clothing. There are three ways of dealing with the risk of hypothermia: diving only in warm water, warming oneself between dives, and wearing

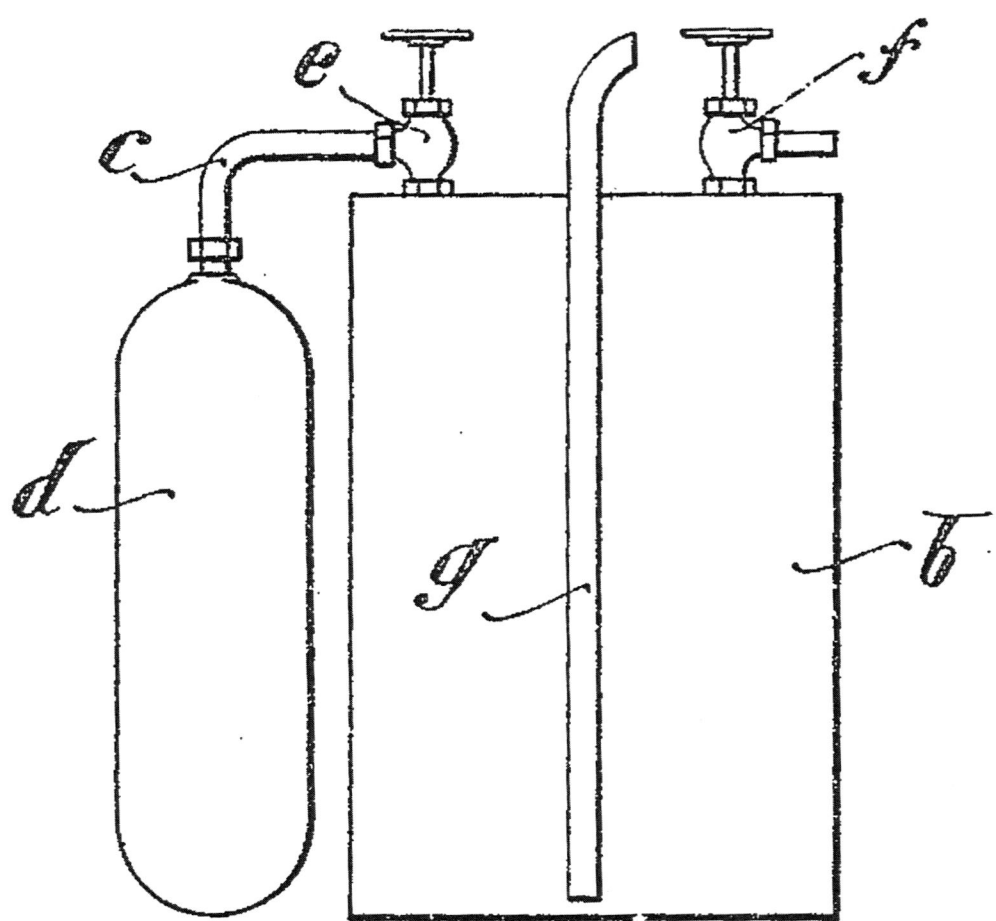

Stolle's control system (Stolle 1917, Fig. 2).

protective clothing. The *ama* didn't start diving until noon and their dive boats were equipped with braziers (Kalland 1995, 165). The *haenyeo* have a traditional cotton swimsuit and spend an average of 30 minutes per day in the water (10°C!) in the winter (Lee and Lee 2014).

Clothing has the disadvantage of absorbing water (making the diver heavier, and keeping the diver cold after emerging from the water) and the advantage of providing thermal insulation. Note that water absorption may reduce the insulating value of the material, because water has a high thermal conductivity. Clothing may be warmed before it is donned.

The water resistance of woven clothing depends on the choice of fiber (hydrophilic versus hydrophobic), the tightness of the weave, and the manner in which seams are joined.

Wool and silk are roughly comparable in thermal conductivity, whereas cotton is substantially higher (Sobuj 2019; Rood 1921). Likewise, wool and silk are roughly comparable in water absorption (~30 percent of own weight), whereas cotton is extremely absorbent (up to 2,500 percent of its own weight). Cotton can be tarred or waxed to make it more water repellent.

The water-repellent properties of wool are attributable to its lanolin content, and lanolin can be added to wool intended for diving use (if you don't mind smelling like a wet sheep). Before synthetics were available, wool was used as insulation inside a drysuit (Stinton 2024).

Oilskin is linen impregnated with linseed oil to make it waterproof. I found instances of oilskins used ad hoc by twentieth-century divers (Perry 2006, 102; Holdridge 2017), but without a way to seal out the water, they weren't too effective. Leather was also an option—it was used in an early eighteenth-century diving suit, the "Old Gentleman of Raahe" (Artifacts 2017).

Natural rubber became available in the nineteenth century, and synthetic rubbers and plastics in the twentieth. Plastics may play a role either as synthetic fibers or as coatings on natural textiles. The plastics most often mentioned in a waterproofing context are polyurethane, polyvinyl chloride, and silicone. Incorporation of polyester fibers has been reported to improve the thermal properties of wet wool (Akçagün et al. 2017).

Sponge rubber and natural rubber were used in some early wetsuits. The modern wetsuit is made of closed-cell foam neoprene (polychloroprene) rubber, lined on both sides with nylon (or later, spandex). The neoprene sheets are preferably joined together by blind-stitch sewing ("Mow" 2007). The closed cells contain nitrogen gas, which is a poor conductor of heat. As you descend, the water pressure compresses the cells, reducing the thermal insulation effect (*Wikipedia*). The thicker the material, the greater is the insulation, but the more restriction there is to movement (United States Navy Diving Manual 1979, 5–13).

A wetsuit, by definition, doesn't keep the body dry, but it does trap water between the suit and the body, and body heat warms up this water. Contrary to popular belief, the trapped water doesn't have much heat insulating value, as it remains a good conductor of heat. However, because it is stagnant, it does not transfer heat outward by convection.

Gloves. These are more to protect the diver from cuts (e.g., by coral) than to keep the hands warm.

Fins. Fins decrease the energy cost of swimming (Zamparo et al. 2002) and therefore permit the diver to swim a greater distance at depth, and to ascend and descend more quickly (subject to health constraints).

Polynesian swimmers used palm fronds attached to their feet (Warshaw 2005, 70). Franklin (1717) made hand paddles out of wood, and de Corlieu (1914) used leather (Geurts 2024). Modern fins are made of rubber, plastic, or carbon fiber (Geurts 2024). Scuba-diving fins are large and wide, whereas those used by free divers are long and narrow (*Wikipedia*/Swimfin).

Goggles and masks. Goggles are used by swimmers to protect the eyes from irritation by saltwater and also to permit clearer vision underwater. In the nineteenth century, the *ama* used goggles adapted for diving by "by attaching small rubber air-filled bulbs to each eyepiece. With descent and increasing ambient pressure, the bulbs collapsed. This increased air pressure in the goggles and prevented a pressure differential from building up between the inside of the goggles and the water outside" (scubakim 2011; Tucker 2015). Absent such an expedient, divers must use masks that cover the nose, too, rather than goggles, to avoid eye barotrauma. (Never use earplugs, either, to avoid ear barotrauma.)

Lights. In clear water there is typically enough light for vision down to a depth of about 300 feet. Even then, judging the size and location of an object can be tricky because light rays are refracted (bend) as they pass from the water to the glass of the faceplate and then to the air inside the mask (United States Navy Diving Manual 1979, 2–25).

Artificial illumination is possible. The first dive lights were battery-powered incandescent lamps. EB1911/Diving refers to use of electric lamps enclosed in a strong glass globe. Light can be produced chemically if you combine a fuel and an oxidizer; the combustion reaction will also produce heat. Glow sticks rely on a chemiluminescent reaction.

Locating a Wreck

The first step in salvaging a shipwreck is locating the wreck. Time is of the essence here; the longer it takes to commence the search, the more likely it is that the wreck will be shifted further from where it sank, be broken up and scattered, or be buried in silt. It is helpful if the ship captain, knowing that the ship was likely to be lost, maneuvered it into shallower waters with protection from wind and wave.

Survivors and observers will be interrogated, but their recollections are not always reliable. If they point out a specific location, divers will be sent down.

From the surface, one can use a "water glass" (aquascope, bathyscope, water-telescope) to see underwater. Basically, it is a tube with a glass bottom. I found an account of the 1845 salvage of a wreck in New Zealand waters. The salvors used buckets having glass bottoms to spy a small box, which was then raised using long poles fitted with screws. It turned out to contain specie (Shipwrecked Mariner 1868, 39).

A more sophisticated version was apparently used by Greek sponge fisherman. In "clear water it is possible to see to a depth of nearly thirty fathoms with this device. It consists merely of a copper or zinc cylinder from two to four feet long by about twelve inches in diameter, treated inside with lamp black, a circular glass being fitted at one end and a pair of handles at the other end. The searcher pushes the cylinder about a foot under water, put his head into the open end" (Davis 1908, 155).

The first glass-bottom boat was reportedly invented in 1878 (Lerner 2018), but I do not know whether any were used in salvage operations.

Another search technique was to drag a line between two small boats, or a grappling line from a single vessel, in the hope of snagging the wreck (Ratcliffe 2011; Marx 1990, 60).

The range of metal detectors is limited, perhaps 6 feet (Bernzweig 2023; Bernzweig 2021). Thus, they are more useful for finding metal concentrations once you have located a wreck than for locating it in the first place (Marx 1990, 67).

Once the wreck is located, it is important to determine the depth of water it is in and the type of bottom. The depth to the wreck, and to nearby underwater obstacles, may be determined by sounding with a lead line. Bottom samples may be taken "by means of snapper or scoopfish type bottom samplers" (Bowditch 1962, 842).

Lifting a Wreck

In 1550, Girolamo Cardano proposed a method of recovering a sunken ship. This involved the use of small lifting boats filled with stones. Divers would tie ropes connecting the wreck to the lifting boats. The ropes would be pulled taut and the stones moved to support boats moored alongside the lifting boats. This would render the latter buoyant, and as they rose, so would the wreck (McManamon 2016; Echeverria 2011).

The following year, his bitter rival Niccolo Tartaglia proposed a similar method (McManamon 2016; Echeverria 2011). Rather than small boats, he suggested two full-size hulls, stripped of guns, anchors, and ballast, and with gunports and hatches sealed. These were connected, with a gap separating them, by at least five thick beams running across the lifting hulls and over the wreck. A work platform was constructed on top of the crossbeams. This catamaran-like assemblage was towed over the wreck.

Tartaglia also devised an oval-shaped lifting cradle of ropes, with a tightening noose. The oval was to have a diameter 1.74 meter greater than the maximum beam of the wreck. This was weighted with anchors and carefully lowered, with divers positioning it so encircled the vessel. The loop of the cradle was pulled under the wreck (its cross section would have narrowed as one moved from the normal waterline to the keel) and the cradle tightened around it, causing the flukes of the anchors to engage the hull of the wreck. The anchor lines ran up to and over the crossbeams and were tightened using hoists or capstans.

Tartaglia filled the lifting hulls with water rather than stones. Their hulls had a hole in the bottom (presumably the hulls were used just for salvage), and Tartaglia designed a removable plug that could be easily manipulated to let water in or keep it out.

The operation preferably began at low tide. The trickiest part was freeing the wreck from the bottom. Tartaglia suggested pumping a small amount of water first from one lifting hull, then from the other, alternating until the wreck was freed. Once that occurred, water would be pumped from both hulls simultaneously. The maximum lift was around 2 meters, so if the wreck were in deeper water, it would be moved to shallower water and the process repeated as needed.

The required size of the lifting hulls depended on the size of the wreck, whether it was embedded in a silty or sandy bottom, and how heavily the wreck was laden. As a rule of thumb, the lifting hulls were to have a total capacity at least twice that of the wreck. If two lifting hulls of the required capacity were not available, Tartaglia proposed using four (or more) hulls, linked in two (or more) pairs, each pair linked stern-to-stern rather than side-to-side.

As for the pump itself, Tartaglia preferred the simple force pump (*tromba*), but a wheel pump or Archimedean screw was suggested as a possible alternative.

Lifting hulls were still used in the twentieth century (Admiralty 1954, 196).

To lift the ship, one must get a good grip on it. Lift lines may be tied around parts of the ship, but one must take care that attachment point is strong enough to resist the pull from above. In 1545, there was an unsuccessful attempt to use the "pontoon" method to lift Henry VIII's flagship, the *Mary Rose*. The lifting hawsers wrenched loose the mainmast (Ratcliffe 2011).

As noted by Tartaglia, anchors may be positioned so their flukes dig into the hull. Lift lines may be reeved through the hull, that is, passed through existing openings. If there are no suitable openings, they may be deliberately created by the divers. The lines may also be swept under the hull by tugs, at least if the bottom is soft. If that fails, divers may tunnel under the ship and then pass the lines through the tunnels (Salvors 2021, 3–43).

Before the *Vasa* was raised in 1959, divers spent two years tunneling through the clay underneath. They used a high-pressure water jet to cut into the bottom and then an induction dredge to suck the debris away. They worked first from one side and then from the other, although aligning the tunnels so they met was tricky (Hocker 2011, 177–8).

Ian Bulmer's contemporary efforts to raise the *Vasa* intact had righted it, a necessary first step, but he was unable to make further progress. In theory, thanks to the buoyancy of the wooden hull, the *Vasa* should have had an effective weight of only 200 tons (compared to 1,200 out of the water). There were frequent reports of broken lines (hemp rope) and lifting gear, so it appears that the hull was embedded in the bottom and thus needed a greater lifting force to free it. There is a suspicion that Bulmer, in righting the ship, also pushed it deeper into the mud and clay bottom (Hocker 2011, 159–61).

In 1959, the *Vasa* began to rise from the bottom when 600 tons of force were applied to the steel cables running through the tunnels. Note that this was the first of five lifts in progressively shallower water.

A lift bag is a waterproof, airtight, inflatable bag that can be inserted into a sunken ship's hull or attached to an item from a shipwreck. Lift bags may be open downward ("parachute") like a diving bell, or closed ("camel"). The uninflated lift bag is lowered, inserted inside the hull or attached to the item of interest, and then inflated. It may be filled with air from a surface compressor or from a compressed gas container carried down with them. The bag should have a spring-loaded relief valve so it doesn't rupture if overpressured.

Lift bags were used in 1863 to raise the 607-gross-ton paddle steamer *Prince Consort*, which broke in two at Aberdeen North Pier that year (Bell 1896; Aberdeen 2024).

Potential problems with lift bags include failing to securely rigging them to the object and experiencing a damaging runaway ascent. If lift bags are used to lift the entire hull, then you have to properly distribute them so you don't stress it to the point of failure. Also, there may be projections that could puncture the bags.

By the late nineteenth century, another method of raising a ship was available: patch-and-pump (Salvors 2021; Deep Sea Diving School 1960; Kinghorn 1896). Patches were applied to all holes below the waterline, making the hull watertight, and then the water is pumped out of the hull. The method required that the dives be able to find and plug all the holes, and if that could not be done from the outside, then the diver had to enter the hull and move cargo or equipment out of the way. As of 1896, the pumps were usually steam-powered (Kinghorn 1896).

When the water is pumped out, it is replaced with air at surface pressure. The result is to create a pressure differential across the hull. A steel hull can handle a greater differential than can wood. For a wooden hull, "8 or 10 feet would probably be the maximum depth for using this method of raising the boat" (Reid 1996, 41).

An alternative method is to pump compressed air into the hull, forcing the water out. However, that requires also patching the holes above the waterline, making the hull gastight. It is probably easier to just use lift bags.

Patches may be oakum, mattresses or blankets, wood (plugs, wedges, and planks), cement, concrete, plastic, fiberglass, or metal. Rather than proceed ad hoc, divers may survey the hull and locate and measure the holes, and then suitable craftspeople on the mother ship prepare customized patches for the divers to install. "Wood's ability to crush, deform and swell in water results in tight seals," and wood patches can be nailed to wooden hulls. There are also underwater sealants that could be used to fill cracks and crevices.

Salvaging Cannon

Cannon, particularly of the larger calibers, would be located on a lower deck for the sake of ship stability. This complicated the process of recovering them from a shipwreck, and in some instances, divers would tear away the deck above in order to provide vertical access to the cannon.

In the mid-seventeenth century, Andreas Peckell developed the "technique of pulling the guns out through the gunports with a giant set of tongs … These were suspended from the dive barge and would be lowered to the bottom, where a diver would place them around the muzzle of a cannon protruding from the side of the ship. The tongs were closed by cables operated from the surface ... Using windlasses at the surface, the tongs could be lifted to pull the gun out of the brackets on its carriage and then maneuvered away from the ship to pull the gun out." Von Treileben, in 1664–1665, used Peckell's tongs and barge, and divers operating out of a diving bell, to bring up about 60 of the *Vasa*'s guns (Hocker 2011, 163–4).

Preservation and Deterioration

There are two threats to the hull and its contents: seawater, which is corrosive, and biological attack (marine borers and wood-degrading microorganisms). There is a complex interaction between them: Organisms may form a protective biofilm, or they may alter the corrosion chemistry so as to accelerate it.

The rate of destruction is reduced if temperature, salinity, and oxygen levels are low. Burial in sediment helps protect the wreck. On the other hand, if the wreck is partially exposed to the surface, wave action assaults it. The corrosion rate falls with increasing depth, presumably because of a decrease in wave action, temperature, and oxygenation.

The wreck will not be uniformly deteriorated. Deterioration of the contents will depend on the local temperature, oxygen level, and water flow rate over the item in question. If the wreck lies at an angle in shallow waters, the sunny side may have greater marine growth than the shaded side (Macleod 2016).

Gold will not be corroded by seawater at all. Corrosion of copper and its alloys brass and bronze is usually minor, although differences in composition affect the corrosion rate. A bronze cascabel on the 340-year-old *Batavia* wreck had corrosion only 1.5 mm thick (Macleod 2016a).

Corrosion is much more pronounced for iron. For a shallow-water wreck, a typical long-term corrosion rate is 0.1 mm/year (Macleod 2016). In general, when iron survived for extended periods, it was because it was buried in mud or silt, protected from oxygen. In the case of the 1954 recoveries from the *Vasa* wreck (1613), cast iron was found to be much more corrosion resistant than wrought iron (Johnson and Francis 1980, 2.13).

Lead usually suffers only surface corrosion because the initial corrosion product (lead carbonate, sulfide, sulfate, chloride, etc.) forms a protective layer. However, if the water is acidic, it suffers active corrosion (Schnitzer 2012, 33). Lead ingots have survived from Roman wrecks (Johnson and Francis 1980).

A further complication is bimetallic (galvanic) corrosion. The seawater acts like the electrolyte in a battery. The less noble metal (lower electrode potential) acts as the anode and corrodes more quickly, and the other becomes the cathode and corrodes more slowly. Thus, copper (or lead) accelerates the corrosion of iron but is protected by it. Copper sheet on the *Batavia* (1629) was less corroded than would have been expected from its shallow depth, most likely because of its proximity to iron cannon and cannon balls (Macleod 2016a).

The threat of deterioration does not necessarily stop when the objects are brought to the surface. If "you recover a heavily corroded cast iron cannon or a cannonball and knock the marine growth from it, oxygen gets direct access to the metal and the liquid corrosion products become solid and you get this essentially massive explosive force just forcing all of the components of the metal apart. The corrosion rate can start taking off so quickly that we have had reports of 400-year-old cannonballs beginning to steam as a result of rapid corrosion, and then shortly after, boomph, they fall apart" (Williams 2013).

Salvage Law and Contracts

According to the Rhodian sea law (circa 800 AD), chapter 47, a salvager received one-third the value of treasure raised from 8 fathoms, and half the value if it came from 15 fathoms. But if the salvaged goods were found on land, or no deeper than a cubit, the salvor's share was one-tenth (Frost; stexboat 2013). The law also required that if part of the cargo was lost or deliberately jettisoned, the value of the remainder was shared ("averaged") among all of the parties according to a formula (chapters 9 and 30)

After the Roman era, the next major maritime law in the Mediterranean was the *Consulat de Mar*. It went through various versions in 1320–1502. Twiss published an English translation under the name "The Good Customs of the Sea," as part of an Appendix to his edition of *The Black Book of the Admiralty*. As best I can tell, it does not really address salvage.

In northern Europe, the major medieval European sea law was the Rolls of Oleron. It was adopted in England and France in the fourteenth century. Article IV (Anonymous c1266) stated that if a ship master "has promised the people who helped him to save the ship the third, or the half part of the goods saved for the danger they ran, the judicatures of the country should consider the pains and trouble they have been at, and reward them accordingly, without any regard to the promises made them by the parties concerned in the time of their distress." Article XXV dealt with the situation of a ship being piloted deliberately to ruin by a local pilot in a land where the local lord and the salvors each received a share of the wreck. These, it said, should be "accursed and excommunicated, and punished as robbers and thieves."

If instead a ship is wrecked on the rocks without such treachery, and any of the merchants or mariners come safe to land, the local lord should aid them "in saving their shipwrecked goods, and that without the least embezzlement, or taking any part thereof from the right owners; but, however, there may be a remuneration or consideration for salvage to such as take pains therein, according to right reason, a good conscience, and as justice shall appoint; notwithstanding what promises may in that case have been made to the salvors by such distressed merchants and mariners" (Article XXIX). Of course, those goods should not be taken from them against their will.

On the other hand, if all on board drowned, the local lord was to send people to save the goods, and "give the relations of the deceased persons who were drowned, notice of it, and to satisfy for the salvage thereof, not out of his own purse, but of the

A. Conventional view of the work of the salvage diver (Hepworth 1904, 160). B. The caption of this tongue-and-cheek cartoon (Bulloch 1899, 205) read "Hard Lines," meaning "unfortunate," implying that the diver was hoping for, say, a chest of gold coins, rather than an unopened bottle of champagne. On the other hand, finding gold was not necessarily a blessing. Siebe (1874, 288) reported that "a diver at work upon a wreck at Valparaiso had fixed the slings around a box containing ingots of gold. In the course of hauling up, the slings broke, and the box of gold descended upon the poor diver, and killed him on the spot."

goods saved, according to the hazards run, and the pains taken therein; and what remains must be kept in safe custody for one year or more; and if in that time they to whom the said goods appertain, do not appear and claim the same, and the said year be fully expired, he may publicly sell and dispose thereof to such as will give most, and with the monies proceeding of the sale thereof, he ought to give among the poor, and for portions to poor maids, and other charitable uses, according to reason and good conscience" (Article XXX).

While thus the goods salvaged from the wreck remained the property of the original owner, a different rule applied to what English law came to call "jetsam": goods deliberately cast overboard for the safety of the ship. These "become his that can first possess himself thereof, and carry them away" (Article XXXII). However this comes with a caveat: "this holds true only in such cases, as when the master, merchant, and mariners have so ejected or cast out the said goods, as that they give over all hope or desire of ever recovering them again, and so leave them as things utterly lost and given over by them, without ever making any enquiry or pursuit after them."

But how would you know? Article XXXIII explains that if the goods "are in chests well locked and made fast …, that they may not be damnified by salt water; in such cases it is to be presumed, that they who did cast such goods overboard, do still retain an intention, hope, and desire of recovering the same."

Another example of the exception would be in the case of what English law came to call "lagan": goods, floating or sunken, attached to a buoy so they can be recovered. The buoy is proof of intent to recover.

In the case of the *Vasa*, Mazalet's 1634 contract provided that "if he succeeded, he would be awarded half of the ship with the exception of the guns and related equipment for which he should be given half the weight of the guns in raw copper." The 1635 contract with Zancke said that "The Admiralty was to supply cables, ships and people at his disposal and if he succeeded he would be rewarded with a sum of 6000 riksdaler" (Bruzelius 2001).

In the English common law, the compensation for salvage varied depending on the peril to the ship or property, the risk to the salvors, the skill required for the salvage, the time invested, and the value of the ship or property.

The modern law of salvage is the International Convention on Salvage (1989). Unlike English common law, it takes into account "the skill and efforts of the salvors in preventing or minimizing damage to the environment" (Article 13[1] [b]; Article 14).

11

Diving Bells, Diving Suits, and Submersibles

There are limits to what can be achieved with breath-hold diving. If you can hold your breath for 2 minutes, and it takes 30 seconds to descend and 30 seconds to ascend, that means you have 1 minute of working time per dive. It is amazing what breath-hold divers have accomplished, but naturally people sought to extend the working time by supplementing the air held in the divers' lungs.

Diving Bell

A diving bell is a device for holding breathable air underwater. The diving bell may be used to lower divers into the water, or just to provide them with an air supply while they are submerged.

Some Early Diving Bells

In 1535, Guglielmo de Lorena and Francesco de Marchi used a diving bell designed by de Lorena to explore Roman barges sunk at the bottom of Lake Nemi. This bell was so small that it actually rested on the diver's shoulders. The bell left the diver's hands free and de Lorena and de Marchi, presumably taking turns, were able to take detailed measurements of the ship and also bring up two mule loads of material. This was apparently not the first use of the de Lorena bell; he claimed to have used it to find a sunken galley and salvage its artillery (McManamon 2016).

Even though it was small (internal volume of about 400 liters), the bell was still too heavy to be carried by a single man on land. Once it was immersed, the buoyancy of the trapped air reduced its apparent weight to about 40 pounds (McManamon 2016).

In 1605, the engineer Jan Adrianszoon Leeghwater demonstrated to Prince Mauritius, in a canal leading to Delft, that he was able to stay underwater for 45 minutes, and he was awarded a Dutch patent with a 10-year term. The patent does not describe his "underwater art," only its utility, but plainly he had a diving bell of some sort (Dekker 2024).

In 1626, Francesco Nunez Melian, Spain's official salvager, used a brass diving bell and slave divers to salvage treasure and cannon from the *Santa Margarita*, which sank in 1622 on a coral reef in the Marquesas Keys, an island group within the Florida Keys. This bronze diving bell, hammered and riveted, was an elaborate one, 680 pounds with seats and windows. One-third of the salvage went to Melian. The wreck lay at a depth of 30 feet. The 13 divers recovered 312 bars of silver, the bronze cannon, and some copper (Viele 1996, 8–9; Fine 2006, 16). Curiously, Melian is a character in the *Uncle Scrooge* comics, appearing in the issue "Treasure Under Glass" (1991).

A more primitive diving bell was the "Bermuda Tub," attributed to Richard Norwood (159?–1675). "This was essentially an oversize wine cask, inverted and weighted around its edges" (Peterson 2020, 163). It was supposedly a spur-of-the-moment invention made while he was in the eastern Mediterranean in 1610–1612, prompted by a gun being dropped into the water during transfer to another ship. Norwood was in Bermuda in 1614–1617 and 1637–1675 (Hamilton 1986; Hamilton 1986a). He brought his diving bell with him to Bermuda for pearl diving and wreck salvage. Unfortunately, his patent (Norwood 1632) provides no useful information about his invention. A similar diving bell was used by Edward Bendall to salvage the *Mary Rose* (not to be confused with the royal warship) in Boston Harbor in 1642 (Peterson 2020).

EB1911/Diving credits Edmund Halley as the inventor, in the early eighteenth century, of "the first really practical diving bell." Its features are described, *en passant*, in the following.

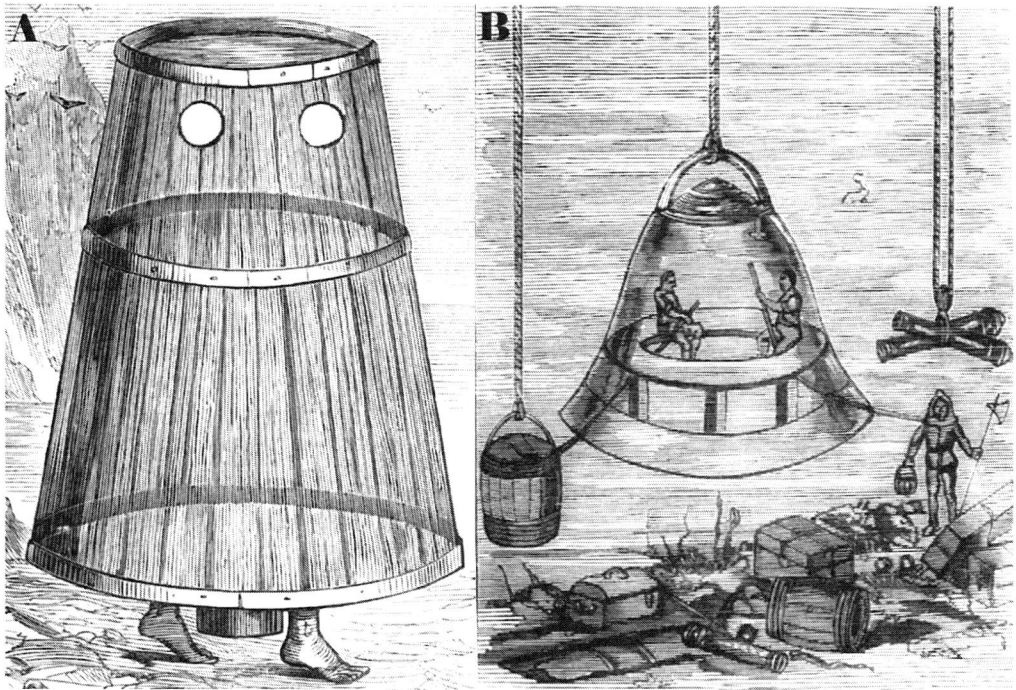

A. Artist's conception of an early diving bell (Siebe 1874, 91). It clearly has two observation windows, but the purpose of the central pedestal seen between the occupant's feet is unclear. B. Artist's conception of Halley's diving bell (Siebe 1874, 83). Note the air tube to the diving bell supplied by the lowered barrel; this system is discussed in Chapter 12. Also note the diver with air supply from the diving bell.

Diving Bell Design and Use Considerations

Diving bells could vary in size from those that barely fit one person to those that hold half a dozen (Halley's 1714 bell held five). Some of the larger bells were equipped with seats, or a suspended stage to stand upon.

The net buoyancy of the diving bell is the weight of the water it displaces, less the weight of the bell and anything it carries (including the air). In order for the diving bell to descend the net buoyancy must be negative, and the more negative it is, the more quickly it will descend.

For this reason, diving bells were often made of metal (usually bronze, but cast and wrought iron bells are known—they were probably painted to prevent rusting). When they were made of wood, they were sheathed in lead to increase their weight (as in the case of Halley's bell [Halley 1714, 496]).

"The bell may be suspended over the side or end of a vessel; or through an opening [a 'moon pool'] in the center of a barge; or from frame-work resting on two barges placed parallel to each other, but at such a distance apart as to allow the bell to descend between them" (Knight 1851, 746).

The machinery for raising the bell had to be sufficiently robust to cope with the weight of the bell in the air, before its effective weight was reduced by submergence. When the bell was suspended over the side or end of the vessel, its weight would have caused the vessel to roll or pitch in that direction, and the smaller the vessel, the greater would have been this the effect. Other shipboard weights could have been shifted to counteract this tendency.

If the bell ever had positive net buoyancy it would have risen, and the cable would have gone slack. There is nothing the surface crew could have done to prevent this; the cable just keeps it from sinking deeper than the length of cable paid out.

When a diver was supported by the bell, the diver's weight would have helped reduce the net buoyancy. But if the diver left the bell, or stood on the sea floor inside the bell, that would no longer have been the case, and the net buoyancy would have increased. Thus, the bell had to be weighted enough so that the bell didn't lurch upward when all the divers left the bell. Also, the cable had to be able to withstand the dynamic load imposed if the divers all let go at once.

The hoist line must also have been chosen with the weight of the bell, and the line itself, in mind. However, in water, tarred hemp rope may be positively buoyant (albeit only slightly once it's waterlogged).

If operations were from a normal sailing ship, tackles suspended from yards, sprits, and stays would have been used to lift the bell, swing it outboard, and lower it. That is indeed how ship's boats were handled before they were hung on custom davits.

The available hoisting equipment would have limited the weight of the bell, but seventeenth- and eighteenth-century standard shipboard equipment seems to have been able to cope with rather heavy bells. Melian's bell (1626) was reportedly 680 pounds. Halley's bell had an internal volume of 60 cubic feet. Seawater weighs about 64 pounds per cubic feet, so for neutral buoyancy, when first immersed, the bell would need to weigh 3,840 pounds (while that could include the divers, they wouldn't have wanted the bell to pop up to the surface when a diver left the bell).

Some early bells may have literally had a bell shape, as one could then go to a bellmaker and say, "I want a bell that is so-big and so-heavy, without a clapper." However, a more rational design is a truncated cone, as in the case of Halley's bell. His had a top diameter of three feet and a bottom diameter of 5 feet, and was 60 cubic feet in volume (Halley 1714, 496).

A diving bell could take the form of a vertical cylinder, but that would have less pitch-and-roll stability than a truncated cone. Since it is critical that the diving bell be kept vertical when it is lowered and raised—tilting would allow air to escape—stability is important.

As previously noted, there were small diving bells ("Bermuda tubs") that were made from some sort of cask. Casks are roughly cylindrical but are wider in the middle (bilge) than at the ends. Box-shaped diving bells were used (e.g., Smeaton's "diving chest") and would have been reasonably stable against pitch and roll if shorter than they were wide. However, the corners and edges would have been weak points.

As one submerges, the water pressure increases (by 1 atmosphere for every 10 meters, or 33 feet), and the volume of air inside the bell decreases (unless there are means for supplying pressurized air to the bell). The level of the water inside the bell would rise accordingly. Thus, one limit on the depth at which a diving bell may be used is the depth at which the level of the water inside would rise to that of the diver's nose.

Since the bell is now displacing less water, the net buoyancy of the bell decreases, placing more of a load on the hoisting equipment (although not as much as when the bell was in the air).

This rise in water level can be avoided if, as the bell descends, pressurized air is delivered to it. Of course, air delivery doesn't just keep your feet dry, it also replenishes the oxygen supply.

Absent some means of oxygen delivery, the ability of a diving bell to sustain the diver is dependent on its internal air volume. At rest, we inhale 0.5 liters of air, 12 times a minute. When exercising, the inhalation increases, to perhaps 3 liters, and also the rate increases to 30 times a minute (BBC 2024).

The work period would be further limited if exhaled air, which contains elevated carbon dioxide levels, weren't removed. One way to expel the exhaled air would be to place an exhaust valve on the roof of the bell, as Halley did, which the diver could periodically open. Eliav (2015) points out that since exhaled air is warmed by passage through the lungs, it would rise to the top of the bell, so the valved air would mostly be exhaled air.

It would be more convenient for the diver if the inlet of a flexible exhalation tube ("reverse snorkel") were held in the diver's mouth. The tube would have to either run down and under the lip of the diving bell, or up through a sealed hole on top. In any event, it doesn't have to run up to the surface, as long as its outlet is above the bell. If so, the water pressure at the outlet is less than the air pressure inside the bell, and water will not enter.

In 1732, Martin Triewald complained that the diver is still mostly breathing the hot air in the upper part of the bell. Triewald suggested furnishing the occupant with a long tube that could draw air from the lower part of the bell (Triewald 1736).

Natural Lighting

Some diving bells, such as Halley's (Halley 1714, 496–7, 498) had thick glass windows built into the roof or upper sides so that the divers inside would have the benefit of natural light. Better yet, these should be planoconvex lenses, with the convex side facing out. The effect would then be to gather light from all directions. Just how much benefit was obtained would depend on the amount of sunlight striking the water surface, the opacity of the water, and the depth.

The windows also have the advantage of letting the occupants see outside the bell. Bear in mind that "locating glass which could combine transparency with sufficient strength was not easy before technical improvements in glass manufacture in the late seventeenth century. It also required very skilled workmen to be able to fit the windows into the bronze or wooden shell so that the bell remained watertight" under pressure (Earle 2008, 35).

With regard to internal illumination (e.g., to study a map), if the diving bell had means for receiving fresh air without surfacing, it might have been possible to light a candle inside (as Halley did; 1714, 499). Also, on Bushnell's *Turtle* submarine, bioluminescent foxfire was attached to the needles of the compass and barometer so they could be seen in the dark (until it got too cold for the fungal luciferase to function).

Communication

Communicating with the surface is necessary, but tricky. One method would have been to tug on a signal line, which would need to be secured so that it remained within reach of the divers in the bell.

A more elaborate form of this idea was a message barrel. A message could be placed inside and the surface crew signaled to bring it to the surface. Halley used weighted barrels to deliver air to the diving bell, and he could send orders—written with an iron pen on a small plate of lead—back with the spent supply barrel (Halley 1714, 499).

A practice I have documented for the nineteenth century, but that may have been used earlier, was to signal the surface crew by striking the bell with a hammer. Supposedly, this could be heard at any depth at which the diving bell was employed. One set of signals was "*one* stroke, more air; *two*, hold on; *three*, raise; *four*, lower; *five*, north; *six*, south, *seven*, east; *eight*, west" (Anglo 1910, 257). I imagine this worked better for the metal bells than for the weighted wooden ones.

EB1911/Diving refers to a diver's helmet equipped with a telephone receiver, from which an insulated wire runs up to the surface. It also says that the forerunner of the telephone was a rubber speaking tube.

The divers may of course attempt to signal the crew to halt, raise, or lower the bell, but in in ordinary practice, the surface crew decides how much of the cable to pay out and thus the depth to which the diving bell descends. Spalding came up with an expedient that gave the occupants some control. As described in the *Edinburgh Encyclopedia* (1830, 4:14), a "balance weight" is suspended from a pulley inside the bell. The relative weight of the bell and the balance weight is such that the bell has negative buoyancy only when the balance weight is suspended below it. If the balance weight comes in contact with the sea floor, it is supported instead by the latter, and the net buoyancy becomes positive. The bell will then rise until either the balance weight is released from the sea floor or it comes to the end of the tether attached to the balance weight, with the latter acting as an anchor. If the occupants haul on the balance weight line, the tether is shortened and the bell descends (cf. Royal Society of Arts 1783).

Another question is what to do if the main cable breaks. If the bell is then relatively close to the surface, the divers could just abandon it and ascend on their own. However, a perhaps better approach would have been to give the divers a means to accord positive buoyancy to the bell until it reaches the surface. This could have been accomplished by making some of the ballast of the bell removable, in increments, as in the case of the modern diving belt.

The Lethbridge Diving Engine and the Armored Diving Suit

EB1911/Diving acknowledges that John Lethbridge, a salvor for the British East India Company, "made a considerable fortune" by his 1715 invention of a diving apparatus. The machine was a wooden barrel with a viewport and armholes and Lethbridge lay prone inside it, according to Lethbridge's 1749 description (Amery 1880). Call it a cross between a diving bell and a diving helmet.

Lethbridge found that "beyond fifty feet the water pressure caused leaks around the armholes, window and entrance hatch" (Amery 1880). Also, since there was no air hose, and only about 60 cubic feet of air was trapped inside, his maximum time at depth was about 30 minutes. Nonetheless, he and his partner Rowe recovered "27 chest of silver, 868 slabs of lead, 64 iron guns and eleven anchors" from the wreck (1719) of the *Vansittart* off the Cape Verde Islands. And that was the first of a series of successful salvage operations.

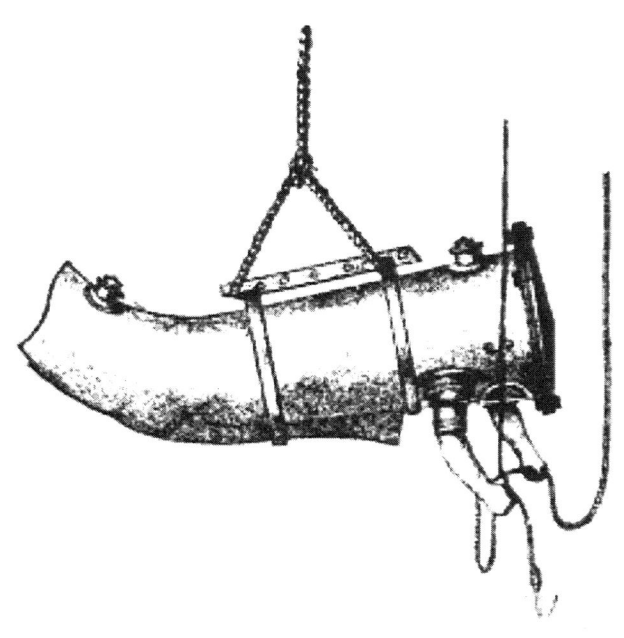

Lethbridge diving engine (Davis 1908, 161).

Rowe even recovered 20,000 pounds of silver coins from the 1728 wreck of the *Adelaar*, in the cold waters of the Outer Hebrides.*

In 1727, Lethbridge was brought to the Dutch Cape colony to recover treasure from the 1722 wrecks of the *Rotterdam* and *Zoetigheid* in Table Bay. He "brought up 200 silver bars and seven cannon from the *Rotterdam* and 2000 silver ducatoons from the *Zoetigheid*." It was not the first salvage attempt on the Cape Coast—a diving team led by Olaf Bergh and including a Malay pearl diver recovered silver from the *Johanna* (1682) wreck—but the first I know to have used a diving apparatus (Loos 2018; Loos 2018a).

Lethbridge's diving engine may be considered the precursor to the armored or atmospheric diving dress. This was intended to maintain an internal pressure of just 1 atmosphere (hence, no decompression sickness risk), but it therefore had to be strong enough to withstand the pressure differential at depth. It was not easy to make it light enough to be mobile, and flexible enough to permit work to be done, while preventing leaks.

Several atmospheric diving suits were designed in the nineteenth century, notably those of Taylor (1838), Philips (1856), and Carmagnole (1882). These showed ever more sophisticated joint designs (Cui et al. 2022, 76). Efforts continued into the twentieth century, with the joints continuing to be an inventive focus (77). A twenty-first-century design, Phil Nuytten's EXOSUIT, could work at a depth of 305 meters for 48 hours (79). While the diver may walk in the atmospheric diving suit (ADS), some had propellers (80), blurring the distinction between them and a submersible.

Diving Helmets and Flexible Diving Suits

The diving helmet in its simplest form is just an open diving bell large enough for the diver's head, equipped with a transparent viewport and connections for an air supply. Since the helmet is open at the bottom, the "hard hat" diver must keep his or her head upright to avoid flooding the helmet. An experiment conducted by Middelaldercentret in 1998 (Lazenby 1998) showed that a viable diving helmet and air compressor could have been built even with fifteenth-century technology, although the depth of operation was limited.

In 1721, Halley pointed out that a diver wearing a helmet ("cap of maintenance") connected by a flexible pipe to the air of a "great" bell would be supplied with air from the latter, as long as the level of water in the helmet was higher than the level of water in the great bell (so the pressure in the helmet was lower). He addressed some practical issues worth noting. First, the supply tube contained a spiral brass wire to keep it from being crushed by external water pressure. Second, the tubes were of leather, waterproofed with a mixture of oil and beeswax. Third, there was a valve at the end near the diver's cap, so he could turn off the supply if he needed

---

* Lethbridge's diving engine was attributed to Rowe (1753) in the *Edinburgh Encyclopedia* (1830, 4:20).

to stoop down. Fourth, the caps were weighted, and the diver also wore a weighted girdle, so he could stand against the current. Since the water was cold, he wore thick woolen "waistcoat and drawers" (*Edinburgh Encyclopedia*/Diving Bell 1830, 8:12–13).

In 1820, John Deane, seeing a stable on fire, grabbed a "helmet from a medieval suit of armor" in a nearby home and had the firefighters on the scene use their hose to pump air into his helmet. With this improvised breathing apparatus to protect him from the smoke, he rescued the horses inside the stable (Veit 2012, 161). He and his brother Charles obtained a patent on a "smoke helmet" in 1823, and in 1828 they marketed "Deane's Patent Diving Dress," in which the helmet, resting "on the diver's shoulders," was strapped "to a waist belt" (Marx 1990, 53). The diving dress covered the body, providing some insulation (reducing heat loss to the water) and protection from some sea life (e.g., jellyfish). The problem was that the helmet had to be kept upright or the air would escape and the helmet would flood with water.

In 1837, Augustus Siebe created a "closed" diving suit; this had a watertight connection between the helmet and the suit. In addition, the diver had control over the helmet's exhaust vent (Marx 1990, 54; Dekker 2024a).

Of course, air now filled the suit as well as the helmet. The advantage was that by adjusting the rate with which air escaped, the diver could adjust his buoyancy. However, there were disadvantages. While the helmet was typically of metal, the suit was not. If the surface crew did not adequately increase the pump pressure when the diver descended, the amount of air in the suit would be reduced and "the external water pressure would squeeze the suit painfully around him," possibly damaging his lungs (Ashcroft 2002, 54). Also, if too much air collected in the legs of the suit, the diver might be turned upside down. In this position, the air escape valve was ineffectual and the diver might be forced to the surface (55).

In the decades that followed, various improvements were made. EB1911/Diving describes the 1911 version in some detail. There are a helmet, a breastplate, and a neck-to-feet suit. The first two are made of tinned copper, and the suit of "two layers of tanned twill with pure rubber in-between." The suit has a vulcanized rubber collar and a set of screws and nuts secure the collar to the breastplate. The helmet is screwed down on the top of the breastplate. The helmet has a non-return air inlet valve to which the air pipe is connected, and which the diver can adjust as needed. The air pipe is made of alternate layers of strong canvas and vulcanized rubber, with metal wire embedded so it is non-collapsible under water pressure.

On the surface, the pump may be human-, steam-, or electric-powered, and it may be possible to switch quickly from one drive to another. The air may be pumped into an intermediate receiver, so there is a reserve supply of air. The pump cylinder may be water-jacketed to cool the supplied air.

In 1898, a light diving apparatus (diving helmet, dress, and air pump), suitable for examining the hull of a ship down to a depth of 40 feet, cost about 60 pounds. A more elaborate one, suitable for an engineer or contractor, cost 100–120 pounds. A two-diver setup cost 150, but the air pumps had to specially designed to be able to simultaneously deliver two streams of air at different pressures. In contrast, a diving bell "complete with air-pumps and tubing, and with three tons of iron weights, costs about 200" pounds (Matheson 1898, 842–3), to which one must add the cost of the hoisting equipment.

Tethered Manned Submersible

This is essentially a closed diving bell with walls strong enough so that internally it can maintain normal atmospheric pressure, like an atmospheric diving suit. Atmospheric quality must still be maintained. Stale air is either exhausted or scrubbed. Fresh air is supplied by an umbilical from the surface, or from a compressed air tank. It does not have a propulsion system so it is held at depth by weights, and moved around and raised by a heavy cable on a crane. It has releasable ballast weights so it can return to the surface even if the cable breaks.

One use for these was as an observation chamber so a supervisory diver could observe underwater operations over a long period of time and give directions to the surface (by telephone link) (divingheritage 2024).

One type of tethered submersible, the submersible diving chamber, was equipped with an airlock so divers could leave and return (divingheritage 2024a).

There are also submersible robots, which may be preprogrammed, guided by artificial intelligence, or remotely operated.

"Salvage diver Frank W. Crilley," wearing a standard diving suit. Crilley received the Medal of Honor in 1929 for lifesaving actions in 1915, and the Navy Cross for his work in 1928 in the recovery of the submarine USS *S-4*. (Naval History and Heritage Command, catalog number UA 57.03.05./).

Early Autonomous Submersibles (Submarines)

The emphasis here is on the development of the propulsion system and propulsion and depth controls.

Cornelius Drebbel (1572–1633) is supposed to have built a submarine in the 1620s, but there are no eyewitness reports, and some doubt that it existed (Goldstone

2017, 12). According to Drebbel's son-in-law, it was propelled by oars, and "that part of the ship where the rowers sit has no bottom." Thus it was, as Tierie puts it, a self-propelled "diving bell." According to De Monconys (1663), Drebbel "could not dive deeper than twelve or fifteen feet." Drebbel's daughter stated in 1690 that her father had a mercury barometer to determine the submarine's depth (Tierie 1932).

A general problem for early submarines was assuring a watertight connection between internal and external controls (rudder, diving plane), and between the drive and the propulsor (propeller, etc.). In Drebbel's submarine, leather gaskets were used (Tierie 1932, 39). Later, rubber became an option. Of course, it is best to have a design that minimizes the number of hull piercings.

Fulton's *Nautilus* (1800), built for the French navy, had a hand-cranked propeller for underwater propulsion (up to 2 knots) and a sail on a hinged mast for use on the surface (up to 4 knots). For depth control it had a ballast tank, hydroplanes, and a vertical propeller (Thomson 1942).

The Confederate *Hunley* made a successful spar torpedo attack on the *Housatonic* in 1864. It was a 25-foot-long converted boiler with a crew of eight, six to eight of whom cranked its propeller, giving it a maximum speed of 4 mph. It had ballast tanks and hand pumps for emptying them, a crude depth gauge, and diving planes (Davis 1908, 167ff) . There were two hatches and also an "air box" (snorkel). (While Chaffin [2008, 111] says this was bellows driven, Lance [2020, 55] doubts its adequacy.)

When a submarine is human powered, one can assume that the crew are breathing more quickly and more deeply than if they are at rest, and consequently the oxygen is consumed more quickly. Whether they would consume oxygen more quickly than the furnace for a boiler supplying steam to a steam engine that turns the propeller as quickly, I do not know. But I do believe that crewmen cranking a propeller will tire faster than those who are merely feeding coal to a boiler.

Monturiol's *Ictineo I* (1859) was a human-powered submersible and intended for coral diving. However, Monturiol's *Ictineo II* is more interesting. He used the reaction of zinc, manganese oxide, and potassium chlorate to heat the boiler of a steam engine. Note that this reaction also produced oxygen, which meant that it did not need a snorkel or oxygen tank (De Decker 2008).

The *Plongeur* (1863) had 23 air tanks, holding compressed air at 180 psi, which could drive an 80-hp reciprocating engine, giving it a maximum speed of 4 knots. Compressed air does have the convenience that it can be used not only for propulsion but also as a breathing air supply, to empty ballast tanks, and to expel torpedoes. The compressed air was generated by its tender, the *Cachalot* (Delgado and Cussler 2011, 59).

The Rev. George Garrett's *Resurgam* (1879) was "fitted with an ingenious adaptation of the Lamm fireless locomotive boiler." Strictly speaking, this is a steam accumulator, not a boiler; it does not produce its own steam. It could travel 10 miles on the impetus of the stored steam (Gray 2006).

Nordenfelt's submarine (1883) had a 100-hp steam engine. This was run only when the submarine was on the surface, but to store steam for underwater use, two large tanks were connected with the boiler. Together, they stored 8 tons of water at 150 psi. The submarine was equipped with vertical propellers, driven by a 6-hp engine, for depth change, which he preferred to the use of diving planes to cause the submarine to angle downward. The submarine had bow "rudders" operated by a plumb weight to keep it horizontal (EB1911/Ship 918). Unfortunately, the heat inside the submarine was intense and it took too long (12 hours) to fill the tanks (Goldstone 2017, 85–6).

Several other late nineteenth- and early twentieth-century submarine designs had interesting features. One Goubet submarine had an automatic trim regulator. The trim regulator used a pendulum to direct the action of a reversible pump, which communicated with fore and aft trim tanks. It also had a universal joint on the drive shaft so that the screw could be set at an angle (so there was no need for separate vertical and horizontal propellers). A later boat added an automatic depth regulator using a pressure gauge to control the movement of a piston and thus the amount of water in a cylinder. The *Gymnote* (1888) had a bronze hull, and this had the advantage that bronze does not affect the compass. The double-hulled *Narval* (1898) had an electric motor (running on storage batteries) for underwater movement and a triple expansion steam engine for surface propulsion (and recharging the batteries). It had fore-and-aft horizontal rudders to maintain a horizontal attitude. Lake's *Argonaut* was intended for sending out divers and therefore was equipped with an airlock (EB1911/Ship 919–920).

By 1910, the "general tendency was undoubtedly to use internal-combustion engines, of which those burning heavy oil are much less expensive in working than those using gasoline" (EB1911/Ship 921). The French continued to build steam-powered submarines, and there were some submarines that used batteries and motors for both surface and submerged propulsion.

For twentieth-century developments, I direct the reader to histories of submarine warfare.

Wheeled Submarines

In 1689, Halley proposed "putting a pair of weighty wheels under a diving bell of considerable capacity," so a diver could explore the bottom and take his air with him, "as a house over his head" (Royal Society Picture Library 2024).

In 1894, Halley's dreams were in part realized when Simon Lake completed his 14-foot-long wooden submarine, the *Argonaut Junior*. It had wooden wheels, powered by a hand crank. Before diving, Lake used a plumber's hand pump to fill a reservoir (a soda water fountain tank) with compressed air (100 psi). Pressurized air would be released from the reservoir during the dive in order to match ambient water pressure (Goldstone 2017, 125–6; Poluhowich 1999, 12).

A "replica" of the *Argonaut Junior* was constructed by an Oklahoma couple in 2010. It is an epoxy-laminated plywood/steel composite with dual propulsion (a trolling motor and a hand crank) and an emergency drop weight system. Scuba tanks were used as the compressed air source (Mone 2010; Seeker 2010).

The later *Argonaut* (1897) was larger, steel-hulled, and powered by a 30-hp gasoline engine. It had an airlock for use by a diver, as it was intended for salvage operations. Air was supplied by a ventilation pipe, supported by a float and driven by a blower within the submarine. This limited its normal operating depth to "less than fifty feet." It had tricycle-type wheels (two large wheels forward, and a small wheel aft) (Goldstone 2017, 135ff; Poluhowich 1999, 52, 67).

The *NR-1* (1969–2008) was a nuclear-powered midget submarine with floodlights, viewing ports, a manipulator arm, bow and stern thrusters, and two retractable rubber-tired extendible bottoming wheels (Sherman 1999).

Unarmed Self-Propelled Submarines

While submarines are primarily warships, there have been some uses of submarines for other purposes.

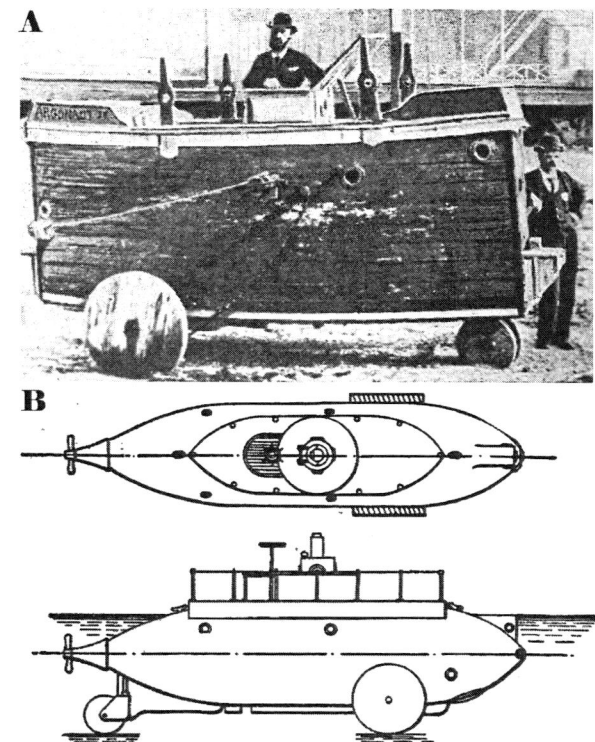

A. Photograph of *Argonaut, Jr.*, built in 1894 (Bishop 1916, 84; cf. Herrman 1961, 4). B. Simon Lake's *Argonaut* as originally built (Bishop 1916, 87). Note that the rudder serves as a steering wheel when on the bottom.

In 1916, Germany operated a "merchant" submarine, the *Deutschland*. As a submarine, she could more easily evade British warships, and as a nominally civilian vessel, she could visit American ports despite the neutrality laws. She was capable of carrying about 1,000 tons of cargo (Messimer 1988, 22). She carried, inter alia, dyestuffs to America, and nickel, rubber, oil, and silver (and possibly gold) back to Germany (62–63, 71–73, 135).

Small submarines have been used for coral fishing in the Pacific, notably in the Formosa Strait and in the Hawaiian Islands. These are towed to the fishery by a mother ship and then forage under battery power. They have a manipulator arm and an external net for storing the corals (Gabriel et al. 2005, 28–29).

12

Fresh-Air Supply for Underwater Exploration

There are essentially three ways of supplying fresh air to divers, diving bells or submersibles: supply it continually from the surface, have the underwater crew take it down with them, or have them generate it themselves underwater.

Breathing 101

To quantify the air supply requirement, we need some data. The data I present in this section are largely taken from the *United States Navy Diving Manual* (USNDM 1979).

The volume of air inhaled with each breath and the rate of breathing depend on exertion. The respiratory minute volume is the total volume of air inhaled over the course of a minute. One may estimate it as follows: resting quietly (8 liters/minute), slow (0.5 knot) swimming or other light work (20 liters/minute), average (0.85 knot) swim (30 liters/minute), swim at 1 knot or moderate work (40 liters/minute), swim at 1.2 knots or heavy work (60 liters/minute), and severe work (80 liters/minute) (USNDM 1979, Fig. 5–28). (Note that there are 28.3 liters to the cubic foot.) Inhaling 60 liters/minute air results in consuming 2.5 liters/minute oxygen.

To sustain life, we must keep the partial pressure of oxygen high enough to avoid hypoxia, and that of carbon dioxide low enough to avoid hypercapnia. The partial pressures are the total air pressure times the percentage of the air the gas contributes.

If the air is not changed, then we are progressively increasing the level of carbon dioxide and decreasing that of oxygen. The U.S. Navy found that in hard hat divers, just 3 minutes of breathing air with a partial pressure of 0.05 atm carbon dioxide induced "distracting discomfort" and "air hunger." At 0.07 atm, the "distracting discomfort" occurred immediately (Lance 2020, 52).

With these data, we can determine how long a diving bell or submarine's internal air can sustain its occupants without fresh air being supplied. For example, Lance (2020, 40) estimates that the internal gas volume of the Confederate submarine *Hunley* was 7.62 cubic meters, of which 1.6 was oxygen when the hatches were first closed. Make assumptions about the crew and their activity level and you can calculate when, say, the oxygen level will reach 0.1 atm or the carbon dioxide level 0.07 atm.

If air is supplied to a diver, whether from the surface or a tank, the pressure of the supplied air should closely match ambient water pressure, which increases by 1 atmosphere for every 33 feet of depth (seawater). If it is lower, more effort is required to inhale, and if it is higher, it is more difficult to exhale (USNDM 1979, 3–13). If the pressure is too low, you can't breathe at all, which is why air supplied to a diver must be pressurized.

The required air supply rate, in air at surface pressure, is the pressure in atmospheres times the respiratory minute volume (subject to slight correction for the air temperature at depth) (5–16).

However, that does not take into account the possibility of buildup of carbon dioxide. The Navy recommends a flow at depth of at least about 57 liters (2 cubic feet) of air per minute, per liter of carbon dioxide production, to sufficiently flush out the carbon dioxide, that is, maintain it below 0.02 atm partial pressure. If one is uncertain about the production rate, the Navy says to provide 6 cubic feet (170 liters) per minute (USNDM 1979, 6–51).

In determining the required pressure at the pump outlet, once must consider not only the diver's depth but also the pressure drop as air flows through the hose (and through the hoses and valves of the mask or helmet). The pressure drop increases with internal hose roughness, hose length, and flow rate, and decreases with increasing internal diameter (Clarke 2003). The compressor must supply an "overbottom pressure" that compensates for this pressure drop. One rule of thumb for overbottom pressure (in psig) is that it should be $25 + 0.71 \times D$ (depth in feet) (USNDM 1979, 6–9). Note that the required bottom pressure is $(14.7/33 =) 0.45 \times D$.

The Fresh-Air Delivery Problem

Consider de Lorena's diving bell. At a depth of 10 meters, the internal volume of air would have been reduced by half, to 200 liters. Based on de Marchi's description of the apparatus, the water level would have risen 61 centimeters above the rim, and this would have been above the diver's nose. (If exhaled air were removed without replacement, then the level would have risen further as the bell was used.) Since the barges were more than 10 meters deep, and they were worked for a couple of hours at a time, it is clear that de Lorena must have also provided means for injecting fresh air into the bell as the depth increased, thus increasing the air pressure to match the water pressure.

But this air supply means was kept a trade secret. It could not have been a simple snorkel. Lung suction is not capable of drawing the air down for any significant depth because the water pressure limits your ability to expand your lungs. It is for this reason that snorkels are typically only 12–15 inches long.

Since De Lorena took his air circulation system to the grave, how was it finagled by subsequent salvors? Essentially, in two ways: by piecemeal delivery and by pumping through a line.

The first solution is ancient and indeed was described in part by Aristotle. Aristotle's *Problemata*, Book XXXII, said that one "can give respiration to divers equally by letting down a cauldron. For this does not fill with water, but retains its air" (Hett and Rackham 1937). (While Aristotle doesn't spell it out, the cauldron is a container with an open end, and if it is inverted, it acts like a miniature diving bell. He does warn against letting it tilt.) As it descends, the air in it is compressed.

However, a large diving bell, like that of Melian, has some advantages. Divers do not have to fight the current. They may meet there and communicate by words rather than signs, or inspect finds and determine whether they are significant, and so on. But how could they transfer the air from the lowered container to the big bell?

Once the container descends below the level of the diver's bell, the air pressure inside it is greater than that in the bell. Now suppose that it is equipped at the opposite end with a plugged supply tube. The diver may pull the supply tube inside his bell and unplug it, thus releasing the compressed air. (Eliav [2015] believes that this is the air supply system used by De Lorena.)

A more elaborate form of this air supply system was devised by Edmund Halley, whose principal bell innovation was the means for supplying compressed air. This relied on the use of two weighted barrels, raised and lowered alternately. The barrels both had holes in the top and bottom, with a leather hose attached to the top hole but weighted so its outlet fell below the bottom of the barrel. Thus, air could not escape unless the lower end of the hose was lifted up. When a barrel was lowered, the diver in the bell grabbed the tube and brought it into the bell, causing the air in the supply barrel to be released. Halley claimed to be able to remain at 9–10 fathoms for 90 minutes, thanks to this design (Halley 1714, 497–98).

His diving bell, with an internal volume of sixty cubic feet and unspecified weight, was suspended from a sprit attached to the mast of a ship. Braces were used to swing it out over the water or to bring it back inboard, as needed. Halley advised that the diving bell be lowered in 12-foot stages. After each drop, he would receive three or four barrels of fresh air to drive out the encroaching water. The top of the diving bell was fitted with "a Cock to let out the hot Air that had been Breathed" (Halley 1714, 497).

The 36-gallon capacity air barrels were cased with lead sufficiently to sink when empty (Halley 1714, 497). They were "fitted ... with tackle proper to make them rise and fall alternately, after the manner of two buckets at a well" (497).

Halley assumes that this linkage is familiar to his readers and hence doesn't explain it. An 1839 frontispiece illustration for *The Saturday Magazine* assumed that the rope connecting the two barrels is passed over the sprit (or a pulley attached to the sprit) and operated by a crewman pulling down on the part leading down to the barrel that is to be lowered (which of course also hauls up the opposite barrel). For that to work, the rope must be within arm's reach of a crewman standing by the taffrail.

The two-barrel hoist could be operated by two men. Descending, the barrels "were directed by lines fastened to the under edge of bell, the which past through rings placed on both sides the leathern hose in each barrel; so that sliding down by those lines they came readily to the hand of a man ... to take up the ends of the hose into the bell" (Halley 1714, 497).

Halley had suspended a weighted "stage" below the bell. The underwater barrel

12. Fresh-Air Supply for Underwater Exploration 245

Artist's conception of Halley diving bell in operation (*Saturday Magazine* 1839; cf. Royal Society Picture Library 2024a).

wrangler stood on that stage. Once the outlet end was brought above the surface of the water in the bell, the air inside the barrel was "blown with great force into the bell," as a result of the water entering at the lower hole (1714, 497–8).

The diver would then signal the surface (presumably by tugging on either the barrel's hoist line or on a separate signal line) and the surface crew would haul the used barrel up. As it ascended, water would drain out of the lower hole. Of course, the second barrel would be descending.

The *Saturday Magazine* interpretation of Halley's contraption accords it a lifting power that is limited to the arm strength of a single person. More effective double-barrel hoisting arrangements are known. For example, a rope is attached to a handle at the top of barrel 1, passes up and then over a right-angle pulley, and goes several times around a vertical drum ("reversing wheel"), then over a second such pulley, and down to the handle at the top of barrel 2. The drum is mounted on a capstan (vertical axle with one or more horizontal lever arms). The crewmen push on these arms, turning the drum and causing one barrel to be lowered and the other to be raised. Then they reverse the direction of turn, causing the barrel movements to be reversed (Lardner 1855, 180).

However, the problem is that both pulleys need to be hanging over the water. If so, how do you avoid positioning the drum and capstan over the water? On a ship, the only way to do that is to have one pulley be on the port side, and the other on the starboard side, so the drum was in between, over the deck. But then where is the hoist for the bell? I suppose the operations could be conducted from the stern. The hoist for the bell is on a sprit pointing rearward, and the pulleys on sprits (or separate structures) are pointing obliquely rearward and to the side.

Pumping Air

The other way of supplying pressurized air to a diving bell (or a hard hat diver) is by some sort of pump (operated as a compressor). The pumping system comprises the pump itself, the prime mover that provides the power to operate the pump, and the air hose that carries the air to its destination.

In large part, the pressure achievable by a pump is determined by the force delivered to its actuator, which in turn depends on the choice of prime mover. Early pumps were driven by human (or animal) muscle power, but more powerful options were developed.

The diver's air may come directly from the pump, or the pump may be used to supply the air to an accumulator (high=pressure tank) and the accumulator later used to supply air to the diver. The latter approach allows the use of a more powerful shore- or large ship-based compressor to charge up accumulators to be carried on a small boat. (This differs from scuba in that the high-pressure tank remains at the surface.)

Let's look more closely at the pumping system components.

Air Hoses

Vegetius, in *De re Militari*, depicts "a diver wearing ... a helmet to which is attached a long leathern pipe leading to the surface" (EB1911/Diving). While this arrangement would be practical only a foot or two below the surface, it does suggest that a leather hose could be used to conduct air. Leather air hoses were used with Deane's diving helmet in 1828 (Ashcroft 2002, 53).

Referring to hard-hat divers, EB1911/Diving says that "the diver's air pipe is …

made of alternating layers of strong canvas and vulcanized india-rubber, with steel or hard-drawn metal wire embedded." The purpose of the latter is to prevent collapse of the hose as a result of water pressure. A hose used in 1875 was able to withstand 100 psi (Bedford 1875, 397). Wire-reinforced India-rubber air pipes date back at least to 1840 (USJ 1840, 333), and vulcanized rubber air pipes were being sold by 1847 (Gardener 1847).

Modern hoses are likely to use neoprene rubber. The air hose may be integrated with electrical lighting and communication wires and even a hot water pipe into a single umbilical cable.

Air Pumps (Compressors)

"One of the major challenges in the development of diving apparatus has been to construct a pump, which could both supply the diver with sufficient fresh air and

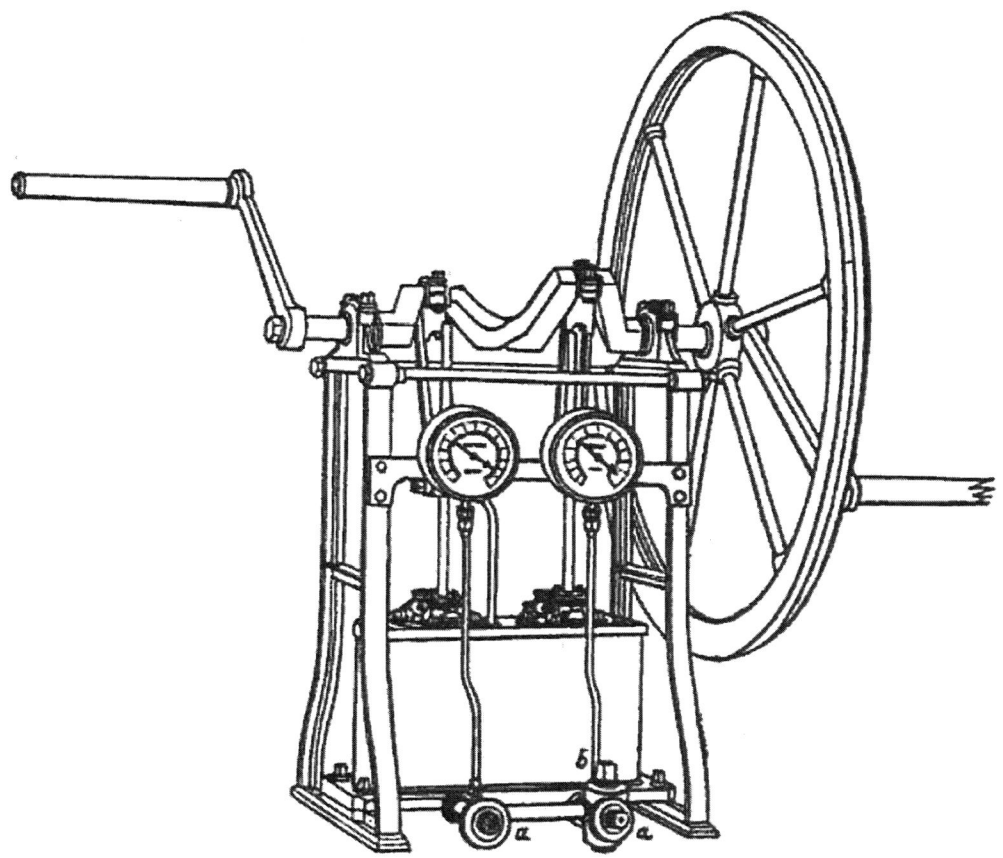

Late-nineteenth-century divers' air pump consisting of "two double-action cylinders, each cylinder capable of supplying about 135 cubic inches per revolution" (Siebe 1874, 149). Each cylinder can supply air to a different diver, so two can work simultaneously yet independently, or both cylinders can supply air to a single diver. The crank-operated pumps can achieve an air pressure of 240 psi, and there is a flywheel to smooth out the air delivery.

deliver a pressure that made it possible to send air divers to greater depths" (Jorgensen 2010).

Strictly speaking, a pump just moves a fluid, whereas a compressor squeezes it into a smaller volume, increasing its pressure, before discharging it. However, if the fluid is compressible, a pump may be operated as a compressor. The simplest way to do this is to compress the fluid in a chamber having a small outlet, like that of a bellows. If so, the escape of the fluid is retarded and hence most of it is compressed before it can exit. Or, more efficiently, one can have a valved outlet that keeps the fluid from escaping until the compression stage is completed.

There are two fundamental requirements for a diver's air supply pump. First it must compress the air sufficiently so that the air that reaches the diver is at the same pressure as the water pressure at the diver's depth. Second, it must supply air quickly enough so the diver breathes in enough oxygen to support the intended activity and also to flush out the carbon dioxide that the diver exhales.

The pump should be equipped with a pressure gauge monitoring the outlet air. This can be calibrated in depth units rather than pressure units for convenience.

Pumps (compressors) may be connected in series or parallel. A series connection increases the delivered pressure, whereas a parallel connection increases the average flow rate. One pump design was a two-cylinder piston pump in which the outlet of the large cylinder was connected to the inlet of the small cylinder, thus increasing pressure at the expense of flow rate (Jorgensen 2010).

A single pump does not produce a continuous flow of compressed air. Parallel pumps could be adjusted so they expelled the compressed air at different times. Also, the pumps could discharge into a buffer tank of relatively large volume to average out the variation in air flow.

"Supplying a diver with air using a hand pump at a great depth required several pumps ... Haldane had a test diver descend to a depth of 60 meters. For this dive three 2-cylinder pumps were coupled together, with three man teams at each handle. That is, it required six men to turn each pump, with the work involved so severe that they had to be replaced every five minutes" (Jorgensen 2010).

Types of Pumps

There are two basic types of pumps. Direct displacement pumps pressurize the fluid by collapsing its volume. Centrifugal pumps first speed up the fluid and then convert the kinetic energy to pressure. I confine my attention here to direct displacement pumps.

Pumps may be single or double acting. If single acting, there is a single pumping chamber, and one movement of the pumping element takes in air and the opposite movement compresses and expels it. If double acting, the pumping chamber is divided by the pumping element into two subchambers, and with each movement, the air in one subchamber is compressed and expelled and in the opposite chamber it is inhaled. The first double-acting pump in European literature is that of Ramelli (1588), although in China it is centuries older (Reti 1970).

Bellows. In the common bellows, two rigid boards are joined by flexible sides (originally leather) to form an air chamber. In the hand bellows, both boards have handles, and the handles are moved together so the boards compress the air, expelling it through a small exit nozzle. The foot bellows is similar except that one board remains flat on the floor and the other is pressed down and released by the operator's foot. Either way, the blast is not continuous. There was also a "double bellows" that provided a more continuous air flow (EB1911/Bellows).

The boards need not be hinged together as in a fireplace bellows. The boards may be parallel and moved apart and together, as if one were playing a concertina, as in the smiths' bellows shown in EB1911.

In 1689, Dennis Papin (Papinus) suggested using a bellows to supply pressurized air to a diver. Marx (1990, 34) states that the bellows of his day were "powerful enough to allow a bell to descend 70 feet" (11.7 fathoms; water pressure about 3 atmospheres). But Halley in 1717 claimed that their practical limit was just 3 fathoms (Hill 1912, 27). Really, it depends on the number, size, and motive force of the bellows.

The Danish Museum Middelaldercentret, which specializes in archaeological reconstructions, decided to investigate whether a diving suit, with a bellows pump air feed, could have been constructed with fifteenth-century technology (Lazenby 1998). The museum's first air supply design was a failure. It featured a pair of eighty liter bellows, linked by a draw rod. It was unable to compress the bellows when the connected helmet was in water deeper than 50 centimeters.

The second was able to permit diving to 5.5 meters. "The pump system comprises three independent bellows of around five liter capacity each, connected by wrought iron pipes to a hand-hewn wooden chest which acts as a pressure reservoir." Each pipe inlet has a simple non-return "clack" valve in the form of a leather flap nailed to the inside of the box. The chest is reinforced with wrought iron straps and sealed within with beeswax. "The lid is fitted with a greased leather gasket and is secured by wooden wedge against the iron bands." A photo shows three blacksmith-type hand bellows connected to the top of the chest. The reconstructors believe that the depth limit of the apparatus could be extended if one used a larger number of smaller bellows.

The hose was semirigid (hollowed-out wooden sections joined by tarred, tubular leather sheaths). The bellows system did not supply air at a constant rate and consequently the helmet periodically flooded up to eye level. The last dive of the day was limited by the exhaustion of the three bellows operators.

Piston (force, reciprocating) pumps. A bellows is not of course the only type of pump available. John Smeaton used a diving bell in conjunction with a force pump when repairing the foundations of Hexham Bridge (1778) and in clearing the foundation of the pier at Ramsgate Harbor (1788). "This bell was never intended to be wholly submerged, and … the pump used for supplying air was fixed on the roof of the bell" (Davis 1908, 168). A windlass was used to lower and raise the diving bell, and the pump was operated by a lever (Smiles 1874, 327).

During the nineteenth century, piston pumps were the dominant form of diving pump.

In a piston pump, a piston rod drives a rigid piston head through a slightly wider chamber of constant cross section (circular, square, rectangular). Piston rings, traveling with the head, seal off the air. The chamber has a fixed internal volume but the head compresses the air within the cylinder into a smaller volume, thus increasing its pressure. A valve opens to discharge the compressed air. Pistons may be single acting or double acting. Cylinders, pistons, and piston rods could be wood, metal, or plastic. Smeaton's 1760 blowing engine was reportedly the first to use cast-iron cylinders and pistons (Westcott 1932, 68).

The efficiency of the piston pump was limited by its clearance volume, that is, the volume remaining in the cylinder after air was squeezed out of the outlet. When the piston reversed, that residual air would expand, which would limit how much new air could be introduced. And for divers' pumps, the power loss increased with the water pressure. "If the [non-]compressible volume came to, for example, 2% of the stroke volume, then loss of effect at 20 meters depth would be 6%" (Jorgensen 2010).

Efficiency is also limited by air loss. With piston pumps, "the primary reasons for air loss were play between the piston and the cylinder, and past the valves ... on a dive to 60 meters, it was not uncommon for only 65% of the compressed air that should have come out of the pump to reach the diver" (Jorgensen 2010).

The air inside the chamber is heated as a result of both the compression by the cylinder and the friction between the gasket and the cylinder. "This heat reduced the capacity of the pump, partly because warm air takes up more space than cold air, and partly because it dried out the leather gaskets and thus increased any tendency for them to leak" (Jorgensen 2010). The heat could also cause mechanical problems (BrightHub Engineering 2022). The pumps could be air or water cooled.

Lubrication is needed to minimize friction and to prevent the leather seals from drying out. In the nineteenth and early twentieth centuries, the "lubricant used was marrow oil, castor oil or acid-free oil." Water could be used while pumping was in progress, but between uses, oiling was needed (Jorgensen 2010). Oil lubrication was dangerous because of the risk that the heat would volatilize the oil and then cause an explosion (Westcott 1932, 71). According to a post on Quora, "A single stage compressor, oil lubricated, is limited to about 150 psi" for this reason (Bangerter 2024). Even if ignition is not an issue, vaporizing your lubricant is not a good idea (BrightHub Engineering 2022).

Apparently, some early compressors used cylinders filled with water, with the water acting "as cooling medium, a safe lubricant, and a filling for the clearance spaces." In these "wet" compressors, the piston moved in a horizontal cylinder, connected at each end to a vertical cylinder in which a column of water moved up and down, drawing in and forcing out air at the top (Westcott 1932, 71). However, these compressors were slow and delivered very moist air (72).

For maximum efficiency, the parts must be manufactured to close tolerances. Also, the pump has to be durable and reliable, since the diver's life depends on it. These considerations increased the price. In 1930, "one had to pay £35–40 for a 2-cylinder lever pump, £56 for a 3-cylinder pump, and £74 for a 2-cylinder double-acting pump," versus a mere £20 for a 12-bolt diving helmet (Jorgensen 2010).

A plunger pump (patented by Morland, 1675) is similar to a piston pump, but the high-pressure seal in a plunger pump is stationary whereas in a piston pump it travels with the piston head. I have also come across a German piston pump used by divers in which it was the cylinder, not the piston, that moved! (Jorgensen 2010).

In 1916, the U.S. Navy's Mark III diving air pump was a two-cylinder, double-acting pump, operated by two hand cranks, one on each side of the pump, set 90 degrees apart. There were two lignum vitae pump wheel handles available, with the small one allowing two men on each handle, and the large one, three. Each side had a 150-pound flywheel and a pressure gauge. The gunmetal cylinders had 4.25 inches internal diameter and the piston stroke was 7.25 inches. A water cistern was provided for cooling the cylinders. The piston–cylinder seal was with leather. The pump's efficiency (E), when new, was required to be 80 percent when operated against a pressure of 100 psi (USNDM 1916, 33, 37ff). The efficiency varied depending on the pressure.

The pump was manufactured for the Navy by Morse. The pump case, without the crank handles, was 41.5 inches wide and 51.5 inches high. The short crank arms were 21.75 inches long, and the long ones 31.25 inches, with a wheel diameter of 35.5 inches (Vallejo 2022).

At the time, the minimum allowable air supply was considered to be 1.5 cubic feet/minute. To achieve this with the Mark III, the required cranking speed (rpm) (on both cranks) was (Depth (ft) × 0.194) + 6.4) × (100/E). The maximum sustained crank rate was 30 rpm (20). If a higher crank rate were used temporarily for operational reasons, the efficiency of the pump was reduced—for example, by 50 percent at 64.8 rpm (USNDM 1916, 50).

The manual comments, "the labor of operating a hand-driven diving air pump increases much faster than the increase in pressure of air being delivered" (USNDM 1916, 51).

At least some modern air compressors designed for surface-supplied diving are piston type. For example, Nuvair's Nomad LP-815-G is a light-duty two-stage, two-cylinder, oil-lubricated compressor with an 11-liter tank, producing 13 cubic feet/minute at 145 psi, weighing 187 pounds and powered by a gasoline engine 4.8 hp (Nuvair 2023).

Modern scuba air compressors, which are used to fill scuba tanks, are three-stage piston compressors—that is, they are three pumps connected in series. For example, in the first stage the air is compressed to 100–140 psi; in the second, to 800–1,000 psi; and in the third, to up to 5,000 psi. They also include coolers, dehumidifiers, lubricating oil separators, and catalysts for converting any carbon monoxide from the compressor engine to carbon dioxide, and activated carbon filters for removing the carbon dioxide (Scuba 2009).

Actuation by Muscle Power

Agricola's *De re Metallica* (1556) depicts bellows operated by a lever (208–9), a foot treadle, a continuous belt, or a capstan (211), and a fan operated by a crank (204).

Use of a piston pump was tiring. If a diver was at 10 meters depth, the water pressure was 2 atmospheres, and to deliver 50 liters a minute at depth one needed to pump 100 liters a minute at the surface. If the pump were 100 percent efficient, and each cylinder of a double pump had 1.5 liter capacity, that would mean 33 double pump strokes per minute. "The effort involved was considerable and a two man pump team had to be regularly changed" (Jorgensen 2010).

Human physiology limits both the maximum force that can be applied and the maximum and sustained power output (work/time). The limits vary depending on age, sex, physical condition, posture, the muscles at work, and the starting point, direction, length, and speed of the movement (CCOHS 2022). The sustained power also depends on the cadence and duration of the activity (McJunkin 1977, 57).

For a piston pump, the piston won't move in the direction of compression unless the driving pressure exceeds the back-pressure. The actual pressure exerted by the pumping element on the air is the applied force (net after friction) divided by the area of the surface to which the force is applied. Thus, we can increase pressure provided by a piston pump by reducing the cylinder diameter. However, that also reduces the swept volume and thus the flow per stroke.

We can increase the force on the air by using an actuator that provides a mechanical advantage. In the case of a lever, there is no mechanical advantage if the input distance (force–fulcrum) equals the output distance (fulcrum to load). We gain a mechanical advantage (output force greater than input force) if we increase the input distance or decrease the output distance. Increasing the input distance means the hand or foot must move a bigger distance, which requires greater movement speed (which is tiring) or greater time per stroke (so a smaller flow rate). Decreasing the output distance means, in the case of a piston pump, a smaller stroke distance, and therefore a smaller outflow per stroke (Kay 2000, 16).

Thus, the force and power constraints mean that there is a trade-off between delivered pressure and flow rate for a muscle-powered pump.

Many piston pumps were hand lever actuated. The Smeaton pump was a single-cylinder pump, operated by a second-degree lever (Smiles 1874). Piston pumps were more commonly manufactured with two cylinders on a seesaw lever, so one piston was lowered as the other was raised.

"For a standing person the maximum force required to repeatedly act on a vertical lever—like a pump—should not exceed 90–100 N. The frequency should not exceed 20 beats per minute" (Engineering Toolbox 2018).

"Lever pumps were only suitable for a maximum of 20 to 25 meters" depth "and demanded 2 to 4 people for operation. For longer dives, these guys had to be refreshed frequently so eight of them were needed. In greater depths two or more pumps were used in parallel" (DivingHeritage 2007). (Note that this increased the rate of air delivery at depth to compensate for the increased pressure, not the pressure delivered.) "In normal operation these pumps would supply [sic, imbibe] about 45 liters of air per minute at the surface" (DivingHeritage 2007) For a depth of 20 meters (3 atm), the effective delivery would 15 liters/minute.

There were also two-stage pumps, in which two pumps were connected in series to increase pressure. The second stage cylinder would be smaller (DivingHeritage 2007).

A crank is really a kind of lever. The mechanical advantage is the ratio of the radius of the crank handle to the radius of the crankshaft. For operating a hand crank, the length of the arm should be about 0.4 meters, the center of the axle at about 1 meter height, the "maximum force should not exceed 130 N," and the speed "should not exceed 25–30 rpm" (Engineering Toolbox 2018).

Hand-cranked diving pumps were popular. There were double-acting pumps with one or two cylinders, and single-acting pumps with two to four cylinders. The crank turned a connecting rod connected to crank pins, which raised or lowered the pistons. A flywheel was mounted on one or both ends of the crank to even out the effort that had to be applied to the crank. These cranked pumps were reportedly much better performers than lever pumps (Jorgensen 2010; DivingHeritage 2005).

In diving literature, these are often referred to as "rotary pumps." However, to a mechanical engineer, a "rotary pump" is one with rotating elements (typically gears, screws, or vanes) that move the fluid (Wärtsilä 2024).

Alternative Prime Movers

In the nineteenth century, all diving pumps were muscle-powered. While steam and later internal combustion engines were used to drive pumps beginning in the early twentieth century, manual pumps remained in use by divers until the 1960s (DivingHeritage 2005).

With any combustion-based prime mover, care must be taken that the exhaust (which can contain carbon monoxide) does not enter the intake for the air compressor.

Rechargeable batteries may be used to power a motor that acts as the prime mover for a compressor. There are modern "hookah diving" air compressors, used to support shallow recreational diving, that run on lithium ion batteries (Brownie 2024).

Self-Contained Underwater Breathing Apparatus (SCUBA)

The sina qua non of scuba is a wearable high-pressure tank containing a breathing gas. This could be air, or some other gas mixture containing sufficient oxygen to support activity and devoid of toxic components.

A typical modern cylinder has an internal volume of 0.39 cubic feet, in which air is held at 3,000 psi (204 atmospheres). Expanded to normal atmospheric pressure, that comes to about 80 cubic feet (2,264 liters).

The cylinders must be thick enough to withstand the internal pressure, which of course makes them heavy. For a thin-walled cylindrical pressure vessel, the hoop stress equals the pressure times the radius, divided by the thickness, and the axial stress is half that (Leckie et al. 2009, 119). The mass of the curved side is proportional to the thickness, the cylinder length, and the radius, and the mass of the ends is proportional to the thickness and the square of the radius.

Diving cylinders are now made either of steel or aluminum. A steel Faber 80 with a service pressure of 2,400 (+10 percent for overfill) has an empty weight of 30 pounds, whereas one with a service pressure of 3,180 (+10%) weighs 32.5 pounds

(HuronScuba 2010). Generally speaking, the steel tanks impart more negative buoyancy than the equivalent aluminum ones.

There are two major types of scuba systems. In open circuit, the diver's exhaled air is discharged into the water. In closed circuit, it passes through a rebreather that scrubs carbon dioxide from it and then adds oxygen from the tank.

There are several advantages to a closed-circuit system. First, it does not produce a trail of bubbles. This is an advantage to a military diver in enemy territory or near an enemy vessel. Second, it avoids wasting the oxygen in exhaled air (16%, vs. 21% in the atmosphere). Hence, for a given duration dive, less oxygen need be provided in the tank.

EB1911/Diving made reference to equipping the diver with "a knapsack consisting of a steel cylinder containing oxygen compressed to a pressure of 120 atmospheres ... and chambers containing caustic soda or caustic potash. The helmet is connected to the chambers by tubes, and the oxygen cylinder is similarly connected to the chambers." This plainly is a closed-circuit system, as the exhaled air is regenerated.

Open-circuit systems may be either continuous-flow or on-demand. Continuous flow is wasteful of air. While a demand regulator was invented in 1838 and another design was commercialized in 1865 (*Wikipedia*/Diving Regulator), there was some early-twentieth-century use of manually adjusted free-flow systems.

The demand regulator has two functions. First, it takes air from the cylinder, reducing its pressure, feeding it ultimately to a low-pressure chamber at ambient water pressure. Second, it detects when the diver starts inhaling and then releases a small amount of the breathing gas from that chamber. Water is on one side of a moveable diaphragm, and the chamber is on the other. If the pressure in the latter equals the ambient pressure, the diaphragm is centered and the valve is closed. If the diver inhales, "the suction produced by his lungs reduces the pressure in the low pressure chamber causing the diaphragm to be pushed inward by the now higher ambient water pressure. The diaphragm actuates the low pressure valve which opens and permits air to flow to the diver" (USNDM 1979, 5–2).

Compressed Air Tanks in Submersibles

The use of compressed air tanks is not necessarily limited to divers. Fulton's *Nautilus* submarine (circa 1800) reportedly carried "flasks of compressed air" for use when submerged (Chaffin 2008, 52). It appears that Fulton actually went through several iterations. In 1801, Fulton proposed a "simple globe or bombe of copper capable of containing one cubic foot," and using a "pneumatic pump" to fill it with "200 cubic feet of common air" (Parsons 1922, 41–42) That would mean a pressure of 200 atmospheres! In contrast, in 1806 Fulton refers (74ff) to an iron or copper box of 27 cubic feet, containing air at a pressure of 20 atmospheres (i.e., 540 cubic feet of "common air"). This compression was to be achieved by "proper pumps." Fulton assumed that air would be consumed at a rate of 20 cubic feet an hour and thus the box's air would support six men for 5 hours. I doubt that either embodiment was ever constructed.

Twentieth-century submarines also used compressed air for blowing water out of ballast tanks and torpedoes out of torpedo tubes, and the 1859 *Plongeur* used compressed air to drive an 80-hp reciprocating engine (EB1911/Ship).

Oxygen Generation in Situ

We discussed Drebbel's 1620s submarine in the preceding chapter. Huyghens (1631) quoted reports that Drebbel was able to stay underwater for 3 hours. Given that in 1604 Drebbel wrote about "saltpetre broken up by the power of the fire and thus changed into the nature of the air," it has been speculated that Drebbel heated saltpeter (potassium nitrate, KNO_3) to generate potassium nitrite (KNO_2) and oxygen (8 grams molecular oxygen per 101 grams of the nitrate). Of course, some oxygen would be consumed in heating the saltpeter to its decomposition temperature.

I have not been able to document any modern use of potassium nitrate as an oxygen generator. The use of ammonium nitrate was proposed. In theory this would decompose to nitrogen, oxygen, and water. In practice, the reaction products included toxic nitrogen oxides (Peters et al. 1964).

Hydrogen peroxide decomposes spontaneously but slowly to water and oxygen. There has been medical use of peroxide in conjunction with a catalyst (catalase) to produce local oxygenation. Other possible catalysts include peroxidase, manganese dioxide, and iron (Ward et al. 2013).

Potassium superoxide has been "used in the mining industry in rescue breathing systems. KO_2 reacts exothermically with water in the exhaled air of a user to produce KOH and oxygen." Unfortunately, KOH is caustic. That said, it may be used to absorb CO_2, reacting with it to form potassium carbonate (Ward et al. 2013).

"U.S. Navy submarines burn sodium chlorate 'candles' in an enclosed furnace for emergency and supplementary oxygen generation" (McCarrick et al. 2011). The reaction is $2\ NaClO_3 \rightarrow 2\ NaCl + 3\ O_2$. In this, 212.88 grams of sodium chlorate nominally generates 48 grams of molecular oxygen. However, the combustion required to raise the chlorate to its decomposition temperature consumes oxygen. The candles are actually composed of sodium chlorate, iron powder, and a binder. The iron reacts with oxygen to generate heat. "Small quantities of water trapped in a chlorate candle can result in the production of significant quantities of chlorine" (Gustafson 1962), and barium peroxide may be incorporated into the candle to act as a chlorine scavenger. Cobalt chloride has been added to act as a catalyst (it reduces the decomposition temperature from 500 to 300°C) (Garcia 2017, 41).

Potassium and lithium chlorate and sodium, potassium, and lithium perchlorate may also be used as chemical sources of oxygen. The chlorate and superchlorate reactions are also exothermic. Exothermic reactions have the advantage that, once started, the heat generated helps sustain the reaction by maintaining the decomposition temperature. However, they have the disadvantage that they can increase the temperature to the point where there is a fire or explosion risk. Even if they don't, insulation or cooling is needed to prevent skin contact and burns. (It may be possible to add an additive that reacts endothermically, such as calcium hydroxide with water, to limit the temperature increase [Ward et al. 2013].)

A submersible could be equipped with apparatus for electrolysis of water, breaking it down into its components, hydrogen and oxygen. Unfortunately, electrolysis requires a lot of energy. A person consumes about 1 kg of oxygen per day. "Commercial electrolysis systems typically require approximately 50 kilowatt-hours of power to produce 1 kg of H_2 and 8 kg of O_2 from 9 kg of water." Because of this power demand, only nuclear submarines have been equipped with electrolysis systems. The hydrogen, by the way, is considered a waste product (Interesting Engineering 2017).

Mixed Gas Diving

The term "mixed gas diving" refers to use of any breathing medium other than air. It may simply have a different ratio of oxygen to nitrogen, or an inert gas other than nitrogen may be included in the mixture.

There is an incentive to reduce the nitrogen content to avoid nitrogen narcosis. It can be replaced in whole or part with oxygen or with an inert gas.

Ordinary air is 21 percent oxygen. In 1879 Fleuss used pure oxygen in an early form of scuba (USNDM 1979, 9–1). The danger with enriched oxygen usage is oxygen toxicity. With normal activity, oxygen toxicity symptoms appear after 30 minutes at an oxygen partial pressure of 1.6 atmospheres. You reach that pressure at 20 feet if you are breathing 100 percent oxygen, or 220 feet if you are breathing air (USNDM 1979, 9–17). Nitrox is about 22–40 percent oxygen, most often 32 percent or 36 percent (Devanney 2017).

It is possible to use a breathing gas with a lower oxygen percentage at depth, because the pressure will increase the partial pressure of oxygen. However, the oxygen percentage would have to be increased when the diver ascended (USNDM 1979, 3–14).

Beginning in 1924, there was experimentation with helium–oxygen (heliox) mixtures, and in the 1940s, with helium–oxygen–nitrogen (trimix). Normoxic trimix is 21 percent oxygen and 35 percent helium, and is used down to about 200 feet. Hypoxic trimix has less oxygen (say, 10 percent oxygen and 70 percent helium) and is used for deeper dives (Devanney 2017).

Heliox is used down to 984 feet, beyond which helium becomes narcotic. For still greater depths there are experimental mixtures—hydrox (hydrogen–oxygen), first used in 1945, or hydreliox (also containing helium). The oxygen level must be not more than 4 percent so the mixture is not explosive—note that if the water pressure is 4 atmospheres, the partial pressure is 0.16 atm (Devanney).

Conclusion

Humans have been working at sea for thousands of years, and over the centuries, they have invented numerous tools, machines, and procedures intended to make that work safer, quicker, easier, or more effective.

Our focus in this book has been on the science underlying the problems they were intended to solve, and the workings of the proposed solutions. Indeed, we only had space to discuss a few, those that were pioneering approaches or that were known to have had a significant impact, or that had features that this author thought were intriguing. Only in passing have we commented on the reception these innovations received.

Many innovations had imperfections ignored by their inventors and were consigned to oblivion, but some were adopted.

That adoption, however, was not necessarily rapid. In 1919, Admiral of the Fleet Lord Fisher declared, "It is a historical fact that the British navy stubbornly resists change" (Gurney 2005, 228). However, it may be added that it sometimes adopted a change that was a questionable improvement, because of the political clout of its inventor or promoter. An example of this would be the Thomson dry-card compass, which was nowhere near as steady as its liquid compass competition (264).

Nor can one fairly assert that these problems were unique to Fisher's navy. A 1976 motor boating article referred to the "innate conservatism of sailors, even if they are neophytes" (Whall 1976, 68).

There is perhaps some justification for this conservatism. "A ship at sea is a small world in itself, in which several hundred people must live and work together for weeks or months at a time, having little or no communication with the outside world … Under such circumstances, it is not surprising that sailors are notoriously conservative about the introduction of new appliances aboard ship; for if these should fail in the middle of a long voyage, when facilities for repairs are not at hand, the lives of the passengers and crew, or the safety of the ship itself might be seriously jeopardised" (Greene 1899, 213).

The risks of injudicious innovation were examined in Arthur C. Clarke's classic science fiction tale, "Superiority" (1951). The more technologically advanced side in an interstellar war was defeated because it replaced its old military hardware with new ones that proved to be unreliable.

A study of pumpkinseed sunfish found that the populations were "mixtures of bold and shy fish" (Dugatkin 2013). The bold fish were willing to take risks—venture into more dangerous areas—in order to eat more food. Among humans, too, there appears to be a "continuum of risk-taking propensities." The risk takers—the inventors and early adopters—are the driving force behind the evolution of work at sea.

Appendix 1
The Grip of Traction-Winches

For the reader who wants to understand the science better, I explain here some of the reasons why (1) the rope doesn't just slip off a traction-winch and (2) the capstan with whelps works just as well as (if not better than) a smooth cylinder of the same effective diameter in gripping the rope.

The tailing force on the rope creates a tension that, because of the curvature of the rope around the barrel, results in a force pressing the rope against the barrel, creating friction (a force here resisting the sliding of the rope along the barrel). The tailing force required to support the load is much less than the load force itself because of the friction between the rope and the winch barrel.

Amonton's Law, which is an empirical relationship discovered by da Vinci and rediscovered by Amonton and Coulomb, says that the frictional force between a sliding object and a surface is linearly proportional to the force exerted by the object normal (perpendicular to) that surface. The constant of proportionality is called the coefficient of sliding friction and has two forms, static and kinetic.

Static friction is the resistive force of an object at rest on a surface to being set into motion thereon, and kinetic friction, of an object in such motion to continued motion. Given the same object and sliding surface, static friction is greater than kinetic friction.

Surface finish, cleanliness, and lubrication can drastically affect the coefficients of sliding friction. Here are some literature values.

Table 1–1

	Static	Source	Kinetic	Source
Hemp rope on wood (dry)	0.5–0.8D 0.87W	Du Bois 1902, 224	0.45D 0.33W	Du Bois 1902, 267
Wood on wood (dry)	0.3–0.7D 0.65–0.71W	"	0.2–0.48D 0.25W	"
Metal on metal	0.15–0.24D 0.11–0.16*	"	0.24D 0.06–0.08*	"
Wood on metal (dry)	0.6D 0.65	"	0.20–0.62D 0.05–0.08W	"
Hemp rope on metal	0.3	Meriam 2020, 230	0.2	Meriam 2020, 230
Wire rope on iron	0.2D	"	0.15D	"
Manila 3 strand**	0.67D	Brown 1977, Table 5	0.23D 0.30W	Brown 1977, Table 5

	Static		Source	Kinetic	Source
Dacron 3 strand**	0.33D 0.30W		Brown 1977, Table 5	0.23D 0.17W	Brown 1977, Table 5
Polypropylene 3 strand**	0.37D 0.30Q		"	0.33D 0.26W	"
Nylon 3 strand**	0.89D 0.76W		"	0.45D 0.5W	"

D, dry; W, wet; *olive oil, **on "standard capstan"

Let's look at the sliding of a rope wrapped tightly around a cylindrical barrel. If the barrel is being rotated (driven by the crew), then the rope is in a "slip" state, and if the rope has negligible stiffness (resistance to bending), then the friction is purely kinetic (Yao Ren 2017). On the other hand, if the barrel is at rest, then we are looking at the question of when a "stick" becomes a "slip," and the friction is static.

Each curved bit of rope, because of the change in angle, converts some of the rope's tangential tensile load into radial contact pressure, creating friction. With calculus, we can add up the contributions of each of these bits. The simplest model of the underlying physics assumes the rope is of negligible thickness, stiffness, and elasticity, and its friction follows Amonton's law.

Based on these assumptions, we may derive the "capstan equation," which shows that the ratio of the frictional force to the tailing force is an exponential function of both the coefficient of friction and the total angular change in slope for the portion of the rope in contact with the barrel.

Table 1–2: Ratio of Frictional Force to Tailing Force, Rope Around Cylinder

Turns	Coefficient of sliding friction			
	0.65	0.45	0.33	0.20
1	59	17	8	3.5
2	3,527	286	63	12
3	209,450	4,829	503	43
3.5	–	19,852	1,418	81
4	–	–	3,999	152

Attaway examined the role of friction experienced by a rope going over the edges of a rock face. Instead of having a smooth angular change, as when a rope slides around a cylinder, there are curved segments in contact with the rock interrupted by straight line segments suspended in the air. He recognized that the capstan equation applies if one sums up the changes in angle at each edge (the tension created by the friction-amplified tailing force is constant in the "free" segments of the rope and increases at each bend).

Harland (2003, 24) realized that Attaway's analysis could be applied to the effect of whelps. It doesn't matter whether the rope is wrapped around a smooth cylinder (continuous contact), one with whelps (partial contact), or even (he says) a star wheel (minimal contact, just at the points), as long as the total angular change is the same. The number of whelps is irrelevant (so far) because the fewer the whelps, the greater the angular change at each one.

However, Attaway's analysis applied calculus, which means it was adding up the effects of infinitesimally small angular changes. That's fine for a barrel with thick whelps. With a star wheel, there would just be a large, abrupt angular change at each point. The proper analysis is then that given by Du Bois (1902, 231–2): The force ratio for one turn about the star wheel is $([1 + k \times \tan(pi/P)]/[1 - k \times \tan(pi/P)])^P$, with coefficient k and number of star points P. This reduces to the capstan equation if applied to the small angular changes about a cylinder (233).

For a single turn around a six-star wheel and a coefficient of 0.4, Harland (2003, 29) calculated 12.3, which also is the result for one turn around a cylinder. However, the correct value for a six-point star wheel, per Du Bois, is 16.8. For a 12-pointer it is 13.22, and for 100 points, 12.36.

Remember that there are a lot of assumptions underlying the capstan equation. One was that the rope is not deformable. In fact, rope is deformable and the whelps will bite into the rope. This creates additional friction because the whelps must "climb out" of the depression they create. (Harland emphasizes, I think mistakenly, the bulge created on the other side of the rope.) In any event, Harland is certainly correct that the whelps bite into the rope and this increases the friction generated. The extent to which the friction is increased by this deformation mechanism will probably depend on the number and shape of the whelps, but I am not aware of any theoretical or experimental analysis of the issue.

I have found an analysis of the effect of deformation and bending rigidity of yarn on textile processes in which the yarn is driven by a rotating object like a capstan. Gao et al. (2015) state that Amonton's law applies to metals, which deform plastically, but not to textiles, which deform viscoelastically. It has been proposed that the frictional force on textile fibers follows a "power law"—that is, rather than be linearly dependent on the normal force, it is dependent on the latter raised to the two-thirds power. Gao developed a modified capstan equation that took into account both power-law friction and bending rigidity. The effect of bending rigidity was complex, sometimes increasing the force ratio and sometimes decreasing it, depending on the exact parameter values.

The derivation of the capstan equation considered the friction of the rope against the barrel, but not the effect of rope-to-rope contact if the coils are in contact with each other. Fulton's analysis (2005, 11) indicated that this would reduce the effectiveness of the tailing force, that is, the ratio of the frictional force opposing the load to the tailing force.

Appendix 2

The Material Properties of Ropes in Nautical Use

Rope Linear Density

Nineteenth-century data frequently express the linear density (D_L) of the rope in pounds per foot of length according to the formula $D_L = K * C^2$, with C being circumference (inches), and the multiplier K depending on the type of rope. The K value for rope depends on the rope construction (lay) and circumference, but the following values are representative:

tarred hemp: 0.0389
manila: 0.0370
iron or steel wire rope: 0.1667 (EB1911/Rope, 717).

Note that the weight of hemp rope will increase by as much as 200 percent if it gets wet (Luce and Ward 1884, 9).

Given the linear density and the length, you may readily calculate the weight of a rope. For example, 120 fathoms of tarred hemp rope of 24 inches circumference should weigh about 16,130 pounds, per the preceding data.

Rope-Specific Strength

Likewise, in the nineteenth century, the breaking load (in pounds) of a rope was expressed as the multiple (K') of the square of the circumference (inches). This value may be converted into a tensile strength in psi and MPa.

Literature values for the strength of hemp, tarred or untarred, vary considerably. The old ropemakers' rule set K' = 448 (Clark 1891, 673), presumably for tarred hemp. However, Luce and Ward (1884, 28) set K' as 1371.4 for untarred hemp and 1044.9 for tarred hemp.

Hemp's strength depends on the geographical source of the fiber (because of differences in soil and climate). Thus, 2-inch ropes had breaking weights of anywhere from 1.97 (Konigsberg) to 2.75 (Neapolitan) tons (Clark 1891, 674). There are also differences in strength and weight depending on how the rope is laid (Rankine 1887, 475ff). Tarred ropes have three-quarters the strength of white ropes (476) but retain strength longer (Anderson 1872, 155) and are more water-resistant. Clark (1891, 673) shows that there are differences in breaking strength depending on whether the rope is made by "the old method," the "cold register," or the "warm register," with the latter being the strongest but also less durable.

With hemp being a natural, inhomogeneous material, "there is much variation in the strength of pieces of ropes even when cut from the same coil … The 6-inch [circumference] ropes [range in breaking load] from 14.25 to 17 tons" (156; ton = 2,240 pounds). Working may reduce the strength of a rope by 50 percent (Anderson 1872, 156).

For manila rope, Luce and Ward (1884, 28) set K' = 783.7. The NBS concluded, based on tests, that its breaking load (pounds) is $5{,}000 \times D \times (D + 1)$, with D the diameter (inches) (0.5–4.5")

Appendix 2

(Stang and Strickenberg 1921); for 1"D, that is equivalent to a K' of 1,013.2, and for 2"D, of 759.9.

For wrought iron wire rope, K' = 3,484, and for nineteenth-century steel, 6,220 (Rankine 1887, 584). Clark indicates that the working load of steel wire rope is nine-fifths that of iron wire rope of the same circumference (1891, 674).

The U.S. Navy rates the strength of various ropes, relative to manila, as follows: nylon (275%), sisal (80%), jute (60%) and agave (60%) (Bureau of Naval Personnel 1964, 27).

Rope-Bending Radius

A rope may needed to be wrapped around the sheave of a pulley, or the barrel of a windlass or capstan. The thicker the rope, the stiffer it is, and sharp bends weaken rope. Harland (2003, 35) says that the 14-inch circumference (4.47-inch diameter) hemp messenger of HMS *Victory* could not be used with a windlass or capstan of less than 48 inches diameter (bending radius 24 inches). Later, Harland (41) says that "the diameter of the barrel is five times that of the largest cable brought to the capstan." But that's misleading, because the rope actually wraps around the whelps of the capstan, and at the bottom those are half as broad as the barrel, thus effectively doubling the capstan diameter.

For wire rope, normally "the diameter of a sheave should never be less than 20 times the diameter of the wire rope" (Bureau of Naval Personnel 1964, 31).

Modern Strength and Linear Density Data

The following table shows the breaking strength and linear density for manila rope, three different synthetic fiber ropes, and steel wire rope, in two different diameters. These should be taken as representative values only. Doubling the diameter quadruples the cross-sectional area. It can be seen that while the linear density is indeed proportional to the cross-sectional area, the breaking strength is not. Thus, the traditional estimation of breaking strength by multiplying the square of the circumference by a factor characteristic of the material overestimates the strength.

Table 2–1: Rope Breaking Strength and Linear Density (Engineering Toolbox 2009)

Material	Breaking strength (lbf)		Linear density (lbm/ft)	
	1"D	2"D	1"D	2"D
Manila	8,100	27,900	0.257	1.02
Nylon	22,230	84,600	0.253	1.00
Polyester polyolefin	13,175	48,050	0.218	0.810
Polyester	19,775	72,000	0.304	0.91
Polypropylene	12,825	46,800	0.18	0.69
Wire rope (6 × 19)	83,600	320,000	2.13	6.72

Note that polypropylene (and polyethylene) has a specific gravity less than that of water, and thus will float.

These data unfortunately do not include polyethylene, Kevlar (para-aramid), the high-molecular-weight polyethylenes Spectra and Dyneema, or the liquid crystal polymer Vectran.

Appendix 3
The Physics of the Ship's Rudder

A ship is normally turned as the result of the action of the rudder (although some would quibble that the rudder doesn't actually turn the ship—it merely initiates the turn). The explanation that follows paraphrases that of Chakraborty (2021).

The rudder acts somewhat like the wing on an airplane. If, while water is moving over the rudder, it is turned to starboard (that is, its trailing edge is moved to the starboard side of the ship), the flow over the rudder creates a hydrodynamic lift force directed to the port side. Since the force is horizontal, but perpendicular to the centerline of the ship, we call it a "sway" force.

Since the rudder is well aft of the center of gravity of the ship, the force also gives rise to a clockwise turning "rudder" moment. The rudder moment is small because the area of the rudder is small, so the result is a slight drift angle (clockwise change in heading).

The sway of the ship exerts a force on the water molecules to port. They resist this force, creating an equal and opposite inertial force in the starboard direction. The inertial force presses against both the bow and stern, but given the drift angle and the shape of the hull, the moment on the bow is greater than that on the stern so, the net turning "inertial" moment is also clockwise. This inertial moment is proportional to the underwater profile area of the ship and hence is much larger than the rudder moment. It is sufficient to turn the ship.

Bibliography

Abbreviations

The abbreviations listed here appear in the parenthetical citations in the body of the book.

AC	Aquatic Community
ACD	Air Compressor Direct
Admiralty	Admiralty, Great Britain
AFP	Agence France-Presse
Anglo	*Anglo-American Encyclopedia*
APS	American Physical Society
Artifacts	Museum of Artifacts
ASA	American Sailing Association
ASME	American Society of Mechanical Engineers
BBC	British Broadcasting Corporation
BHE	BrightHub Engineering
BMIN	Bureau of Maritime Inspection and Navigation, U.S.
BNP	Bureau of Naval Personnel, U.S.
BOIMA	British Optical Instrument Manufacturers' Association
CCOHS	Canadian Centre for Occupational Health and Safety
CSO	Chief Signal Officer
DMS	Datawave Marine Systems
DP	Dreadnought Project
DPA	DPA Microphones
DSDS	Deep Sea Diving School, U.S. Navy
EB	*Encyclopaedia Britannica*
EE	*Electrical Engineer, The*
EMMS	*English Mechanic and Mirror of Science*
ER	*Electrical Review*
ETB	Engineering Toolbox, The
FAA	Federal Aviation Administration, U.S.
FPL	Forest Products Laboratory, U.S. Dept. of Agriculture
Franklin	Franklin Institute, The
Gardener	*Gardeners' Chronicle*
GBBT	Great Britain Board of Trade
GBNS	Great Britain Naval Staff
GBPO	Great Britain Patent Office
Harper's	*Harper's Weekly*
IWM	Imperial War Museums, Great Britain
JRUSI	*Journal of the Royal United Service Institution*
Lenard	Lenard Audio
Life	*Life* magazine

LOC	Library of Congress
ME	*Marine Engineer*
MEL	*Marine Engineering/Marine Log*
MINN	Marine Insight News Network
MJ	*Marine Journal*
MM	*Mechanics' Magazine*
MMC	Merchant Marine Council
MSB	Maritime Subsidy Board
NAI	*Navy and Army Illustrated, The*
NAVCOMM	Naval Communications
NAVEDTRA	Naval Education and Training Support Command, United States
NCEI	National Centers for Environmental Information
NGIA	National Geospatial-Intelligence Agency
NM	*Nautical Magazine*
NMA	Naval Marine Archive
NMAH	National Museum of American History
NMM	National Maritime Museum, Greenwich
NMPMO	Navy Music Program Management Office
NOAA	National Oceanic and Atmospheric Administration
NWC	Naval War College
Parliament	Parliament, Great Britain
PBE	*Pacific Builder & Engineer*
PopSci	*Popular Science*
Riviera	Riviera Newsletters
RMG	Royal Museums Greenwich
RSA	Royal Society of Arts
RSPL	Royal Society Picture Library
SBF	Short Blue Fleet, The
SciAm	*Scientific American*
SM	*Shipwrecked Mariner, The*
Soane	(Sir John) Soane's Museum London
USCG	United States Coast Guard
USHO	United States Hydrographic Office
USNDM	*United States Navy Diving Manual*
USJ	*United Services Journal*
Vallejo	Vallejo Gallery
VAM	Victoria and Albert Museum
WAM	Western Australian Museum
War	War Department, United States

Journal Title Abbreviations

Amer.	American
Bull.	Bulletin
Eng'g	Engineering
Eur.	European
Hist.	History
Inst.	Institution (or Institute)
Int.	International
J.	Journal
Pop.	Popular
Proc.	Proceedings
Rev.	Review
Sci.	Science (or Scientific)

Technol.	Technology (or Technological)
Transac.	Transactions
U.S.	United States

"A.B." (author's initials). 1865, April. "British Sea-Fish, Fishermen, and Fisheries." *Fraser's Magazine* 71: 507–525.

Aberdeen Maritime Museum. 2024, March 26 (access date). "The Wreck of the 'Prince Consort.'" (Painting by George Reid, accession ABDAG001280.) https://artuk.org/discover/artworks/the-wreck-of-the-prince-consort-106145.

Abridgments. 1872. *Abridgments of Specifications Relating to Raising, Lowering, and Weighing, AD 1617–1866* (2d ed.).

Ackerley, Charles Henry. 1834. *A Plan for the Better Security of Vessels Navigating the River Thames.* Longman, Rees, Orme, Brown, Green & Longman.

Aczel, Amir D. 2001. *Riddle of the Compass; The Invention That Changed the World.* Harcourt.

Admiralty, Great Britain. 1883. Manual of Seamanship for Boys' Training Ships of the Royal Navy. 1883. HM Stationery Office.

_____. 1951. *Manual of Seamanship, Vol. 1.* HM Stationery Office.

_____. 1954. *Manual of Seamanship Vol. III.* B.R. 67 (3/51). HM Stationery Office.

_____. 1995. *BR 67 Admiralty Manual of Seamanship.* HM Stationery Office. https://www.amphion.ca/wp-content/uploads/2019/02/br-67-admiralty-manual-of-seamanship-1995-05-01.pdf.

ageFotoStock. 2024, January 17 (access date). "Stock Photo—HMS Victory's Signal. The flaghoist signals used by Admiral Horatio Nelson (1758–1805) at the Battle of Trafalgar (1805), England." https://www.agefotostock.com/age/en/details-photo/hms-victory-s-signal-the-flaghoist-signals-used-by-admiral-horatio-nelson-1758-1805-at-the-battle-of-trafalgar-1805/MEV-11057950.

Agence France-Presse. 2018, August 7. "Cast from the Past: World's Oldest Fishing Net Sinkers Found in South Korea." https://phys.org/news/2018-08-world-oldest-fishing-net-sinkers.html.

Agricola, Georgius. 1556. *De re Metallica.* Transl. by Herbert Clark Hoover and Lou Henry Hoover, 1912. Upload of 1950 Dover edition. https://www.gutenberg.org/files/38015/38015-h/38015-h.htm

Agullella-Arzo, Marcel, et al. 2003, April. "Heat Loss and Hypothermia in Free Diving: Estimation of Survival Time Unde Water." *Am. J. Physics* 71(4): 333–7. https://www.researchgate.net/publication/233951484_Heat_loss_and_hypothermia_in_free_diving_Estimation_of_survival_time_under_water

Air Compressor Direct ("Melissa C."). 2024, January 17 (access date). "The Ultimate Air Compressor Guide: How to Pick the Perfect Air Compressor." https://www.aircompressorsdirect.com/stories/156-How-to-Pick-the-Perfect-Air-Compressor.html

Aitken, George Atherton. 1889. *The Life of Richard Steele, Volume 2.* W. Isbister, Ltd.

Akçagün, Engin, et al. 2017, December 19. "Thermal Insulation and Thermal Contact Properties of Wool and Wool/PES Fabrics in Wet State." *J. Natural Fibers* 16: 199–208. https://doi.org/10.1080/15440478.2017.1414650

Alam, Akm Nowsad. 2007, June. "Chilling and Icing of Fish." In *Participatory Training of Trainers—A New Approach Applied in Fish Processing,* 93–112. Bangladesh Fisheries Research Forum.

Allan, Philip K. 2022, June. "How the Propeller Displaced the Paddle Wheel." *Naval History Magazine.* https://www.usni.org/magazines/naval-history-magazine/2022/june/how-propeller-displaced-paddle-wheel.

Allen, Thomas W. 2021. "A History of the Development of Mathematics Teaching in England from Before the Sixteenth Century to the Beginning of the Twentieth Century." Master's thesis, Loughborough University. https://doi.org/10.26174/thesis.lboro.14572488.v1.

Alston, Lt. Alfred Henry. 1860. *Seamanship, and Its Associated Duties in the Royal Navy.* Routledge, Warne, & Routledge.

American Angler. 1899, October. "Keeping Fish Alive." *American Angler* 29: 293–94.

American Physical Society. 2012, July. "This Month in Physics History: July 24, 1954: Operation Moon Bounce." *APS News* 21(7). https://www.aps.org/publications/apsnews/201207/physicshistory.cfm.

American Sailing Association. 2014. "Bareboat Cruising Made Easy." *The Official Manual For The ASA 104 Bareboat Cruising Course.* American Sailing Association.

American Society of Mechanical Engineers, History and Heritage Committee. 2020, November. "Perkins Vapor-Compression Cycle for Refrigeration." https://www.asme.org/getmedia/cb9bea09-6d23-425e-bfe5-5f6d786919fb/274-Perkins-Vapor-Comp-Refrig.pdf.

Amery, John S. 1880, July. "John Lethbridge and His Diving-Machine (1880)." https://devonassoc.org.uk/devoninfo/john-lethbridge-and-his-diving-machine-1880/

Amplifon. 2021, June 29. "The Human Hearing Range." https://www.amplifon.com/au/blog/human-hearing-range.

Anderson, John. 1872. *The Strength of Materials and Structures.* D. Appleton and Co.

Anderson, Ray C., et al. 2012a. *The Future of Sustainability.* (The Berkshire Encyclopedia of Sustainability, Volume 10.) Berkshire Publishing Group LLC.

Anderson, Romola, and R.C. Anderson. 2012. *A Short History of the Sailing Ship.* Dover.

Angell, John. 1879. *Elements of Magnetism and Electricity.* Collins.

Anglo-American Encyclopedia. 1910. "Diving." 13: 255–260.

Aniker, Alpertunga. 2021. *Shiphandling with Azumuthing Podded Propellers.* Alpertunga Aniker.

Anonymous. c1266. *The Rules of Oleron.* Source of text: 30 Fed. Cases 1171–1187. http://www.admiraltylawguide.com/documents/oleron.html.

Anonymous. 1839, April 20. "The Diving Bell." *The Saturday Magazine*, no. 436, 145.

Anschütz & Co. 1910. *The Anschütz Gyro Compass: History, Description, Theory, Practical Use.* Anschütz & Co.

Aquatic Community. 2024, February 19 (access date). "History of Trawling; Not a Modern Problem." https://www.aquaticcommunity.com/history-of-trawling-not-a-modern-problem.

Arevalo, Wendy. 2021, September 22. "The Navy's Use of Carrier Pigeons." *Naval History and Heritage Command.* https://www.history.navy.mil/browse-by-topic/exploration-and-innovation/navy-pigeons.html.

Army. 1966, July 13. *Marine Salvage and Hull Repair.* TM 55–503. U.S. Army.

_____. 1979, April 9. *Soldier's Manual, Construction Equipment Supervisor, MOS 62N, Skill Level Three.* Field Manual 5–62N3. U.S. Army.

_____. 1979a, July 18. *Terminal Operations Coordinator's Handbook.* Field Manual 55–17. U.S. Army.

Arnold, Charles D. 1989. "Arctic Harpoons." *Arctic* 42(1): 80–81. https://journalhosting.ucalgary.ca/index.php/arctic/article/view/64696.

Artizan, The. 1862, October 1. "Trial of the 'Black Prince.'" *The Artizan* 20(238): 230–31.

Ash, Eric C. 2007. "Navigation Techniques and Practice in the Renaissance." In David Woodward (ed.), *The History of Cartography, vol. 3, Cartography in the European Renaissance* 509–27. University of Chicago Press.

Ashcroft, Frances. 2002. *Life at the Extremes: The Science of Survival.* University of California Press.

Åström, Karl Johan. 2012, February 21 (pdf creation date). "Ships and Aerospace." https://archive.control.lth.se/media/Education/Doctorate Program/2012/HistoryOfControl/L06ShipsAndAerospaceeight.pdf.

Athiyaman, N., and K. Rajan. 2018, November 29. "An Age-Old Practice: Pearl and Chank Diving in South India." https://www.uw360.asia/an-age-old-practice.

Attaway, Stephen W. 1999. "The Mechanics of Friction in Rope Rescue, International Technical Rescue Symposium." https://itrsonline.org/tproduct/1-969924084631-the-mechanics-of-friction-in-rope-rescue.

Avis, Robert. 2012. *Superyacht Master: Navigation and Radar for the Master (Yachts) Certificate.* Bloomsbury Publishing.

Bachhuber, Christoph. 2011, October 15 (pdf creation date). "Seafaring During the Mesolithic and Neolithic in the Mediterranean Region." https://www.brown.edu/Departments/Joukowsky_Institute/courses/maritimearchaeology11/files/18306900.pdf.

Bacon, Sir Francis. 2000 [1626]. *The New Atlantis.* Project Gutenberg ebook edition, ebook #2434. https://www.gutenberg.org/files/2434/2434-h/2434-h.htm.

Baird, Spencer F. (Commissioner, Fish and Fisheries). 1879. *Report of the Commissioner for 1877.* GPO.

Bakels, Pieter. 2024, "Interior Communications." http://www.navsource.org/archives/01/57v.htm.

Baker, William Avery, and Tre Tryckare. 1965. *The Engine Powered Vessel: From Paddle Wheeler to Nuclear Ship.* Crescent Books.

Baldino, Lexi. 2020, October 26. "From Bellows to Beyond—A Brief History of the Air Compressor." https://europe.sullair.com/en/blog/bellows-beyond-brief-history-air-compressor.

Baldwin, Bert L. 1897, December. "Pittsburg and Cincinnati Packet Line Stern Wheel Steamer Queen City." *Marine Engineering* 1: 19–22.

Ball, Nick, and Simon Stephens. 2018. *Navy Board Ship Models.* Pen & Sword Books.

Bangerter, Ken. 2024, January 25 (access date). Post in response to "What is the effect of clearance volume in reciprocating air compressor?" https://www.quora.com/What-is-the-effect-of-clearance-volume-in-reciprocating-air-compressor.

Barbieri, Patrizio. 2004. "The Speaking Trumpet: Developments of Della Porta's Ear Spectacles (1589–1967)." *Studi Musicali* 33(1): 205–47. https://www.academia.edu/5116175/The_speaking_trumpet_developments_of_Della_Portas_ear_spectacles_1589_1967.

Barker, Thomas. 1651 (printed 1653). *The Art of Angling ....* (Early English Books Online: Text Creation Partnership.) http://quod.lib.umich.edu/e/eebo/A30936.0001.001.

Bart, Ryan M., and Henry Lau. 2023, January 30. "Shallow Water Blackout." NIH StatPearls. https://www.ncbi.nlm.nih.gov/books/NBK554620.

BCW Project. 2010, March 11, "Fighting Instructions, March 1653." https://web.archive.org/web/20220816050120/http://bcw-project.org/texts/fighting-instructions.

Beckwith, L. 1962. "Ships' Cargo Handling Gear." *Lloyd's Register Technical Association Session 1962–1963.* Lloyd's Register.

Bedford, Frederick George D. 1875. *The Sailor's Pocket Book.* J. Griffin & Co.

Bell, Alexander Graham. 1877, January 30 (issue date). U.S. Patent No. 186,787, Improvement in Electric Telegraphy (applied for January 15, 1877).

Bell, Captain James. 1896, October 23. "Salvage Appliances." *The Practical Engineer* 14: 409–10.

Bell, John. 1763. *Travels from St. Petersburg in Russia to Diverse Parts of Asia.* Foulis.

Ben-Yami, M. 1988. "Attracting Fish with Light." FAO Training Series, no. 14. https://www.researchgate.net/publication/39005660_Attracting_fish_with_light.

Bennett, John R., and Susan Rowley. 2004. *Uqalurait: An Oral History of Nunavut.* McGill-Queen's University Press.

Bennett, Stuart. 1984, November. "Nicolas Minorsky and the Automatic Steering of Ships." *IEEE Control Systems Magazine* 4(4): 10–15. https://ieeexplore.ieee.org/document/1104827.

———. 1993. *A History of Control Engineering, 1930–1955.* Peter Peregrinus Ltd.

Bergman, Albert. 1890. *On Board the "Pensacola": The Eclipse Expedition to the West Coast of Africa.* No publisher identified.

Bernstein, William J. 2009. *A Splendid Exchange: How Trade Shaped the World.* Grove Atlantic.

Bernzweig, Daniel. 2023, June 30. "Treasure You Can Find in up to Six Feet of Water." https://www.metaldetector.com/learn/buying-guide-articles/beach-water-hunting/treasure-you-can-find-in-six-feet-of-water.

Bernzweig, Michael. 2021, November. "What are the Best Metal Detectors for Underwater Search and Recovery?" https://www.metaldetector.com/learn/buying-guide-articles/marine-salvage-search-recovery/best-metal-detectors-for-underwater-search-recovery.

Bertin, Louise Emile, and Leslie Stephen Robertson. 1906. *Marine Boilers: Their Construction and Working Dealing More Especially with Tubulous Boilers.* John Murray.

Bingeman, John, et al. 2021. *The Wrecks of HM Frigates* Assurance *(1753) and* Pomone *(1811).* Oxbow Books.

Bishop, Farnham. 1916. *The Story of the Submarine.* Century.

Biss, Gerald. 1921, June 1. "Many Inventions." *The Sketch* 114(1470): 342.

Blackmore, Howard L. 2000 [1971]. *Hunting Weapons: From the Middle Ages to the Twentieth Century.* Dover.

Blake, William P. 1870. *Report of the United States Commissioners to the Paris Universal Exposition, 1867.* GPO.

Blanckley, Thomas Riley. 1750. *A Naval Expositor.* Project Gutenburg online edition released August 27, 2016. https://www.gutenberg.org/cache/epub/52902/pg52902-images.html.

Bloxham, Jeremy. 1992. "The Steady Part of the Secular Variation of the Earth's Magnetic Field." *J. Geophys. Res.* 97:19565–79. https://ntrs.nasa.gov/citations/19930037356.

Boetto, Giulia. 2010. "9. Fishing Vessels in Antiquity: The Archaeological Evidence from Ostia." In Tønnes Bekker-Nielsen and Darío Bernal Cassala (eds.), *Ancient Nets and Fishing Gear, Proceedings of the International Workshop on "Nets and Fishing Gear in Classical Antiquity: A First Approach*, 243–55 (Cadiz, 2007, November 15–17). Servicio de Publicaciones de la Universidad de Caìdiz.

Borg, John E. 1986, March 28 (DTIC date stamp). "Magnus Effect—An Overview of Its Past and Future Practical Applications." 2 Vols. Contract N00024–83-C-5350. Naval Sea Systems Command.

Borwick, John. 2012. *Loudspeaker and Headphone Handbook.* Taylor & Francis.

Boston, City of. 2017, May 9. "Seaport Shipwreck." https://www.boston.gov/departments/archaeology/seaport-shipwreck.

Boudriot, Jean. 1986. *The Seventy-Four Gun Ship: Fitting Out the Hull.* Naval Institute Press.

Boughton, Claudius V. 1898, January 16 (issue date). US Patent 597,636, "Signal-Telegraph."

Bourne, John. 1855. *A Treatise on the Screw Propeller, with Various Suggestions of Improvement* (2nd ed., rev.). Longman, Brown, Green, and Longmans.

Bowditch, Nathaniel. 1826. *The New American Practical Navigator* (6th stereotype ed.). Edmund M. Blunt.

———. 1962 (revised by Hydrographic Office). *American Practical Navigator: An Epitome of Navigation.* Hydrographic Office Publ. No. 9. US Navy Hydrographic Office.

Boyd, John McNeill. 1857. *A Manual for Naval Cadets.* Longman & Co.

Bradford, Sir Edward Eden. 1923. *Life of Admiral of the Fleet Sir Arthur Knyvet Wilson.* E.P. Dutton & Co.

Brady, William N. 1864. *The Kedge-Anchor, Or, Young Sailors' Assistant,* etc. (18th ed.) Published by the author.

Braly, Edmund. 1973, July 24 (issue date). U.S. Patent 3,747,139, "Buoyancy Compensation."

Branson, Edward J. 2008. *Fish Welfare.* Wiley.

Breckenridge, F.C. 1967, April 3. *Colors of Signal Lights: Their Selection, Definition, Measurement, Production and Use.* NBS Monograph 75. National Bureau of Standards.

Bremner, David. 1869. *The Industries of Scotland: Their Rise, Progress, and Present Condition.* A. and C. Black.

Bretagnon, P., and G. Francou. 1988. "Planetary Theories in Rectangular and Spherical Variables. VSOP Solutions." *Astronomy and Astrophysics* 202: 309–15.

Brewer, Benjamin B., and Barton B. Ward. 1879, June 3 (issue date). U.S. Patent 216,140, "Improvement in Hydraulic Propellers. "

Brewster, Hugh. 2012. *Gilded Lives, Fatal Voyage: The* Titanic's *First-Class Passengers and Their World.* Crown.

Bridger, Capt. Herbert H., and Oswald Martin Watts. 1927. *The Stowage of Cargo: A Practical Treatise on the Description, Handling and Stowage of Cargo.* Imrae, Laurie, Norie & Wilson, Ltd.

BrightHub Engineering. 2022. "The Limitations of Single Stage Air Compressors." https://www.brighthubengineering.com/hvac/63720-the-operation-of-air-compressors-part-two.

British Broadcasting Corporation. 2024. "Lung

Volumes and Vital Capacity." https://www.bbc.co.uk/bitesize/guides/z3xq6fr/revision/2.
British Optical Instrument Manufacturers' Association. 1921. *Dictionary of British Scientific Instruments.* Constable and Company Ltd.
Brooks, Randall C. 1999, Fall. "A Problem of Provenance: A Technical Analysis of the 'Champlain' Astrolabe." *Cartographica* 36(3): 1–16.
Brown, Lloyd A. 1977a [1949]. *The Story of Maps.* Dover.
Brown, W.E. 1977, February. *Friction Coefficients of Synthetic Ropes.* Naval Undersea Center.
Brownie's Third Lung. 2024, January 18 (access date). "Sea LiOn™ 3+ Hours Battery System." https://www.browniedive.com/sea-lion-battery-powered-tankless-diving-system.
Brownlie, David. 1923, November. "Marine Steam Generation." *Combustion* 9(5): 375–382, 387.
Brunel, Marc Isambard. 1801, March 9. British Patent 2478 of 1801,"A New and Useful Machine for Cutting One or More Mortices Forming the Sides of and Cutting the Pin-Hole of the Shells of Blocks, and for Turning and Boring the Shivers, and Fitting and Fixing the Coak."
Bruzelius, Lars. 1995a, March 5. "The Tiller and Whipstaff." http://www.bruzelius.info/Nautica/Shipbuilding/Whipstaff-Use.html
Bruzelius, Lars. 1995, May 13. "Dockyard Models." http://www.bruzelius.info/Nautica/Shipmodels/Dockyard_Models/dock1600x.html.
_____. 1998, June 26. "17th Century Navy Board Style Models." http://www.bruzelius.info/Nautica/Shipmodels/dock1700.html.
_____. 2001, April 5. "Wasa." http://www.bruzelius.info/Nautica/Ships/War/SE/Wasa(1627).html.
Buffington, Charles W. 1873, December 9. US Patent 145,394. "Steering Apparatus for Vessels."
Bulloch, J.M. 1899. "The Romance of a Diver's Life." *The New Illustrated Magazine,* May 21, 195–205.
Bunker, John. 1982, November. "Sailing the Magnetic Fields." *Surveyor.* https://publicationsonline.carnegiescience.edu/legacy/exhibits/ault_exhibition/bibl_sailing.html
Burch, Ernest S. 2006. *Social Life in Northwest Alaska: the Structure of Inupiaq Eskimo Nations.* University of Alaska Press.
Bureau of Maritime Inspection and Navigation. 1936, July. "Equipment Approved." *Merchant Marine Bulletin* 1(1): 20–21.
Bureau of Naval Personnel. 1946, June. *Submarine Electrical Installations.* NAVPERS 16162. https://maritime.org/doc/fleetsub/elect/chap17.php.
_____. 1953. *Manual for Buglers, U.S. Navy.* Navy Training Courses, NAVPERS 10137-B (Original ed. 1919; reprinted with changes 1953).
_____. 1958. "Machinist's Mate 3." Nav Training Course NAVPERS 10522. GPO.
_____. 1961. *Naval Communications.* Navy Training Course NAVPERS 10898. GPO.
_____. 1962. *Military Sea Transportation Service* (major revision). Navy Training Course NAVPERS 10829-A, -B. GPO.
_____. 1964. "Boatswain's Mate 3 & 2." Navy Training Course NAVPERS 10121-D (3rd ed.). GPO.
_____. 1965. "Tradevman 3 & 2." Navy Training Course NAVPERS 10376-B. GPO.
_____. 1966. "A Navigation Compendium." Navy Training Course NAVPERS 10494. GPO.
_____. 1970. *Principles of Naval Engineering.* NAVPERS 10788-B. GPO.
Burger, W., and A.G. Corbet. 2016 [1963]. *Marine Gyro-Compasses and Automatic Pilots: A Handbook for Merchant Navy Officers.* Volume 1, Gyro Compasses. Elsevier Science.
Burney, William. 1815. *A New Universal Dictionary of the Marine.* (Also titled *Falconer's Marine Dictionary.*) T. Cadell and W. Davies; J. Murray.
Burningham, Nick. 2001. "Learning to Sail the Duyfken Replica." *Intl. J. Nautical Archaeology* 30(1): 74–85.
Burns, R.W. 2004. *Communications: An International History of the Formative Years.* Institution of Engineering and Technology.
Buxton, Ian Lyon, et al. 2012. *Cargo Access Equipment for Merchant Ships.* Springer US.
Caiella, J.M. 2019, August. "Technological Dead End." *Naval History Magazine* 33(4). https://www.usni.org/magazines/naval-history-magazine/2019/august/technological-dead-end.
Calahan, H.A. 1952. *The Sky and the Sailor: A History of Celestial Navigation.* Harper.
Canadian Centre for Occupational Health and Safety. 2022, December 20. "Pushing and Pulling: General." https://www.ccohs.ca/oshanswers/ergonomics/push1.html.
Canas, Antonio Costa. 2017. "The Astronomical Navigation in Portugal in the Age of Discoveries." *Cahiers François Viète [Online]* II-3. https://doi.org/10.4000/cahierscfv.752
Carlton, John S. 2011. *Marine Propellers and Propulsion* (2nd ed.). Elsevier Science.
Casey, Emily Clare. 2016, Summer. "Mapping Ocean Currents." *Open Rivers Journal—Rethinking Water, Place & Community.* https://libpubsdss.lib.umn.edu/openrivers/article/mapping-ocean-currents.
Casson, Lionel. 1965. "Harbour and River Boats of Ancient Rome." *J. Roman Studies* 55: 31–39.
_____. 1991. *The Ancient Mariners: Seafarers and Sea Fighters of the Mediterranean in Ancient Times* (2nd ed.). Princeton University Press.
Cayley, Arthur. 1806. *The Life of Sir Walter Raleigh, Knight,* Vol. 2. Cadell and Davies.
Cerbule, Kristine, et al. 2022, May. "Comparison of the Efficiency and Modes of Capture of Biodegradable Versus Nylon Gillnets in the Northeast Atlantic Cod (*Gadus morhua*) Fishery." *Marine Pollution Bull.* 178: article 113618. https://www.sciencedirect.com/science/article/pii/S0025326X22003009
Chaffin, Tom. 2008. *The H.L. Hunley: The Secret Hope of the Confederacy.* Hill and Wang.
Chakraborty, Soumya. 2021, August 13. "How

Does a Rudder Help in Turning a Ship." https://www.marineinsight.com/naval-architecture/rudder-ship-turning.

Chalmers, Thomas Wightman. 1920. *The Gyroscopic Compass: A Non-Mathematical Treatment.* Constable.

Chatham 1847. Chatham Committee of Naval Architects. 1847. *Reports on Naval Construction, 1842–44.* W. Clowes and Sons.

Chatterton, Edward Keble. 1913. *Ships & Ways of Other Days.* Sidgewick & Jackson, Ltd.

Chief Signal Officer. 1890. *Annual Report of the Chief Signal Officer of the Army to the Secretary of War for the Year 1889, Part I.* GPO.

_____. 1938, December 5 [reprinted July 15, 1940]. *Instruction Book for Telephone TP-3-T1 (Sound Powered).* GPO.

Chilton, Frederick Ernest. 1923. *Electric Cranes and Hauling Machines.* Pitman & Sons, Limited.

Choudhury, Asim Kumar Roy. 2014. *Principles of Colour and Appearance Measurement.* Woodhead Publishing.

Christensen, Rob. 2010, June 22. "What Is the Average Jaw Opening?." https://www.craniorehab.com/average-jaw-mouth-opening.

Church, Irving Porter. 1889. *Mechanics of Engineering (Fluids): A Treatise on Hydraulics and Pneumatics, for Use in Technical Schools.* Wiley.

Clark, Daniel Kinnear. 1891. *A Manual of Rules, Tables and Data for Mechanical Engineers* (6th ed.). D. Van Nostrand.

Clark, John R., et al. 1959. "Development and Operation of Television for Studying Fish Behavior in Otter Trawls." Special Scientific Report—Fisheries No. 320. U.S. Fish and Wildlife.

Clarke, Arthur C. 1951, August. "Superiority." *The Magazine of Fantasy and Science Fiction* 2(4): 3–11. https://archive.org/details/Fantasy_Science_Fiction_v002n04_1951-08_AK/page/n11/mode/2up?view=theater.

Clarke, John R. 2003, May. "Gas Flows Supporting Umbilical Diving—Requirements and Measurements." https://www.researchgate.net/publication/235125932_Gas_Flows_Supporting_Umbilical_Diving_-_Requirements_and_Measurements.

Cline, Duane A. 1999, October 28. "The Back-Staff." www.rootsweb.com/~mosmd/backstaf.htm.

Collins, J.W. 1890. *Report on the Construction and Equipment of the Schooner Grampus.* GPO.

Colomb, Captain, and Major Bolton. 1870. *The System of Flashing Signals Adopted in Her Majesty's Army and Navy* (alternatively titled *Colomb & Bolton's Flashing Signals*). W. Nunn & Co.

Colton, Tim. 2010, March 12. "C2 Cargo Ships." http://shipbuildinghistory.com/merchantships/2c2cargoships.htm.

Conover, Emily. 2023, May 11. "Science Explains Why Shouting into the Wind Seems Futile." *Science News.* https://www.sciencenews.org/article/science-explains-shout-wind-acoustics.

Cookman, Mark S. 2001. "Navigational Cross-Staff." https://web.archive.org/web/20130513222432/http://www.florilegium.org/files/TRAVEL/Nav-Crosstaff-art/Nav-Crosstaff-art.html

Cooper, Carolyn C. 1984, April. "The Portsmouth System of Manufacture." *Technol. & Culture* 25(2): 182–225.

Cooper, David J., et al. 1991. *By Fire, Storm, and Ice: Underwater Archeological Investigations in the Apostle Islands.* State Historical Society of Wisconsin.

Cooper, Iver P. 2024. *Poseidon's Progress: The Quest to Improve Life at Sea.* McFarland.

_____. 2024a. *Arming the Warship: Naval Weapons Technology and Gunnery from the Spanish Armada to the Cold War.* McFarland.

Corbett, Julian, Stafford. 1908. *Signals and Instructions 1776–1794.* Navy Records Society.

Corbett, Julian Stafford. 1889, *Drake and the Tudor Navy,* Vol. 2. Longmans.

_____. 1905. *Fighting Instructions 1530–1816.* Navy Records Society.

Coston, Martha. 2024, January 18 (access date). "Night Signals." http://www.civilwarsignals.org/pages/signal/signalpages/flare/coston.html.

Coston, Martha (administratrix). 1859, April 5 (issue date). US Patent 23,536, "Pyrotechnic Night-Signals."

Countiss, K.W. 1898. "Naval Marine Electric Lighting." *American Electrician* 10(6): 282–7.

Cousins, Tom. 2022. "The Rigging of HMS Invincible." (Master of Research, Bournemouth University.) https://eprints.bournemouth.ac.uk/37484/1/COUSINS%252C%2520Thomas_M.Res_2022.pdf.

Cowles, William S. 1907. "Report of the Chief of the Bureau of Equipment." (October 1, 1906), in *Annual Reports of the Navy Department for the Year 1906,* 337–384. GPO.

Craddock, David. 2021. *What Ship, Where Bound?: A History of Visual Communication at Sea.* Seaforth Publishing.

Craik, A.D.D. 2018, April 11. "The Hydrostatical Works of George Sinclair (c1630–1696): Their Neglect and Criticism." *Notes and Records* 72: 239–273.

Crane, Nicholas. 2010. *Coast: Our Island Story.* Ebury Publishing.

Cresy, Edward. 1865. *An Encyclopaedia of Civil Engineering, Historical, Theoretical, and Practical,* Volume 1. Longman, Brown, Green, and Longman.

Crowhurst, N.H. 2010, November 2. "Horn Shapes." http://www.vias.org/crowhurstba/crowhurst_basic_audio_vol1_049.html.

Cui, Weicheng, et al. 2022. *Encyclopedia of Ocean Engineering.* Springer Nature Singapore.

Culver, Henry B. 1924. *The Book of Old Ships.* Garden City Publishing.

Cunliffe, Barry W., et al. 2002. *Archaeology: The Widening Debate.* British Academy.

Cunningham, Brysson. 1924. *Cargo Handling at Ports: A Survey of the Various Systems in Vogue, with a Consideration of Their Respective Merits.* Chapman & Hall, Ltd.

Cunningham, H.D.P. 1862, "On the Rig and Sails of Steam Ships of War." *Trans. Institution Naval Architects* 3: 98–109.

Cunningham, Henry Duncan Preston. 1850, November 30. British Patent 13,368 of 1850. "Reefing Sails."

———. 1853. *Cunningham's Patent Mode of Reefing Topsails.* W. Woodward.

———. 1855, July 20. British Patent 1,640 of 1855. "Reefing Sails." https://www.google.com/books/edition/English_Patents_of_Inventions_Specificat/nQcADAPqbO0C?hl=en&gbpv=1&dq=cunningham++reefing+patent&pg=RA11-PA4&printsec=frontcover.

———. 1856, January 15 (issue date). U.S. Patent 14,094. "Reefing Sails."

D'Antonio, Steve. 2018, July 12. "The Rudimentaries of Rudders." https://www.proboat.com/2018/07/the-rudimentaries-of-rudders.

Dahl, Cmdr. Erik J. 2001. "Naval Innovation. From Coal to Oil." https://apps.dtic.mil/sti/citations/ADA524799.

Dana, Richard Henry. 1841. *The Seaman's Manual, Containing a Treatise on Practical Seamanship.* Edward Moxon.

Danton, Graham. 1996. *Theory and Practice of Seamanship.* Routledge.

Das, Biman, and Yangqing Wang. 2004. "Isometric Pull-Push Strengths in Workspace: 1. Strength Profiles." *Intl. J. Occupational Safety & Ergonomics* 10(1): 43–58. DOI: 10.1080/10803548.2004.11076594

Datawave Marine Systems. 2021. "Strength and Stiffness: Design of Structural Foundations." https://www.dmsonline.us/strength-and-stiffness-design-of-structural-foundations.

David, Elizabeth, and Jill Norman. 1995. *Harvest of the Cold Months: The Social History of Ice and Ices.* Viking.

Davids, Karel. 2022. *The Republic of Skill, Artisan Mobility, Innovation, and the Circulation of Knowledge in Premodern Europe.* Brill.

Davies, J.D. 2008. *Pepy's Navy: Ships, Men & Warfare, 1649–1689.* Seaforth Publishing.

———. 2017. *Kings of the Sea: Charles II, James II and the Royal Navy.* Pen & Sword Books.

Davies, Stephen. 2014. *East Sails West: The Voyage of the Keying, 1846–1855.* Hong Kong University Press.

Davis, Charles G. 2012 [1926]. *The Ship Model Builder's Assistant.* Dover.

Davis, Gary, and Ralph Jones. 1989. *The Sound Reinforcement Handbook.* Hal Leonard.

Davis, John. 1595. *The Seamans Secrets.* Thomas Dawson. Transcription by the Early English Books Text Creation Partnership. https://quod.lib.umich.edu/e/eebo/A19937.0001.001?view=toc.

———. 1595a. "Of the Crosse Staff and His Demonstration." Id., https://quod.lib.umich.edu/e/eebo/A19937.0001.001/1:7.57?rgn=div2;view=fulltext.

———. 1595b. "How Is the Use of This Staffe?" https://quod.lib.umich.edu/e/eebo/A19937.0001.001/1:7.58?rgn=div2;view=fulltext.

———. 1595c. "What Is Great Circle Navigation?" https://quod.lib.umich.edu/e/eebo/A19937.0001.001/1:7.41?rgn=div2;view=fulltext.

———. 1626. *The Seamans Secrets* ("newly corrected and amended"). John Dawson. https://www.google.com/books/edition/The_Seamans_Secrets_deuided_into_2_parte/kMtlAAAAcAAJ?hl=en&gbpv=1&dq=davis+seaman+secrets&printsec=frontcover.

Davis, P. 2024, January 18 (access date). "HMS Agincourt (1865)." https://www.pdavis.nl/ShowShip.php?id=1026

Davis, Sir Robert Henry. 1908. *Diving Scientifically and Practically Considered: Being a Manual and Handbook of Submarine Appliances.* Siebe.

Dawson, Kevin. 2006, March. "Enslaved Swimmers and Divers in the Atlantic World." *J. American History* 92(4): 1327–55. https://www.jstor.org/stable/4485894.

DeBlois, Captain E.T. 1881. "The Origin of the Menhaden Industry." *Bull. U.S. Fish Commission* 1: 46–51.

De Decker, Kris. 2008, August 24. "A Steam Powered Submarine: The Ictíneo." https://solar.lowtechmagazine.com/2008/08/a-steam-powered-submarine-the-ictineo.

Deep Sea Diving School, US Naval Weapons Plant. 1960, September. "Ship Salvage Notes: Used in the Course of Instruction for Salvage Officers and First Class Divers." https://books.google.com/books/download/Ship_Salvage_Notes.pdf.

DeFrangesco, Ralph, and Stephanie DeFrangesco. 2022. *The Big Book of Drones.* CRC Press.

De Hilster, Nicolàs. 2011. "The Early Development of the Davis Quadrant." *Bull. Sci. Instrument Soc'y* 110: 14–22.

———. 2014, August 7. "Observational Methods and Procedures for the Mariner's Astrolabe." *The Mariner's Mirror* 100(3): 261–81. DOI: 10.1080/00253359.2014.935141.

Dekker, David L. 2024, January 18 (access date). "1605. Jan Adriaansz. Leeghwater." https://www.divinghelmet.nl/divinghelmet/1605_Jan_Adriaansz._Leeghwater.html

———. 2024a, January 18 (access date). "1839. Augustus Siebe." https://www.divinghelmet.nl/divinghelmet/1839_Augustus_Siebe.html

Del Rosso, Pietro. 2024, February 16 (access date). *Maritime English: A Comprehensive & Updated Maritime English C.L.I.L. (Content and Language Integrated Learning) Handbook for Deck Officers in Conformity with I.M.O.* Vincenzo Cappai.

Delgado, James P., and Clive Cussler. 2011. *Silent Killers: Submarines and Underwater Warfare.* Bloomsbury Publishing.

De Moor, Janny. 1998. "In the Beginning Was Fish. Fish in the Ancient Near East." In Harlan Walker (ed.), *Fish: Food from the Waters,* 84–93. Prospect Books.

Dempsey, Lt. Cmdr. J.K. 2018, January 31 (pdf creation date). "The Evolution of Signalling at Sea by Flags." www.commsmuseum.co.uk/publications/evolution.pdf.

De Russett, Edwin William, 1899. "Recent Developments in Cargo-Steamers." *Minutes Proc. Inst. Civil Engineers* 138: 450–7.

Desmond, Charles. 1915. *Naval Architecture Simplified: Explained in Non-Technical Terms.* Rudder Publishing Company.

Desurvire, Emmanuel. 2009. *Classical and Quantum Information Theory: An Introduction for the Telecom Scientist.* Cambridge University Press.

Deutsches Historic Museum. 2024, February 4 (access date). https://www.dhm.de/mediathek/en/ship-types/milestones-in-the-history-of-european-shipbuilding/09-fluyt.

Devanney, Richard. 2017, Janurary 21. "The Difference Between Scuba Diving Gas Mixes." https://scubadiverlife.com/difference-scuba-diving-gas-mixes.

DeVoy, Louise. 2023, October 30. "What Is a Nocturnal?" https://www.rmg.co.uk/stories/blog/astronomy/what-nocturnal.

Dinsdale, J. 1974, May. "Horn Loudspeaker Design—2." *Wireless World.* https:/www.volvotreter.de/downloads/Dinsdale_Horns_2.pdf.

dipndive. 2023. "Heat Loss and Hypothermia When Diving." https://dipndive.com/blogs/scuba-health-and-safety/heat-loss-and-hypothermia-when-diving.

divingheritage.com, 2005, January 11. "Rotary Pumps." https://www.divingheritage.com/rotarypumps.htm.

_____. 2007, June 18. "Lever Pumps." https://www.divingheritage.com/leverpumps.htm.

_____. 2024, April 11 (access date). "Diving Bells and Observation Chambers." https://www.divingheritage.com/chambers.htm.

_____. 2024a, April 11 (access date). "Modern Diving Bells and Chambers." https://www.divingheritage.com/bells.htm.

Dowell, Vice-Admiral William Montagu. 1882. "On Naval Tactics." *J. Royal United Service Institution* 25:350–73.

Downing, John A. 2024. "Hypothermia: Understanding and Prevention." Minnesota Sea Grant. https://seagrant.umn.edu/programs/recreation-and-water-safety-program/hypothermia#graphics.

DPA Microphones. 2021, March 3. "Facts About Speech Intelligibility." https://www.dpamicrophones.com/mic-university/facts-about-speech-intelligibility.

Dreadnought Project. 2013, November 21. "Evershed Helm Indicator." http://www.dreadnoughtproject.org/tfs/index.php/Evershed_Helm_Indicator.

Dubiago, Alexander D. 1961. *The Determination of Orbits.* MacMillan.

Du Bois, Augustus Jay, 1902. *Kinematics, Statics, Kinetics, Statics of Rigid Bodies and of Elastic Solids.* Wiley.

Duffett-Smith, Peter. 1988. *Practical Astronomy With Your Calculator.* Cambridge University Press.

Dugatkin, Lee Alan. 2013, January–February. "The Evolution of Risk-Taking." Cerebrum. https://www.ncbi.nlm.nih.gov/pmc/articles/PMC3600861.

Dunkerley, Stanley. 1905. *Mechanism.* Longmans, Green, and Co.

Dunn, Ambrose C. 1889, June 4 (issue date). US Patent 404,472. "Steering Apparatus." (Application filed July 27, 1887.)

Earle, Peter. 2008. *Treasure Hunt: Shipwreck, Diving, and the Quest for Treasure in an Age of Heroes.* St. Martin's Publishing Group.

Earle, Ralph. 1912. "The Origin of Our Signal Book." *US Naval Institute Proceedings* 38: 1037–40.

Echeverria, Virginia Iommi. 2011, December. "Hydrostatics on the Fray: Tartaglia, Cardano and the Recovering of Sunken Ships." *Brit. J. Hist. Sci.* 44(4): 479–91. https://philpapers.org/rec/ECHHOT-2.

Eckerton, E.A., et al. 1926, November 8. "Transmission of Sound Through Voice Tubes." Technologic Papers of the Bureau of Standards No. 333. GPO.

Edgerton, William F. 1927, July. "Ancient Egyptian Steering Gear." *Amer. J. Semitic Languages* 43(4): 255–265.

Edinburgh Encyclopedia. 1830. 18 Vols. William Blackwood et al.

Edles, Henry. 1799. *Signal-Book for the Ships of War.* University of Rhode Island, Special Collections (Miscellaneous Paper 15). http://digitalcommons.uri.edu/sc_pubs/15.

Ehrhart, Josef. 1935, September 24 (issue date). U.S. Patent 2,015,514, "Device for the Control of Motion of Movable Blades on Blade Wheels." (Application filed March 23, 1935.)

Ekirch, Roger. *At Day's Close: Night in Times Past.* W.W. Norton and Company.

El-Hawary, Ferial. 2000. *The Ocean Engineering Handbook.* CRC Press.

Electrical Engineer, The. 1895, November 20. "The Boughton Telephotos for Long-Distance Visual Signalling." *The Electrical Engineer* 20 (394): 489–90.

_____. 1896, September 4. "Some Electrical Features of the United States Battleship 'Indiana.'" *The Electrical Engineer* 18 (New Series), 263–65.

Electrical Review. 1919, September 6. "Interesting Electrical Features of the Battleship New Mexico," *Electrical Review,* 75 (10): 398–99.

Eliav, Joseph. 2015, January. "Guglielmo's Secret: The Enigma of the First Diving Bell Used in Underwater Archaeology." *Int. J. Hist. Eng'g & Technol.* 85: 60–69. https://www.researchgate.net/publication/273293818.

Elliot, Vice-Admiral G. 1868. "The Hydraulic Propeller as a Motive Power for Ships." *J. Royal United Service Institution,* 9(47): 589–611.

Emerson, James. April 17, 1855, US Patent 12,718. "Ship Windlass."

Emerson, James. August 28, 1855. US Patent 13,506, "Windlass."

Emerson, James. June 17, 1856. US Patent 15,123, "Windlass."

Emerson, James. 1894. *Treatise Relative to the Testing of Water-wheels and Machinery* (6th ed.). James Emerson.

Encyclopaedia Britannica. 1911 (11th ed.). "Bellows and Blowing Machines." *Encyclopaedia Britannica* 3: 705–10.

———. 1911 (11th ed.). "Compass." *Encyclopaedia Britannica* 6: 804–9.

———. 1911 (11th ed.). "Divers and Diving Apparatus." *Encyclopaedia Britannica* 8: 326–31.

———. 1911 (11th ed.). "Magnetism, Terrestrial." *Encyclopaedia Britannica* 17: 353–85.

———. 1911 (11th ed.). "Navigation." *Encyclopaedia Britannica* 19: 284–98.

———. 1911 (11th ed.). "Planet." *Encyclopaedia Britannica*.

———. 1911 (11th ed.). "Refraction." *Encyclopaedia Britannica* 23: 25–29.

———. 1911 (11th ed.). "Rope and Rope-Making." *Encyclopaedia Britannica* 23: 713–18.

———. 1911 (11th ed.). "Ship." *Encyclopaedia Britannica* 24: 860–922.

———. 1911 (11th ed.). "Sounding." *Encyclopaedia Britannica* 25: 460–62.

Endsor, Richard. 2020. *The Master Shipwright's Secrets: How Charles II Built the Restoration Navy.* Bloomsbury Publishing.

Engelard, Georg H. 2005, August. "Catalogue of Defra Historical Catch and Effort Charts: Six Decades of Detailed Spatial Statistics for British Fisheries." Science Series Technical Report No. 128. https://www.researchgate.net/publication/242189545_Catalogue_of_Defra_historical_catch_and_effort_charts_six_decades_of_detailed_spatial_statistics_for_British_fisheries.

Engineering Toolbox, The. 2004. "Illuminance—Recommended Light Levels." https://www.engineeringtoolbox.com/light-level-rooms-d_708.html.

———. 2009."Manila Ropes—Strengths." https://www.engineeringtoolbox.com/manila-rope-strength-d_1512.html; "Nylon Ropes—Strengths." https://www.engineeringtoolbox.com/nylon-rope-strength-d_1513.html; "Polyester Polyolefin Ropes—Strengths." https://www.engineeringtoolbox.com/polyester-polyolefin-rope-strength-d_1515.html; "Polyester Ropes—Strengths." https://www.engineeringtoolbox.com/polyester-rope-strength-d_1514.html; "Polypropylene Fiber Ropes—Strengths." https://www.engineeringtoolbox.com/polypropylene-rope-strength-d_1516.html; "Wire Ropes—Strengths." https://www.engineeringtoolbox.com/wire-rope-strength-d_1518.html.

———. 2018. "Maximum Force or Power Available for Human Operated Machines." https://www.engineeringtoolbox.com/human-force-power-d_2086.html.

English Mechanic and Mirror of Science. 1868, July 31. "Improvement in Steering Apparatus for Vessels." *English Mechanic and Mirror of Science* 7(175): 397.

Eros, Terri. 2017. "Kalmar Nyckel: Using a 17th Century Dutch Pinnace to Teach Physics and More." https://www.dti.udel.edu/content-subsite/Documents/2016-units/T. Eros—Kalmar Nyckel Using a 17th century—Unit.pdf.

Espenak, Fred, and Jean Meeus. 2011, May 23. "Lunar Eclipses: 1601–1700." https://eclipse.gsfc.nasa.gov/LEcat5/LE1601-1700.html.

Evans, John M. 2006, June 15. "Standards for Visual Acuity." Report prepared for Elena Messina, Intelligent Systems Division, National Institute for Standards and Technology under Contract 2006-02-13-01 through KT Consulting. https://www.nist.gov/system/files/documents/el/isd/ks/Visual_Acuity_Standards_1.pdf.

Evelpidou, N., et al. 2018. "A Tentative Methodology of Sea level Change Based on Fish Tanks from Hellenistic Alexandria, Vis-a-Vis, the Submerged El Hassan Rock Provide a New Look for Subsidence Estimates." In C.S. Zerefos and M.V. Vardinoyannis (eds.), *Hellenistic Alexandria: Celebrating 24 Centuries*, 289–96. Archaeopress.

Eyres, David J. 2001. *Ship Construction.* Elsevier.

Eyton, Thomas Campbell. *A History of the Oyster and the Oyster Fisheries.* J. Van Voorst.

Faber, Henry Burnell. 1919. *Military Pyrotechnics: The Manufacture of Military Pyrotechnics*, Vol. 2. GPO.

Falconer, William. 1769. *An Universal Dictionary of the Marine.* T. Cadell.

———. 1784. *An Universal Dictionary of the Marine.* T. Cadell.

———. 1822. *The Shipwreck.* John Sharpe.

Fearne, Charles, et al. 1746. *Minutes of the Proceedings of a Court-martial, Assembled on the 23d of September, 1745 …* (publisher not stated).

Federal Aviation Administration. 2022, June 24. "Satellite Navigation—GPS—How It Works." https://www.faa.gov/about/office_org/headquarters_offices/ato/service_units/techops/navservices/gnss/gps/howitworks.

Feenker, Thomas M. 1985. *Naval Safety Supervisor*, NAVEDTRA 10808. GPO.

Feldman, S., and T. I. Monahan. 1962, February. "New Signaling Beacon." *Naval Research Revs.* 12–14.

Felkin, William. 1867. *A History of the Machine-wrought Hosiery and Lace Manufactures.* Longmans, Green, and Company.

Fernö, Anders, et al. 2020. *The Welfare of Fish.* Springer International.

Fessel, H.E. 1858, December 28. US Patent 22,417, "Steering-Propeller."

Ficken, N.L., and Mary C. Dickerson. 1969, July. "Experimental Performance and Steering Characteristics of Cycloidal Propellers." Report 2983. Naval Ship Research and Development Center, Navy Dept.

Fieldphones. 1922, April 5. "U.S. Signal Corps

Field Telephones Timeline." https://fieldphones.org/004-us-fieldphones.

Fillingane, Walter B., and Warren C. Williams. 1993. *Seaman.* NAVEDTRA 12016. GPO.

Fincham, John. 1851. *A History of Naval Architecture.* Whittaker & Co.

Findlay, Gordon D. 2005. *My Hand on the Tiller.* AuthorHouse UK.

Fine, John Christopher. 2006. *Treasures of the Spanish Main: Shipwrecked Galleons In The New World.* Lyons Press.

Fitch, Charles H. 1881. *Report on Marine Engines and Steam Vessels in the United States Merchant Service.* Publisher not identified.

Flettner, Anton. 1926. *The Story of the Rotor.* F.O. Willhöfft.

_____. 1928, February 28 (issue date). U.S. Patent 1,661,115, "Rudder."

Foerster, Ernst. 1923, October. "A Rudder That Turns Itself." *Sci. Am.* 129(4): 237. https://www.jstor.org/stable/10.2307/24974674.

Foley, Michael. 2021. *Britain's Railways in the First World War.* Pen & Sword Books.

Ford, Guy Stanton. 1922. "Compass." In *Compton's Pictured Encyclopedia.* F.E. Compton Company.

Forest Products Laboratory, Forest Service, U.S. Dept. Agriculture. 2010, April. *Wood Handbook: Wood as an Engineering Material.* Agriculture Handbook No. 72. Forest Service.

Fork, Werner, and Birgit Jürgens. 2002. "How the VSP Works." In *The Fascination of the Voith-Schneider Propeller: History and Engineering,* 116–123. Koehler.

Fornshell, John A., and Alessandra Tesei. 2012. "The Development of SONAR as a Tool in Marine Biological Research in the Twentieth Century." *Int. J. Oceanography.* https://www.hindawi.com/journals/ijocean/2013/678621.

Forrester, John M. 2013 [Cardano, 1550]. *The De Subtilitate of Girolamo Cardano.* Arizona Center for Medieval and Renaissance Studies.

Franklin, John. 1989. *Navy Board Ship Models, 1650–1750.* Naval Institute Press.

Franklin Institute, The. 1866, December. "The Water-Jet Propeller." *J. Franklin Institute* 52 (3rd Ser.)(Whole No. 81): 392–96, and "H.M.S. Water-Witch." *J. Franklin Institute* 52 (3rd Ser.) (Whole No. 81): 396–98. (Both reprinted from *London Engineering,* No. 43.)

Freiesleben, H.C. 1950. "Investigations into the Dip of the Horizon." 270–79. fer3.com/arc/imgx/Dip-of-the-horizon-Freiesleb.pdf.

Friedman, Norman. 2014. *Naval Anti-Aircraft Guns and Gunnery.* Naval Institute Press.

Frost, Scyllias. 1968, October. "Diving in Antiquity." *Greece & Rome* 15(2): 180–85.

Fulton, Alasdair. 2005. "Investigation into the Effect of Rope to Rope Contact for a Winch in a Multi-Turn Configuration." https://personal.strath.ac.uk/andrew.mclaren/AlasdairFulton2005.pdf.

Furuno Electric Co., Ltd. 2014. "How Does Sonar Work." https://www.furuno.com/en/technology/sonar/basic.

Gabriel, Otto, et al. 2005 (4th ed.). *Von Brandt's Fish Catching Methods of the World.* Wiley.

Gakuran, Michael. 2013, November 5. "Ama—The Pearl Diving Mermaids of Japan." http://gakuran.com/ama-the-pearl-diving-mermaids-of-japan.

Gao, Xiaoping, et al. 2015. "An Improved Capstan Equation Including Power-Law Friction and Bending Rigidity for High Performance Yarn." *Mechanism Machine Theory* 90: 84–94.

Garcia, Jorge David. 2017, June. "Sodium Chlorate Oxygen Generation for Fuel Cell Power Systems." MS thesis, Massachusetts Institute of Technology. https://dspace.mit.edu/handle/1721.1/112489.

Gardener's Chronicle. 1847, June 12, p. 386.

Garry, R.C. 1930. "The Factors Determining the Most Effective Push or Pull Which Can Be Exerted by a Human Being on a Straight Lever Moving in a Vertical Plane." *Arbeitsphysiologie* 3, 330–346. https://doi.org/10.1007/BF02015064.

Geurts, Collin. 2024, January 23 (access date). "Evolution of Swim Flippers." https://www.sutori.com/story/evolution-of-swim-flippers—gTLgeav2xNSYFjCQVPbTjKLw.

Gibbons, Lt. J.H. 1897. "Operations of the Naval Militia." In Theodore Roosevelt, *Annual Reports of the Navy Department for the Year 1897.* GPO.

Gibbs, D.D. 1971. "The Physician's Pulse Watch," 187–90. https:/www.cambridge.org/core/services/aop-cambridge-core/content/view/A336EA45F583AEAFC6DDADBF2C8A5172/S0025727300016409a.pdf/the-physicians-pulse-watch.pdf.

Giesen, Jorgen. 2019, April 25. "Astronomical Algorithms." http://www.jgiesen.de/elevaz/basics/meeus.htm.

Gilbert, K.R. 1965. *The Portsmouth Blockmaking Machinery: A Pioneering Enterprise in Mass Production.* HM Stationery Office.

Gingerich, Owen, and Barbara L. Welther. 1983. *Planetary, Lunar, and Solar Positions: New and Full Moons, A.D. 1650–1805.* American Philosophical Society.

Glynn, Joseph. 1867. *On the Construction of Cranes and Other Hoisting Machinery* (4th ed.). Virtue Brothers & Co.

Goldstone, Lawrence. 2017. *Going Deep: John Philip Holland and the Invention of the Attack Submarine.* Pegasus Books.

Goodeve, Thomas Minchin. 1880. *The Elements of Mechanism.* Longmans, Green, and Company.

Goodwin, Peter. 2002. *Nelson's Ships: A History of the Vessels in Which He Served, 1771–1805.* Stackpole Books.

Gordon, Amelie. 2020, November 18. "The Centuries-Long Saga of the 'Oyster Wars.'" https://boundarystones.weta.org/2020/11/18/centuries-long-saga-oyster-wars.

Gordon, Donald A. 1957, November. "A Survey of Human Factors in Military Night Operations (with Special Application to Armor)." George Washington University, Special Report 11. https://apps.dtic.mil/sti/citations/trecms/AD0149357.

Göteborg of Sweden. 2024, February 4 (access date). "About the Ship." https://www.gotheborg.se/about-gotheborg/about-the-ship; "The Sails." https://www.gotheborg.se/about-gotheborg/about-the-ship/#tab—the-sails.

Gougeon, Mede, and Ty Knoy. 1973. *Evolution of Modern Sailboat Design*. Winchester Press.

Gower, Richard Hall. 1796. *A Treatise on the Theory and Practice of Seamanship* (2nd ed.). G.G. & J. Robinson. https://archive.org/details/bim_eighteenth-century_a-treatise-on-the-theory_gower-richard-hall_1796.

_____. 1808. *A Treatise on the Theory and Practice of Seamanship* (3rd ed.). Wilkie and Robinson.

Graham, Bob. 1999, January [archive date December 26, 2019]. "Determination of Latitude by Francis Drake on the Coast of California in 1579." https://web.archive.org/web/20191226133619/www.longcamp.com/nav.html.

Graham, Joseph J., and Dorothy D. Stewart. 1958, December. "Estimating Maximum Fishing Depth of Longline Gear with Chemical Sounding Tubes." Special Scientific Report—Fisheries No. 285, U.S. Fish and Wildlife Service. https://spo.nmfs.noaa.gov/content/estimating-maximum-fishing-depth-longline-gear-chemical-sounding-tubes.

Gray, Edwyn. 2006 [2003]. *Disasters of the Deep: A Comprehensive Survey of Submarine Accidents & Disasters*. Pen & Sword Books.

Great Britain Board of Trade. 1894. "The International Code of Signals for the Use of All Nations."

Great Britain Naval Staff, Signal Division. 1918, November 21. *Handbook of Signalling*. O.U. 5041. The Division.

Great Britain Patent Office. 1903. Patent 7545 (Gensichen, E., and E. Ehrke. 1900, April 24) in *Patents for Inventions. Abridgments of Specifications. Class 113, Ships, Boats, and Rafts, Division I … Period—A.D. 1897–1900*. HM Stationery Office.

Greene, S. Dana. 1899. "Electricity in Marine Work: Its Growing Application in the United States." *Cassier's Magazine* 16: 207–22.

Greenhood, David. 1964. *Mapping*. University of Chicago Press.

Greenway, Ambrose. 2011 [2009]. *Cargo Liners: An Illustrated History*. Pen & Sword Books Ltd.

Greenwood, Jonathan. 1715. *The Sailing and Fighting Instructions Or Signals as They are Observed in the Royal Navy of Great Britain*. J. Greenwood.

Gregg, Thomas M. 1999, February 15. "The Commercial Code of Signals • 1857–1902." https://tmg110.tripod.com/sigf_2.htm.

_____. 2023. "Captain Frederick Marryat's Code of Signals for the Merchant Service, 1817–90." https://tmg110.tripod.com/sigf_1.htm.

_____. 2023a. "England Expects: Nelson's Trafalgar Signals, 21 October 1805." https://tmg110.tripod.com/british21.htm.

Gu, Weilie. 2022. *A General Introduction to Chinese Culture*. American Academic Press.

Guillemin, Amédée. 1877. *The Applications of Physical Forces*. Macmillan and Company.

Guilmartin, John Francis, Jr. 2003. *Gunpowder & Galleys: Changing Technology & Mediterranean Warfare at Sea in the 16th Century*. Conway Maritime Press.

Gunther, Robert Theodore. 1923. *Early Science in Oxford*. Oxford Historical Society.

Gurney, Alan. 2004. *Compass: A Story of Exploration and Innovation*. W.W. Norton.

Gustafson, P.R. 1962, January 23. "Chlorate-Candle Fabrication by Hot Pressing." NRL Report 5732. https://apps.dtic.mil/sti.tr/pdf/AD0272580.pdf.

Hall, William. 1904. *Modern Navigation*. W.B. Clive.

Hallen, Alf Elskil. 1965, November 23. U.S. Patent 3,219,197, "Derricks for Ships." See also Hallen, Alf Elskil. 1966, August 17. British Patent Specification 1,039,178, "Improvements in Derricks for Ships." Both Swedish priority, November 9, 1962.

_____. 1969, April 1. U.S. Patent 3,435,960, "Ship's Derrick for Container Handling." Swedish priority, September 2, 1966.

Halley, Edmund. 1714. "The Art of Living Under Water: Or, a Discourse Concerning the Means of Furnishing Air at the Bottom of the Sea, in Any Ordinary Depths." *Philosophical Transactions* 29: 492–99. https://royalsocietypublishing.org/doi/pdf/10.1098/rstl.1714.0063.

Hamilton, Alan M. 1986, July. "Richard Norwood, Part 1." *The Bermuda Beacon* 3(3): 24–27 http://homepages.rootsweb.com/~norwood/id50.htm.

_____. 1986a, October. "Richard Norwood, Part 2." *The Bermuda Beacon* 3(4): 35–39. http://homepages.rootsweb.com/~norwood/id51.htm.

Hammond, N. 1968, May. "A Study of the Factors in Cargo Gear Selection." *Ship and Boat* 21: 18–19.

Hannah, Lindsay. 2007. "Wind and Temperature Effects on Sound Propagation." *New Zealand Acoustics* 20(2): 22–29.

Hansen, Viveka. 2022, October 6. "Fishing Nets and Lines—Textile Observations by 18th Century Travelling Naturalists." https://www.ikfoundation.org/itextilis/fishing-nets-and-lines.html.

Happer, Richard, and Mark Steward. 2015. *River Forth: From Source to Sea*. Amberley Publishing.

Harbord, John Bradley, and H.B. Goodwin. 1897. *Glossary of Navigation: A Vade Mecum for Practical Navigators*. Griffin.

Harland, John. 2013a. "Capstans Handling Chain: Gordon and Barbotin." *The Mariner's Mirror* 99(3): 338–41. DOI: 10.1080/00253359.2013.816022.

Harland, John. 1985. *Seamanship in the Age of Sail*. Naval Institute Press.

_____. 1991. "The Design of Winches Used at Sea in the 1800s." *The Mariner's Mirror* 77(2): 151–165. DOI: 10.1080/00253359.1991.10656346.

———. 2003. *Capstans and Windlasses: An Illustrated History of Their Use at Sea.* Pier Books.

———. 2011. "The Whipstaff." *The Mariner's Mirror* 97(1): 103–26. DOI: 10.1080/00253359.2011.10709035.

———. 2013. "The Transition from Hemp to Chain Cable: Innovations and Innovators." *The Mariner's Mirror* 99:1, 72–85. DOI: 10.1080/00253359.2013.767000.

———. 2015, "The Evolution of the Windlass in the Nineteenth Century." *The Mariner's Mirror,* 101:1, 38–62 (2015). DOI: 10.1080/00253359.2015.994874.

Harper's Weekly. 1860, March 10. "The Steamer 'Peytona.'" *Harper's Weekly* 4:156.

Harrington, Purnell F. 1882. *Notes on Navigation and the Determination of Meridian Distances for the Use of Naval Cadets at the U.S. Naval Academy.* GPO.

Harris, Mike. 2010. *The Compass Book: Maintain, Repair and Adjust Your Own Compass.* Paradise Cay Publications.

Hayase, Shinzo. 2018, March. "Manila Hemp in World, Regional, National and Local History." *J. Asia-Pacific Studies (Waseda Univ.)* 31: 171–88. https://core.ac.uk/download/pdf/286959049.pdf.

Hayes, Isaac I. 1892. *Cast Away in the Cold: An Old Man's Story of a Young Man's Adventures As Related by Captain John Hardy, Mariner.* Lee and Shepard.

Hayes, Matthew. 2020, March 12. "Going Fishing: A Year in the Life of a Tuna Boat Helicopter Pilot." https://verticalmag.com/features/going-fishing-a-year-in-the-life-of-a-tuna-boat-helicopter-pilot.

He, Pingguo, et al. 2021. *Classification and Illustrated Definition of Fishing Gears.* FAO Fisheries and Aquaculture Technical Paper 672. Food and Agriculture Organization of the United Nations. https://doi.org/10.4060/cb4966en.

Hebert, Luke. 1836. *The Engineer's and Mechanic's Encyclopaedia,* Volume 1. T Kelly.

Hedderwick, Peter. 1830. *A Treatise on Marine Architecture.* Printed for the author.

Heilbron, J.L. 1999. *The Sun in the Church: Cathedrals as Solar Observatories.* Harvard University Press.

Heinz, J. 1894. "Electric Search Lights at Sea." *Proc. US Naval Inst.* 21(4): 763–82.

Hellenkamp, Johann F. 1967, October 31 (issue date). U.S. Patent 3,349,680, "Underwater Scanning Device." (Application filed June 28, 1966.)

Heller, Eric J. 2013. *Why You Hear What You Hear.* Princeton University Press.

Henderson, J.B. 1922, August 14 (publication date). British Published Application 185,612. "Improvements in the Automatic Control of the Steering of Ships, Aircraft and the Like." (Application filed May 13, 1921.)

Henderson, Wilfred. 1907. *Seamanship.* J. Griffin.

Hepworth, T.C. 1904. "How Men Work Under Water." *Cassell's Popular Science* 2: 159–68.

Hermann, E.M. 1961, April. "Historical Development of the Modern Submarine, Part II." *Bureau Ships J.* 10(4): 4–7.

Hett, W.S., and H. Rackham, eds. 1937. *Aristotle: Problems II, Books XXII–XXXVIII, Rhetorica ad Aldexandrum* (The Loeb Classical Library). Harvard University Press.

Heyboer, Marvin III, et al. 2017, June 1. "Hyperbaric Oxygen Therapy: Side Effects Defined and Quantified." *Advanced Wound Care (New Rochelle)* 6(6): 210–24. https://www.ncbi.nlm.nih.gov/pmc/articles/PMC5467109.

Hill, Mary. 2000. *Gold: The California Story.* University of California Press.

Hill, Sir Leonard. 1912. *Caisson Sickness and the Physiology of Work in Compressed Air.* Arnold.

Hines, Frank Thomas, and Franklin Wilmer. 1910. *The Service of Coast Artillery.* Goodenough & Woglom.

Hirsh, Aaron. 2013. *Telling Our Way to the Sea: A Voyage of Discovery in the Sea of Cortez.* Farrar, Straus and Giroux.

Hiscox, Gardner Dexter. 1899. *Mechanical Movements, Powers, Devices and Appliances, Used in Constructive and Operative Machinery and the Mechanical Arts ....* N.W. Henley.

History of Tynemouth, Its Priory and Castle. 1789. https://archive.org/details/fisherchapbook342/page/n3/mode/2up?q=tyzack.

Hocker, Fred. 2011. *Vasa: A Swedish Warship.* Oxbow Books.

Hocker, Fred (ed.). 2023. *Vasa II: Rigging and Sailing a Swedish Warship of 1628. Part 1. The Material Remains and Archaeological Context.* Nordic Academic Press.

Hodgins, George Sherwood. 1911, December. "Sounding the Sea." *The Marine Review* 41: 453–54.

Holdridge, Desmond. 2017. *Northern Lights.* Papamoa Press.

Holzmann, Gerard. 1994, January. "Data Communications: The First 2500 Years." *IFIP 13th World Computer Congress* 52: 271–78. https://www.researchgate.net/publication/221330073.

Homer, Lt. William Bradford, 1895. *Notes and Problems on the Elements of Mechanism and the Transmission of Power.* Artillery School Press.

Hooper, Henry N. 1850. *The Sixth Exhibition of the Massachusetts Charitable Mechanic Association.* Eastburn's Press.

Hornish, Ashley. 2024, March 26 (access date). "LORAN Day and Night Coverage, 1945." https://timeandnavigation.si.edu/multimedia-asset/loran-day-and-night-coverage-1945.

Hose-McCann. 2010. "About Hose-McCann Communications." https://www.hose-mccann.com/about.cfm.

House, David J. 2018 [2013]. *Seamanship Techniques: Shipboard and Marine Operations* (5th ed.). Taylor & Francis.

———. 2024 [2005]. *Cargo Work: For Maritime Operations.* Taylor & Francis.

Hoving, Ab, and Cor Emke. 2000. *The Ships of Abel Tasman, Volume 1.* Verloren.

Howeth, Linwood S. 1963. *History of Communications Electronics in the United States Navy.* GPO.

HuronScuba. 2010. "SCUBA Cylinder Specifications." https://huronscuba.com/equipment/scuba-cylinder-specifications.

Huss, Hans H. 1995. "Quality and Quality Changes in Fresh Fish." FAO Fisheries Tech. Paper 348. https://www.fao.org/3/V7180E/v7180e01.htm.

Hyperphysics. 2024. "Moment of Inertia." http://hyperphysics.phy-astr.gsu.edu/hbase/mi.html.

———. 2024a. "Moment of Inertia: Sphere. http://hyperphysics.phy-astr.gsu.edu/hbase/isph.html.

Iitaka, Yunosuke. 1971. "History and Global Review of Purse Seines," 179–84. https://kindai.repo.nii.ac.jp%2Frecord%2F5234%2Ffiles%-2FAN00064044-19710315-0179.pdf.

Imperial War Museums. 2024. "The Royal Navy on the Home Front, 1914–1918." Photograph Catalogue No. Q 18620. https://www.iwm.org.uk/collections/item/object/205253098.

Inglefield, Edward Fitzmaurice, and Arthur David. 1894, July 28. GB 189316908, "Improvements in Hooks and their Attachments, suitable for Signal Flags, etc." https://worldwide.espacenet.com/patent/search/family/035242867/publication/GB189316908A?q=GB189316908.

Inglefield, Edward. 1940, March 2. "Letter to [Commander] Mead." www.commsmuseum.co.uk/tactical/flags/inglefield clip.pdf.

Interesting Engineering. 2017, July 19. "The Interesting Engineering Behind Submarines." https://interestingengineering.com/interesting-engineering-behind-submarines.

InterlogUSA. 2024. "What Is Bulk Cargo?" https://www.interlogusa.com/answers/blog/supplychain-what-is-bulk-cargo.

Isherwood, B.F., et al. 1882, January 31. *Report of a Board of United States Naval Engineers on the Mallory Steering and Propelling Screw, as Applied to the United States Torpedo Boat Alarm, and on the Experiments with it in That Vessel....* GPO.

Jackson, A., et al. 2000, March 15. "Four Centuries of Geomagnetic Secular Variation from Historical Records." *Philosophical Transactions Royal Soc'y London* 358 (1768): 957–90. https://royalsocietypublishing.org/doi/10.1098/rsta.2000.0569.

Jamieson, Alexander. 1829. *A Dictionary of Mechanical Science, Arts, Manufactures, and Miscellaneous Knowledge,* Volume 2. H. Fisher, Son & Company.

Jarvis, John Charles Barron. 1890, September 26. British Provisional 15,267. "Improved Means for Bracing the Yards in Square-Rigged Ships." http://www.bruzelius.info/Nautica/Rigging/Jarvis(patent).html.

Jeffers, William Nicholson. 1850. *A Concise Treatise on the Theory and Practice of Naval Gunnery.* D. Appleton.

Jenkins, James Travis. 1921. *A History of the Whale Fisheries: From the Basque Fisheries of the Tenth Century to the Hunting of the Finner Whale at the Present Date.* H.F. & G. Witherby.

Jenkins, Wallace T. 1976. *A Guide to Polar Diving.* Naval Coastal System Laboratory.

Johansen, Herbert O. 1957, January. "TV Robot Roams Ocean Bottom." *Pop. Sci.* 170(1): 140–43.

Johnson, C.A., and E.J. Casson. 1995. "Effects of Luminance, Contrast, and Blur on Visual Acuity." *Optometry Vision Science* 72(12): 864–69.

Johnson, Capt. Edward J. 1852. *Practical Illustrations of the Necessity for Ascertaining the Deviations of the Compass,* etc. (2nd ed.) J.D. Potter.

Johnson, A.B., Jr., and B. Francis. 1980, January. "Durability of Metals from Archaeological Objects, Metal Meteorites, and Native Metals." https://www.osti.gov/biblio/5406419.

Jolly, Thomas William, and Robert Beatty. 1779a, April 22. British Patent 1220. "Machine for Steering Ships by a Horizontal Wheel, Quadrants, Pinions and Spindles."

Jolly, Thomas William, and Robert Beatty. 1779. *Description of the New Patent Steering Machine.* No publisher identified.

Jones, Eric Lionel. 2010. *Locating the Industrial Revolution: Inducement and Response.* World Scientific.

Jorgensen, Sven Erik. 2010, June. "Manually-Operated Air Pumps." *Int. J. Diving Hist.* 3(1): 18.

Journal of the Royal United Service Institution. 1899. "Naval Notes." *J. Royal United Service Institution,* 43(2): 1359–75.

Jürgens, Dirk. 2011. "Voith Schneider Propeller—Current Applications and New Developments." Voith Turbo, Brochure G 1949. https://www.semanticscholar.org/paper/Voith-Turbo-The-Voith-Schneider-Propeller-Current-J%C3%BCrgens/b2d78bedbdff5893120de12a7a1d128045e8712d.

Kalland, Arne. 1995. *Fishing Villages in Tokugawa Japan.* University of Hawaii Press.

Kasten, Michael. 2023. "What About the Junk Rig ...?" http://www.kastenmarine.com/junk_rig.htm.

Kay, Melvyn. 2000, October. "Treadle Pumps for Irrigation in Africa." FAO. https:/www.fao.org/3/x8293e/x8293e00.pdf.

Kellett, William P. 1942, June 30 (issue date). U.S. Patent 2,287,886. "Container Ship." (Filed May 17, 1940.)

Kellogg, James Lawrence. 1910. *Shell-Fish Industries.* H. Holt.

Kenninger, Madeline. 2019, August 26. "Difference Between Braided and Twisted Rope." https://ropeandcord.com/guides-ideas/difference-between-braided-and-twisted-rope.

Kent, Barrie. 2001, July 23–27. "Flag Signalling at Sea." *Proc. XIX Int'l Congr. Vexillology* 187–92 (York).

Kidd, Lt. Alexander Campbell. 1924. *Notes on Naval Communications, A Text Book for the Instruction of Midshipmen in the Department of Seamanship, U.S. Naval Academy* (rev. ed.). United States Naval Institute.

Kimball, John. 2015. *Physics Curiosities, Oddities, and Novelties*. CRC Press.

King, James Wilson. 1878. *Report of Chief Engineer J.W. King, United States Navy on European Ships of War and Their Armament, Naval Administration and Economy, Marine Constructions and Appliances, Dockyards, Etc.* (2nd ed.). GPO.

Kinghorn, J.G. 1896. "Salvage Appliances." *Transactions Inst. Naval Architects* 37: 250–56.

Kipfer, Barbara Ann. 2008. *Dictionary of Artifacts*. Wiley.

Kipping, Robert. 1857. *Rudimentary Treatise on Masting, Mast-making, and Rigging of Ships*. John Weale.

Knecht, Heidi. 2013 [1997]. *Projectile Technology*. Springer US.

Knight, Austin Melvin, and Harry Alexander Baldridge. 1921. *Modern Seamanship*. D. Van Nostrand, Inc.

Knight, Charles. 1851. *Cyclopaedia of the Industry of All Nations*. Charles Knight.

_____. 1861.*The English Cyclopedia: A New Dictionary of Universal Knowledge*. Bradbury & Evans.

Knight, Edward H. 1881. *Knight's American Mechanical Dictionary*, 3 Vols. Houghton, Mifflin & Co.

Knox-Johnston, Robin. 2013. "Practical Assessment of the Accuracy of the Astrolabe." *The Mariner's Mirror* 99(1): 67–71. DOI: 10.1080/00253 359.2013.766699.

Kolbrek, Bjorn. 2008. "Horn Theory: An Introduction, Part 1." audioXpress 1–8. https://www.grc.com/acoustics/an-introduction-to-horn-theory.pdf.

Kollerstrom, N. 1995. "Flamsteed's Lunar Data, 1692–95, Sent to Newton." *J. Hist. Astron.* 26: 237–46. http://www.ucl.ac.uk/nk/flamsteedmoon.htm.

Koutsoumanis, Konstantinos, et al., 2021, January 28. "The Use of the So-Called 'Superchilling' Technique for the Transport of Fresh Fishery Products." *EFSA J.* https://efsa.onlinelibrary.wiley.com/doi/full/10.2903/j.efsa.2021.6378.

Krishnamurti, Chandrasekhar. 2019, April 18. "Historical Aspects of Hyperbaric Physiology and Medicine." https://www.intechopen.com/chapters/66258.

Kristjonsson, Hilmar, ed. 1959. *Modern Fishing Gear of the World*. Fishing News (Books) Ltd.

Kulweic, Raymond A. 1985. *Materials Handling Handbook*. Wiley.

Kuntz, Jerry. 2012. *A Pair of Shootists: The Wild West Story of SF Cody and Maud Lee*. University of Oklahoma Press.

Kurlansky, Mark. 2011 [1997]. *Cod: A Biography of the Fish That Changed the World*. Knopf Canada.

Lacey, George E. 1921, March. "Steering Gear for Warships." *The Draughtsman* 4(3): 18–20.

LaGrone, Sam. 2017, July 20. "Office of Naval Research Set to Upgrade the 200-Year-Old Signal Lamp for Modern Stealth Communication." https://news.usni.org/2017/07/19/office-of-naval-research-project-set-to-give-21st-century-relevance-to-the-200-year-old-signal-lamp.

Lance, Rachel. 2020. *In the Waves: My Quest to Solve the Mystery of a Civil War Submarine*. Penguin Publishing Group.

Lang, Oliver. 1853. *Improvements in Naval Architecture*. Edwards.

Lardas, Mark. 2016. *The Battleship Texas*. Arcadia Publishing.

Lardner, Dionysius. 1851. *Hand-books of Natural Philosophy and Astronomy: Mechanics. Hydrostatics, Hydraulics, Pneumatics, and Sound*. Blanchard and Lea.

Lardner, Dionysius, ed. 1855. *The Museum of Science and Art*, Volume V. Walton and Maberly.

Lavery, Brian. 1987. *The Arming and Fitting of English Ships of War, 1600–1815*. Conway Maritime Press.

Lavery, H.I. 2013 [1984]. *Shipboard Operations*. Taylor & Francis.

Lazenby, David. 1998, Winter. "From the Depths of the Dark Age! A Danish Museum's Reconstruction of a Mediaeval Diving Dress." *Historical Diving Times* 23: 9–12. https://forum.historischetauchergesellschaft.de/download/file.php.

Leckie, Frederick A., et al. 2009. *Strength and Stiffness of Engineering Systems*. Springer US.

Ledyard, John O., ed. 2012. *The Economics of Informational Decentralization: Complexity, Efficiency, and Stability: Essays in Honor of Stanley Reiter*. Springer.

Lee, Joo-Young, and Hyo Hyun Lee. 2014. "Korean Women Divers 'Haenyeo': Bathing Suits and Acclimitization to Cold." *J. Human-Environment System*, 17(1):1–11. https://www.jstage.jst.go.jp/article/jhes/17/1/17_001/_pdf.

Leggett, Don. 2016. *Re-inventing the Ship: Science, Technology and the Maritime World, 1800–1918*. Taylor & Francis.

Lekang, Odd-Ivar. *Aquaculture Engineering*. Wiley.

Lenard Audio. 2021. "Horns." http://education.lenardaudio.com/en/07_horns.html.

Leniman, Daniel J., ed. 2001, April 27. "The USS *Utah*: Construction and Operational History." In "Submerged Cultural Resources Study, USS *Arizona* and Pearl Harbor National Historic Landmark, Chapter II: Historical Record." https://www.nps.gov/parkhistory/online_books/usar/scrs/scrs2m.htm.

Lerner, Andrew. 2018, August 20. "Do You Know the History of Glass Bottom Boats?" https://www.procurveglass.com/do-you-know-the-history-of-glass-bottom-boats.

Lewis, David. 1994. *We, the Navigators: The Ancient Art of Landfinding in the Pacific*. University of Hawai'i Press.

Lewis, William H. 1986. *Underground Coal Mine Lighting Handbook, Part 1. Background*. Information Circular 9073. Bureau of Mines, U.S. Dept. Interior.

Library of Congress. 1988 (drawing date). "1–54 Clarke, Chapman's Patent Windlass—Ship Balclutha." Drawings from Survey HAER CA-54. Historic American Engineering Record, National Park Service. Delineator Mark T. Bittle. https://www.loc.gov/resource/hhh.ca1493.sheet/?sp=54&st=image.

_____. 1989 (drawing date). "1-55 Capstans—Ship Balclutha." Drawings from Survey HAER CA-54. Historic American Engineering Record, National Park Service. Delineators Chang K. Yi and Marta M. Cubina Jackson. https://www.loc.gov/resource/hhh.ca1493.sheet/?sp=55&st=image.

Liburn, Frank. 1866, April 10. U.S. Patent 53,840. "Improved Steering-Screw."

Life Magazine. 1944, September 4. "Nerve Systems for Battle Wagons." *Life* 17(10): 16.

Lin, Chin-Chiuan. 2003, July. "Effects of Illumination, Viewing Distance, and Lighting Color on Perception Time." *Perceptual and Motor Skills*, 96(3 Pt 1):817–26. https://www.researchgate.net/publication/10687276_Effects_of_Illumination_Viewing_Distance_and_Lighting_Color_on_Perception_Time.

Littleshales, George Washington. 1899. *The Development of Great Circle Sailing*. GPO.

Liversidge, John George. 1899. *Engine-room Practice: A Handbook for Young Marine Engineers, Treating of the Management of the Main and Auxiliary Engines on Board Ship*. Griffin.

London, Jack. 1911. *The Cruise of the Snark*. Macmillan.

London Encyclopædia. 1845. *London Encyclopædia, Or, Universal Dictionary of Science, Art, Literature, and Practical Mechanics*. T. Tegg.

Loos, Jackie. 2018, August 2. "Diving Barrel Upped Bounty for Treasure Divers." https://www.pressreader.com/south-africa/cape-argus/20180802/281745565197341

_____. 2018a, August 15. "Devon Diver Discovers Silver Ducatoons in 1727." https://web.archive.org/web/20180815130046/https://www.iol.co.za/capeargus/opinion/devon-diver-discovers-silver-ducatoons-in-1727-16491064.

Luce, Cmdr. Stephen Bleecker, and Lt. Aaron Ward. 1884. *Text-book of Seamanship: The Equipping and Handling of Vessels Under Sail Or Steam. For the Use of the United States Naval Academy*. D. Van Nostrand.

Luckiesh, Matthew. 1920. *Artificial Light: Its Influence on Civilization*. Century Company.

Lukens, Ronald R., and Carrie Selberg. 2004, January. *Guidelines for Marine Artificial Reef Materials* (2nd ed.). Atlantic and Gulf States Marine Fisheries Commissions.

Lyon, Frank, and Alfred Walton Hinds. 1912. *Marine and Naval Boilers*. United States Naval Institute.

Ma, Qing-Ping. 2024. *Economics and Politics in the Robotic Age: The Future of Human Society*. Cambridge Scholars Publishing.

Machovec. 2020. "Rope Comparison Charts." https://www.machovec.com/rope/compare.htm#strength.

Mackenzie, Thomas. 1920 [1896]. *Practical Mechanics Applied to the Requirements of the Sailor*. Charles Griffin.

Mackrow, Clement. 1879. *The Naval Architect's and Shipbuilder's Pocket-Book of Formulae, Rules, and Tables, etc.* Crosby Lockwood and Co.

Macleod, Ian D. 2016, March 30. "In-Situ Corrosion Measurements of WWII Shipwrecks in Chuuk Lagoon, Quantification of Decay Mechanisms and Rates of Deterioration." *Frontiers in Marine Sci.* 3: Article 38. https://www.frontiersin.org/articles/10.3389/fmars.2016.00038/full.

_____. 2016a, November 15. "Corrosion of Copper Alloys on Historic Shipwrecks and Materials Performance." Corrosion and Prevention Australasian Corrosion Association Conference 2016 (Auckland, New Zealand). https://www.researchgate.net/publication/311515419_CORROSION_OF_COPPER_ALLOYS_ON_HISTORIC_SHIPWRECKS_AND_MATERIALS_PERFORMANCE.

Maggs, Colin. 2018. *Great Britain's Railways: A New History*. Amberley Publishing.

Maguire, Garreth. 2023. *Advanced Pike Fishing Tactics and Strategies*. Self-published.

Mallory, William H. 1874, June 23. U.S. Patent 152,238, "Improvement in Steering-Propellers."

_____. 1881, November 8 (issue date). U.S. Patent 249,191, "Steering-Propeller."

Maloney, Elbert S. 1985. *Dutton's Navigation and Piloting* (14th ed.). Naval Institute Press.

_____. 2006 [1917]. *Chapman Piloting and Seamanship* (65th ed.). Hearst.

Manwayring, 1644. *Sea-Mans Dictionary*. https://quod.lib.umich.edu/e/eebo2/A51871.0001.001/1:25.29?rgn=div2;view=fulltext '

Marine Digital. 2024. "Top 10 Biggest RoRo Ships In The World." https://marine-digital.com/article_roro.

Marine Engineer. 1893, October 1. "The Loading and Discharging of Ships." *The Marine Engineer* 15: 285.

Marine Engineering Log. 1902, November. "Mechanical Equipment of a Seven-Masted Schooner." *Marine Engineering Log* 7: 560–65.

Marine Insight News Network. 2019, February 2. "Selandia: The First Motor Ship in the World." https://www.marineinsight.com/maritime-history/selandia-the-first-motor-ship-in-the-world.

Marine Journal. 1921, August 6. "The Largest Capstan Ever Built in the Port of New York." *Marine Journal* 43: 2. (Advertisement for Morse Dry Dock & Repair Co.)

Maritime Subsidy Board. 1965. American President Lines, Ltd. v. Bethlehem Steel Company. No. CA-2, in *Decisions of the Maritime Subsidy Board* 1: 590–608.

Marryat, Captain Frederick. 1841. *A Code of Signals for the Use of Vessels Employed in the Merchant Service* (8th ed.). J.M. Richardson.

Marsh, Charles L. 1887, December 6 (issue date). U.S. Patent 374,196, "Oyster Tongs." (Application filed October 7, 1887.)

Marshall, Michael. 1990. *Ocean Traders from the Portuguese Discoveries to the Present Day.* Facts on File.

Martin, Commander Tyrone G. 2017, February. "Steering a Ship with a Pole." *Naval History Magazine* 31(1). https://www.usni.org/magazines/naval-history-magazine/2017/february/armaments-innovations.

Martin, Jay C. 2014. "Strands That Stand: Using Wire Rope to Date and Identify Archaeological Sites." *Int. J. Naut. Archaeol.* 43(1): 151–61. doi: 10.1111/1095-9270.12038.

Martin, Robert E. 1926, August. "We Can Trick the Wind into Saving Billions!" *Pop. Sci. Monthly* 105, 38–39. [No volume number.]

Marx, Robert F. 1977. *The Capture of the Treasure Fleet: The Story of Piet Heyn.* McKay.

———. 1990 [1978]. *The History of Underwater Exploration.* Dover Publications.

Mason, Henry W., and Patrick Cunningham. 1882, December 12 (issue date). U.S. Patent 269,080, "Mounting Boat Guns." (Application filed January 12, 1880.)

Massa, Frank. 1985, April. "Some Personal Recollections of Early Experiences on the New Frontier of Electroacoustics during the Late 1920s and Early 1930s." *J. Acoustical Soc'y Amer.* 77(4): 1296–302.

Massey, Edward. 1820. "A Statement of the Case of Mr. Edward Massey … most respectfully offered to the Notice of every Member of Parliament." A.T. Ducker.

Matheson, Ewing. 1898. *Aid Book to Engineering Enterprise.* E. & F.N. Spon, Ltd.

Mattes [Wikimedia User Name]. 2007, August 23. Photograph of work by unknown 19th century artist. Said to be exhibited in Deutsches Museum Verkehrszentrum, Munich, Germany, with title "Maschine zum Übersetzen der Diligencen auf Eisenbahnwaggons." https://commons.wikimedia.org/wiki/File:Maschine_zum_%C3%9Cbersetzen_der_Diligencen_auf_Eisenbahnwaggons.JPG.

Matthews, Commander E.O., and Lt. R.M.G. Brown. 1879, October 16. "The U.S.S. Alarm." *Proc. U.S. Naval Institute* 5(10): 499–512.

Mayr, Otto. 1970. *The Origins of Feedback Control.* MIT Press.

Mazur, Allan. 2022. *Ice Ages: Their Social and Natural History.* Cambridge University Press.

McCarrick, Alan, et al. 2011, July 17–21. "U.S. Navy Sodium Chlorate Oxygen Candle Safety." AIAA 2011-5045, 41st Int'l. Conf. Environmental Systems (Portland, Oregon). https://arc.aiaa.org/doi/abs/10.2514/6.2011-5045

McCourt, Mark. 2017, December 29. "A Brief History of Mathematics Education in England." https://markmccourt.blogspot.com/2017/12/a-brief-history-of-mathematics.html.

McJunkin, F.E. 1977, July. "Hand Pumps for Use In Drinking Water Supplies in Developing Countries." Technical Paper No. 10. https://www.ircwash.org/resources/hand-pumps-use-drinking-water-supplies-developing-countries.

McKay, John. 2020. *Sovereign of the Seas, 1637: A Reconstruction of the Most Powerful Warship of Its Day.* Pen & Sword.

McKay, Richard C. 2013 [1995]. *Donald McKay and His Famous Sailing Ships.* Dover Publications.

McManamon, John M. 2016. *Caligula's Barges and the Renaissance Origins of Nautical Archaeology Under Water.* Texas A&M University Press.

———. 2018, Spring/Summer. "The World's First Archaeological Divers." *INA (Institute of Nautical Archaeology) Quarterly* 45 (1/2): 15–21.

———. 2021. *"Neither Letters Nor Swimming": The Rebirth of Swimming and Free-Diving.* Brill.

McMillan, Joseph. 2001, "Signaling at Sea." https://www.seaflags.us/signals/Signals.html.

McMillin, John S. 1866, February 20 (issue date). US Patent 52,730, "Capstan."

McNeil, Gomer T. 1964, July 21 (issue date). U.S. Patent 3,141,397, "Underwater 360° Panoramic Camera." (Application filed October 19, 1962.)

McPhee, John, and Tappan Adney. 1975. *The Survival of the Bark Canoe.* Farrar, Straus & Giroux).

Meade, Richard Worsam. 1869. *A Treatise on Naval Architecture and Ship-building.* J.B. Lippincott & Co.

Mechanics' Magazine. 1836, March 5. "Rapson's Patent Steering Apparatus." *Mechanics' Magazine,* 24(656): 449–451.

Mechanisms. Club Videos https://mechanisms.club/video/ratchet_mechanism_18; https://mechanisms.club/video/ratchet_mechanism_19; https://mechanisms.club/video/ratchet_mechanism_20; https://mechanisms.club/video/ratchet_mechanism_22; https://mechanisms.club/video/ratchet_mechanism_23; https://mechanisms.club/video/ratchet_mechanism_24.

Mentzer, Robert. 2002, May. "Jupiter's Moons and the Longitude Problem." *Mercury* 31(3): 34–39. https://web.archive.org/web/20120724234557/http://www.lawrencehallofscience.org/pass/passv07/jupmoons.html

Merchant Marine Council. 1947, October. "Cargo Handling." *Proc. Merchant Marine Council* 4(10): 171–72.

Meriam, James L., et al. 2020. *Engineering Mechanics: Statics, Volume I.* Wiley.

Messimer, Dwight R. 1988. *The Merchant U-Boat: Adventures of the Deutschland 1916–1918.* Naval Institute Press.

Meyer, Ralph. 2020, December. "Bell, Watson, Soft Iron, and the Insight That Commercialized the Magneto Telephone." *Proc. IEEE,* 108(12): 2311–20. https://ieeexplore.ieee.org/document/9264835.

Miettinen, Arto, et al. 2008, January. "The Palaeoenvironment of the 'Antrea Net Find.'" *Iskos* 16: 71–87. https://www.researchgate.net/publication/256115911_The_palaeoenvironment_of_the_%27Antrea_net_find%27.

Milham, Willis Isbister. 1923. *Time & Timekeepers:*

Including the History, Construction, Care, and Accuracy of Clocks and Watches. Macmillan.

Miller, Jeffrey. 2024, "Tools for Navigating the High Seas." http://www.immigrantships.net/newcompass/ships/ship_files/navigation_tips.html.

Miller, William Davis. 1944. "The Background and Development of Naval Signal Flags." *Proc. Massachusetts Hist. Soc'y, Third Series* 68: 60–71, 563 (October 1944–May 1947).

Mills, A.A., et al. "Mechanics of the Sandglass." *Eur. J. Physics* 17: 97–109.

Mills, Chester B. 1928, December 18 (issue date). U.S. Patent 1,695,601, "Automatic Steering for Dirigible Craft." (Application filed June 19, 1922.)

Milton, James H., and Roy M. Leach. 2013. *Marine Steam Boilers* (4th ed.). Elsevier Science.

Milton, James Tayler, et al. 1899. *Water-Tube Boilers for Marine Engines*. Institution of Civil Engineers (London).

Mindell, David A. 2000. *War, Technology, and Experience Aboard the USS* Monitor. Johns Hopkins University Press.

_____. 2002. *Between Human and Machine: Feedback, Control and Computing Before Cybernetics*. Johns Hopkins University Press.

Misra, Suresh Sandra. 2015. *Design Principles of Ships and Marine Structures*. CRC Press.

Mixter, George W. 1967. *Primer of Navigation: with Problems in Practical Work and Complete Tables*. Van Nostrand Reinhold.

Molland, Anthony F., and Stephen R. Turnock. 2011. *Marine Rudders and Control Surfaces: Principles, Data, Design and Applications*. Elsevier Science.

Mone, Gregory. 2010, September 12. "A Husband and Wife Build a 19th-Century Wooden Submarine." https://www.popsci.com/diy/article/2010-09/submobile.

Monmonier, Mark. 2010. *Rhumb Lines and Map Wars: A Social History of the Mercator Projection*. University of Chicago Press.

Moon, Robert E. 2023, April. "Decompression Sickness (Caisson Disease; The Bends)." https://www.merckmanuals.com/home/injuries-and-poisoning/diving-and-compressed-air-injuries/decompression-sickness.

Moore, Frank. 1862. *Rebellion Record: a Diary of American Events: Documents and Narratives, Volume 8*. Van Nostrand.

Moore, John Hamilton, and Joseph Dessiou. 1810 [1794]. *The New Practical Navigator* (17th ed.). Multiple publishers.

Morland, Samuel. 1672, January 2. "An Account of the Speaking Trumpet etc." *Philosophical Transactions Royal Soc'y London* 6(79): 3056–58. https://royalsocietypublishing.org/doi/10.1098/rstl.1671.0067.

Morrison-Low, A.D. 2017 [2007]. *Making Scientific Instruments in the Industrial Revolution*. Taylor & Francis.

Morton, Vanda. 2019. *Brass from the Past*. Archaeopress Publishing.

Mott, Lawrence V. 1999. *The Development of the Rudder: A Technological Tale*. Gerald Duckworth & Co.

"Mow" (pseud.). 2007, January 15. "Wetsuit History—From Wool Sweaters To Heated Wetsuits." https://360guide.info/wetsuits/wetsuit-history.html.

Moyer, Richard H., and Susan A. Everett. 2013, December. "Twisting and Braiding—From Thread to Rope." *Science Scope* 37(4): 72–79.

Muckle, W. 2013 [1975]. *Naval Architecture for Marine Engineers*. Elsevier Science.

Muller, W.J. 1923. "A Few Results from Practical Experience with Mechanical Stokers Applied to Marine Water-Tube Boilers." *Transactions Royal Institute Naval Architects* 65: 310–20.

Mungan, Carl E. 2012. "Rolling Friction on a Wheeled Laboratory Cart." *Phys. Ed.*, 47(3): 288–92.

_____. 2025. "Rolling a Barrel Up a Ramp," *Eur. J. Physics*, 46(5): 055008.

Murray, Athole James. 1916. *Strength of Ships*. Longmans, Green and Company.

Murray, Carl D., and Stanley F. Dermott. 1999. *Solar System Dynamics*. Cambridge University Press.

Museum of Artifacts. 2017, August 17 (Wayback Machine archive date). "One of the Oldest Diving Suits in Existence—Called Wanha Herra." http://web.archive.org/web/20170810115810/https://museum-of-artifacts.blogspot.com/2017/07/one-of-oldest-diving-suits-in-existence.html.

mvsmith (pseud.). 2007, September 22. "Re: The Voith-Schneider Propeller." Ship Simulator Forum, http://forum.shipsim.com/index.php?topic=2550.0.

Myer, Albert James. 1864. *A Manual of Signals for the Use of Signal Officers in the Field*. (Excerpts selected by Signal Corps Association.) https:/www.civilwarsignals.org/pdf/lgmanualofsignals.pdf.

_____. 1872. *A Manual of Signals for the Use of Signal Officers in the Field*. D. Van Nostrand.

Napier, James Robert. 1871, October 18. British Patent 2775 of 1871. "Water Pressure Speed Indicators." In *English Patents of Inventions, Specifications, 871, 2748–2818* (1872). H.M. Stationery Office.

Nares, Captain George Strong. 1876. *Seamanship* (5th ed.). Griffin & Co.

National Centers for Environmental Information. 2024, March 28 (access date). "Magnetic Declination." https://www.ncei.noaa.gov/products/magnetic-declination#:~:text=Magnetic%20declination%20(sometimes%20called%20magnetic,over%20time%2C%20and%20with%20location.

National Geospatial-Intelligence Agency. 2004. *Handbook of Magnetic Compass Adjustment*. ww.nga.mil/MSISiteContent/StaticFiles/Files/HoMCA.pdf.

_____. 2019. *American Practical Navigator: An*

Epitome of Navigation, Volume 1. Pub. No. 9. https://thenauticalalmanac.com/2019_Bowditch-_American_Practical_Navigator.html.

National Maritime Museum, Greenwich. 1570 (putative manufacture date). "Mariner's Compass." Catalog No. NAV0276 https://www.rmg.co.uk/collections/objects/rmgc-object-42488.

———. 1702 (date model made). "Warship; First Rate; 90 Guns." ID: SLR0387. https://www.rmg.co.uk/collections/objects/rmgc-object-66348.

———. 1703 (model date). "Warship (1703); Fourth Rate; 50 Guns." ID SLR0218. https://www.rmg.co.uk/collections/objects/rmgc-object-66179.

National Museum of American History. 1775 (manufacture date). "Ramsden Dividing Engine." Catalog No. 215518. https://americanhistory.si.edu/collections/nmah_694508.

National Oceanic and Atmospheric Administration. 2005, November 25. "Map of Magnetic Elements from the WMM 2005." https://web.archive.org/web/20061209141011/http://www.ngdc.noaa.gov/seg/WMM/image.shtml.

National Research Council. 1992. *Intermodal Marine Container Transportation: Impediments and Opportunities.* NRC.

Natural Resources Canada. 2008, May 1. "Geomagnetism: Early Concept of the North Magnetic Pole." https://web.archive.org/web/20111207104412/http://gsc.nrcan.gc.ca/geomag/nmp/early_nmp_e.php.

Nature. 1878, October 24. "Edison's Inventions." *Nature,* 18: 674–76.

Nautical Magazine. 1907, November. "America on the Ocean." *Nautical Magazine,* 78(5): 361–71.

Naval Communications 2006. "Chronological History U.S. Naval Communications." https://www.navy-radio.com/manuals/NAVCOMM-history-2006.pdf.

Naval Education and Training Support Command. 1973. *IC [Interior Communications] Electrician 3 & 2.* NAVEDTRA 10558-B. https://www.navy-radio.com/manuals/ic.htm.

———. 1993, January. *Seaman.* Nonresident Training Course. NAVEDTRA 14067. GPO.

———. 1996, July. *Signalman 1 & C.* Nonresident Training Course. NAVEDTRA 14243. GPO.

———. 2010, November. *Boatswain's Mate.* Nonresident Training Course. NAVEDTRA14343A. Center for Surface Combat Systems.

Naval Marine Archive. 2022, October 10. The Canadian Collection. "England Expects That Every Man Will Do His Duty." http://navalmarinearchive.com/research/england_expects_signalflags.html.

Naval War College. 1901. "Rules for the Conduct of the War Games." https://dnnlgwick.blob.core.windows.net/portals/0/NWCDepartments/Wargaming Department/War-Games-1901.pdf.

Navy and Army Illustrated. 1899, September 9. "A Mediterranean Training-Ship." *Navy and Army Illustrated* 8(136): 590–92.

Navy, Bureau of Supplies and Accounts. 1953. *Cargo Handling.* NAVPERS 10124. GPO.

Navy, Secretary of. 1883. *Annual Report of the Secretary of the Navy for the Year 1883,* Volume I. GPO.

Navy Department. 1822. *Proceedings of the General Court Martial Convened for the Trial of Commodore James Barron, etc.* Jacob Gideon, Jr.

Navy Music Program Management Office, Navy Personnel Command. 2003. *Manual for Buglers.* GPO.

Nelson, August T. 1912, January 16 (issue date). U.S. Patent 1,015,061, "Automatic Steering Device." (Application filed February 12, 1906.)

Nelson, Jack. 1981, August. "Sailing Great Circle Routes." *Cruising World* 7(8): 38–41.

New Haven Reef Conservation Program. 2016. "Artificial Reefs: What Works and What Doesn't." https://newheavenreefconservation.org/marine-blog/147-artificial-reefs-what-works-and-what-doesn-t.

Niblack, Albert Parker. 1892. "Naval Signaling." *U.S. Naval Institute Proceedings,* 18(4): 431–89.

Nicolls, Bruce. 1991. "The Talking Flags from Trafalgar Onwards." XIV International Congress of Vexillology (Barcelona). https://fiav.org/wp-content/uploads/2021/04/14-20-Nicolls.pdf.

Nilsson, Jan Erik. 2023. "The Historical Background Sweden and the City of Gothenburg: 5. Götheborg." https://gotheborg.com/project/ships/skepp5.shtml.

Noack, Thomas. 2017, March 7. "Danish Seine—Ecosystems Effect of Fishing." Ph.D. Thesis, Technical University of Denmark. https://orbit.dtu.dk/files/132792467/Publishers_version.pdf.

Noel, Sir Gerard Henry Uchtred. 1874. *The Gun, Ram, and Torpedo.* J. Griffin & Co.

North, Albert H. 1965, June 6. U.S. Patent 48,087, "Steering Apparatus."

North, Mike F., and Simon J. Lovett. 2005. "Freezing Methods and Equipment." In Da-Wen Sun (ed.), *Handbook of Frozen Food Processing and Packaging,* Chapter 9. CRC Press.

Norwood, Richard. 1632, April 2. British Patent 56, "Making and Using Engines or Instruments for Diving, and for Raising or Bringing out of the Sea or Other Deep Waters, Any Goods Lost or Cast Away by Shipwreck or Otherwise." https://ryanfb.github.io/loebolus-data/L317.pdf.

Noye, R.J. 1998, June 22. "Sources of Light for Magic Lanterns." https://noye.agsa.sa.gov.au/Lantern/Lighting.htm.

Nuvair. 2023. "Nomad Low Pressure Gas Powered Compressor." https://www.nuvair.com/nuvair-nomad-gas-compressor.html.

O'Dwyer, Aidan. 2005, January 1. "PID Control: The Early Years." *Control in the IT Sector Seminar,* Cork Institute of Technology, Cork, May. doi:10.21427/q287-3p17.

Oestmann, Günther. 2001. "On the History of the Nocturnal." *Bull. Sci. Instrument Soc'y* 69: 5–9.

Oldknow, R.C. 1893. "Marine Engineering." In T.A. Brassey (ed.), *The Naval Annual,* Chapter VII. J. Griffin & Co.

Oleron. 1266 (circa). "The Rules of Oleron." http://

www.admiraltylawguide.com/documents/oleron.html.

Oliver, John. 2000. "Lunar Eclipse Longitude Observation." https://web.archive.org/web/20061206001555/http://www.astro.ufl.edu/~oliver/lelo/

10.3390/heritage6050219.

Ozanam, Jacques. 1708. *Recreations Mathematical and Physical.* R. Bonwick et al.

Pacific Builder & Engineer. 1918, April 5. "New Cargo Hook." 24(14): 1.

Paddle, Muddy (pseud.). 2020, June 26. "Building & Sailing the 'Half Moon.' Part 5." https://www.hrmm.org/history-blog/category/half-moon

Pagenstecher, Albert. 1864, October 4. U.S. Patent 44,584, "Improved Hydraulic Propeller."

Papinus, Dionysus [Papin, Denis]. 1689. "De instrumentis ad flammam sub aqua conservandam." *Acta Eruditorum* 1689: 485. https://web.archive.org/web/20221211185035/http://www.izwtalt.uni-wuppertal.de/Acta/AE1689.pdf#page=501.

Park, Robert W., and Douglas R. Stenton. 1998. "Ancient Harpoon Heads of Nunavut: An Illustrated Guide." Parks Canada. http://parkscanadahistory.com/publications/north/nunavut-harpoon-heads.pdf.

Parker, ICCS Bert A. *Interior Communications Electrician*, Volume 1, NAVEDTRA 14120 (original release September 1994, administrative update October 2003). GPO.

Parliament, Great Britain. 1803. *Journals of the House of Commons,* Volume 35 (Nov. 29, 1774, to Oct. 15, 1776). Parliament.

Parsons, William Barclay. 1922. *Robert Fulton and the Submarine.* Columbia University Press.

Pasachoff, Jay M., and Alex Filippenko. 2019. *The Cosmos: Astronomy in the New Millennium* (5th ed.). Cambridge University Press.

Pasley, Sir Charles William. 1823. *Description of the Universal Telegraph, for Day and Night Signals.* T. Egerton.

Patarino, Vincent, Jr. 2012. "The Religious Shipboard Culture of Sixteenth and Seventeenth-Century English Sailors." In Cheryl A. Fury (ed.), *The Social History of English Seamen, 1485–1649,* 141–192. Boydell Press.

Patowary, Kaushik. 2018, December 12. "John Lethbridge's Diving Machine." https://www.amusingplanet.com/2018/12/john-lethbridges-diving-machine.html.

Payne, Brian. 2022. *Eating the Ocean: Seafood and Consumer Culture in Canada.* McGill-Queen's University Press.

Peake, James. 1851. *Rudiments of Naval Architecture.* John Weale.

_____. 1867. *The Elementary Principles of Naval Architecture.* Crosby Lockwood (1897 reprint).

Peddie, John. 1996. *The Roman War Machine.* Combined Books, Inc.

Pelios, Nick. 2023, April 24. "The Techniques and Tools of Skandalopetra Diving—A Look at a Unique Practice." https://alchemy.gr/post/692/skandalopetra-the-ancient-art-of-kalymnian-sponge-diving.

Pellett, C. Roger. 2018. *Whaleback Ships and the American Steel Barge Company.* Wayne State University Press.

Perez-Mallaina, Pablo E. 1998. *Spain's Men of the Sea: Daily Life on the Indies Fleets in the Sixteenth Century.* Johns Hopkins University Press.

Perry, Felix. 2006. *In Deep Water.* Breakwater Books.

Peters, Gavin H., et al. 1964. *A Solid Chemical Air Generator.* Research Report No. AG-1. Amoco Chemicals Corp.

Peterson, Mark. 2020. *The City-State of Boston: The Rise and Fall of an Atlantic Power, 1630–1865.* Princeton University Press.

Phillipps, H. Cranmer. 1835. *A Code of Universal Naval Signals.* Longman, Rees, Orme, Brown, Green, and Longman.

Phillips, Charles. 1819. English Patent 4394. Specification duplicated in *London J. Arts Sci.* 2:1 (1821).

_____. 1827. English Patent 5504. Invention described in *London J. Arts Sci.* 6:88–90 (1831); *Repertory of Patent Inventions* 7: 115–17 (1829).

Phillips, Edward. 1720. *The New World of Words, or, Universal English Dictionary* (7th ed.). J. Philips.

Phillips-Birt, Douglas. 1971. *A History of Seamanship.* Doubleday.

Pickering, Keith A. 2004, September 11. "The Transatlantic Tracks of Columbus." http://columbuslandfall.com/ccnav/shd2004.shtml.

Pilato, Denise E. 2024, January 25 (access date). "Martha Coston: A Woman, a War, and a Signal to the World." *Signal Corps Association.* http://www.civilwarsignals.org/pages/signal/signalpages/flare/coston2.html

Pinto, Marcia Freire, et al. 2017. "Chapter 8: People and Fishery Resources: A Multidimensional Approach, in Alves, Romulo Romeu Nobrega, and Ulysses Paulino Albuquerque, *Ethnozoology: Animals in Our Lives.* Elsevier Science.

Pipping, Olof. 2000. "Whipstaff and Helmsman: An Account of the Steering-Gear of the *Vasa.*" *The Mariner's Mirror,* 86: 1, 19–36.

Plimpton, George. 1984. *Fireworks: A History and Celebration.* Doubleday & Co.

Poluhowich, John. 1999. *Argonaut: The Submarine Legacy of Simon Lake.* Texas A&M University Press.

Pope, Dudley. *Life in Nelson's Navy.* House of Stratus.

Popham, Sir Home. 1803. *Telegraphic Signals; or Marine Vocabulary.* T. Egerton.

Popular Science. 1944, August. "Sound-Powered Phones Carry Battle Orders." 124.

Post, Natasha. 2021, June 29. "What is Roll On/Roll Off Services?" https://www.heavyhaulers.com/blog/what-is-roll-on-roll-off-services.

Preston, Richard S. 2000, June. "The Accuracy of the Astronomical Observations of Lewis and Clark." *Proc. Amer. Philos. Soc'y* 144: 168–91. https://www.jstor.org/stable/1515630.

Pritchett, Robert H. III, and William K. Seliger.

2010. *ARS Shipwreck Projects Dominican Republic Volume I*. Never Mind Publishing.

Proc, Jerry. 2022. "Visual Signalling in the RCN: Flags Section." http://www.jproc.ca/rrp/rrp2/visual_flags.html.

_____. 2022a. "Visual Signalling in the RCN: Light Signalling." www.jproc.ca/rrp/rrp2/visual_lights.html.

Proctor, Richard Anthony. 1888. *Great-Circle Sailing*. Longmans, Green, and Company.

Puretic, Mario J. 1956, February 7 (Issue date). U.S. Patent 2,733,530, "Method of Operating a Purse Seine with a Power Block Unit." (Filed August 16, 1954.)

_____. 1956, February 7 (Issue date). U.S. Patent 2,733,531, "Net Handling Apparatus."

"QSVC" (pseud.). 2012, December 17. "Begbie Signal Lamp." Post to Victorian Wars Forum. https://web.archive.org/web/20130709142723/http://www.victorianwars.com/viewtopic.php?f=19&t=7839.

Qualtrough, Edward Francis. 1881. *The Sailor's Handy Book and Yachtsman's Manual Adapted for the Use of the Navy, Merchant Service, Revenue Marine, and Yachtsmen*. C. Scribner's Sons.

Quin, Theophilus. 1814. *The Biographical Exemplar; Comprising Memoirs of Persons who Have Risen to Eminence by Industry and Perseverance in the Beneficial Occupations of Life, Etc*. Sharpe and Hailes, etc.

Radau, Rodolphe. 1872. *Wonders of Acoustics; or; the Phenomena of Sound*. Scribner's, Armstrong & Company.

Radcliffe, William. 1921. *Fishing from the Earliest Times*. J. Murray.

Radhalekshmy, K., and S. Gopalan Nayar. 1973. "Synthetic Fibres for Fishing Gear." *Fishery Technology* 10(2): 142–65. http://hdl.handle.net/1834/33629.

Radiocommunications Bureau. 2014. *Handbook on Ground Wave Propagation*. International Telecommunications Union. https://www.itu.int/pub/R-HDB-59-2014.

Raeside, Bob. 2011, December 24. "Popham's Signal Flag." https://www.fotw.info/flags/xf~psf.html.

_____. 2018, February 10. "England Expects That Every Man Will Do His Duty." https://www.fotw.info/flags/gb%5Etraf.html.

_____. 2020a, April 25. "Maritime Signal Flag History." https://www.crwflags.com/fotw/flags/xf~sfh.html.

_____. 2020b, April 19. "International Code of Signals (Overview)." https://www.fotw.info/flags/xf-ics.html.

_____. 2024, February 2 (access date). "NATO Naval Signal Flags." http://www.loeser.us/flags/nato_note_1.html.

Raeside, Rob. 2020a, May 23. "Signal-book for the Ships of War, 1799." https://www.crwflags.com/fotw/flags/xf~1799.html.

Raible, Daniel Edward. 2011. "Free Space Optical Communications with High Intensity Laser Free Space Optical Communications with High Intensity Laser Power Beaming." ETD Archive 251 (Cleveland State University). https://engagedscholarship.csuohio.edu/etdarchive/251.

Raines, James Orvill. 2019. *Good Night Officially: The Pacific War Letters of a Destroyer Sailor*. Taylor & Francis.

Raines, Rebecca R. 1996. *Getting the Message Through: A Branch History of the U.S. Army Signal Corps*. GPO.

Randles, William Graham Lister. 1989. "Pedro Nunes' Discovery of the Loxodromic Curve …." *Revista da Universidade de Coimbra*, 35: 119–30.

_____. 1993. "The Alleged Nautical School Founded in the Fifteenth Century at Sagres by Prince Henry of Portugal, Called the 'Navigator.'" *Imago Mundi*, 45: 20–28.

Rankine, William John MacQuorn. 1887. *A Manual of Machinery and Millwork* (6th ed.). Charles Griffin and Company.

Raper, Henry. 1828. *A New System of Signals by Which Colours May be Wholly Dispensed With*. Saunders and Otley.

Rapson, John. 1834, August 23. British Patent 6665, "Facilitating the Steering of Vessels." Abstract in Commissioner of Patents. 1875. *Patents for Inventions: Abridgments of Specifications Relating to Steering and Maneuvering Vessels*, AS 1763–1866, 18. George E. Eyre and William Spottiswoode. https://www.google.com/books/edition/Patents_for_inventions_Abridgments_of_sp/YIur7JGqJSAC?hl=en&gbpv=1&dq=rapson+6665+1834&pg=PA18&printsec=frontcover. The latter also cites *Mechanics' Magazine*, vol. 24, 449, and *London J*. (Newton's), vol. 20 (conjoined series): 374 (no figures provided).

_____. 1839, September 9. British Patent 8214, "Steering Ships and Vessels." Abstract in Commissioner of Patents. 1875. *Patents for Inventions: Abridgments of Specifications Relating to Steering and Maneuvering Vessels*, AS 1763–1866, 20–21. George E. Eyre and William Spottiswoode. The latter also cites *Repertory of Arts*, vol. 13 (new series), 326, and *Inventors Advocate*, vol. 2, 195.

Ratcliffe, John. 2011, Spring. "Bells, Barrels and Bullion: Diving and Salvage in the Atlantic World, 1500 to 1800." *Nautical Research J*. 56(1): 35–56. https://www.academia.edu/1522075/Bells_Barrels_and_Bullion_Diving_and_Salvage_in_the_Atlantic_World_1500_to_1800.

Raupp, Jason Thomas. 2015, June. "'And So Ends this Day's Work': Industrial Perspectives on Early Nineteenth-Century American Whaleships Wrecked in the Northwestern Hawaiian Islands." Ph.D. Thesis, Archaeology, Flinders University. https://theses.flinders.edu.au/view/0c79bee2-53dd-40c3-91fe-8fc4d09d7fca/1.

Rawlins, Dennis. 1993, October. "Tycho's Star Catalog: The First Critical Edition." DIO 3:3–106, www.dioi.org/vols/w30.pdf.

Raymond, Keith A., et al. 2023, June 26. "Diving

Buoyancy." https://www.ncbi.nlm.nih.gov/books/NBK470245.
RCA Sound Products. 2016, April 4. (pdf last modified date). Catalog 218. https:/www.worldradiohistory.com/Archive-Catalogs/RCA/Theatre/RCA-Catalog-218.pdf
Read, Bob. 2021, September 12 (pdf creation date). "Titanic's Kelvin Sounding Machines." https://rms-titanic.fr/otb/armement/pontemb/ot_pont_emb_sonar.pdf.
Reed, Sir Edward James. 1869. *Shipbuilding in Iron and Steel*. Murray.
Rees, Abraham. 1819. *The Cyclopaedia: Or, Universal Dictionary of Arts, Sciences, and Literature.* Longman, Hurst.
Reid, George H. 1996. *Marine Salvage: A Guide for Boaters and Divers*. Sheridan House.
Reid, Phillip Frank. 2017, February. "'A Very Good Sailer': Merchant Ship Technology and the Development of the British Atlantic Empire 1600-1800." Ph.D. Thesis, Dept. of History, Memorial University of Newfoundland. https://research.library.mun.ca/12577/1/thesis.pdf.
Reis, Omar. 2004, August. "Technical Note—The Proper Motion of the Stars." http://www.tecepe.com.br/nav/ProperMotion.htm.
Rektorys, Karel. 2013 [1994]. *Survey of Applicable Mathematics*, Vol. 2. Springer Netherlands.
Renwick, James. 1832. *The Elements of Mechanics*. Cary & Lea.
Reti, Ladislao. 1970, April. "The Double-Acting Principle in East and West." *Technology and Culture* 11(2): 178–200.
Richardson, G.B. 1864. *The Universal Code of Signals for the Merchant Marine of All Nations, by the Late Captain Marryat, R.N., etc.* Richardson & Co.
Riess, Warren Curtis. 1987. "The Ronson Ship: The Study of an Eighteenth Century Merchantman Excavated in Manhattan, New York in 1982." Dissertation, University of New Hampshire. https://core.ac.uk/download/pdf/215519477.pdf.
Rijksmuseum. 1771 (artifact date). "Model of a Capstan (1771)." https://www.rijksmuseum.nl/en/collection/NG-MC-187/catalogue-entry.
Ritchie, Edward S. 1862, September 9 (issue date). "Mariner's Compass." U.S. Patent 36,422.
Rivera, Flor Trejo, and Roberto Junco Sánchez. 2023. "The Remains of a Manila Galleon Compass: 16th-Century Nautical Material Culture." *Heritage* 6: 4173–4186. https://doi.org.
Riviera Newsletters. 2008, September 26. "An Azimuth Thruster Ahead of Its Time." https://www.rivieramm.com/news-content-hub/news-content-hub/an-azimuth-thruster-ahead-of-its-time-51852.
Roberts, Callum. 2009 [2007]. *The Unnatural History of the Sea*. Island Press.
Roberts, G.N., and Robert Sutton. 2006. *Advances in Unmanned Marine Vehicles*. Institute of Engineering and Technology.
Robinson, Rowan. 1996. *The Great Book of Hemp*. Inner Traditions/Bear.

Robison, S.S. 1932, April. "Commodore Thomas Truxtun." *US Naval Institute Proceedings* 58(4): 350. https://www.usni.org/magazines/proceedings/1932/april/commodore-thomas-truxtun-u-s-navy.
Roche, John J. 1981, November. "Harriot's 'Regiment of the Sun' and Its Background in Sixteenth-Century Navigation." *British J. Hist. Sci.* 14(3): 245–61.
Rogers, Woode. 1712. *A Cruising Voyage Round the World.* Reprinted 1928, Cassell and Company.
Rohn, Arthur C. 1946. "Cargo Handling and Its Relation to Ocean Commerce." *The Log* 41(4): 73–78, 92.
Ronny, Ang Toon Yiam. 2012, March. "Link Performance Analysis of a Ship-to-Ship Laser Communication System." Naval Postgraduate School. https://faculty.nps.edu/thuynh/Theses/12Mar_Ang.pdf.
Rood, E.S. 1921, November 1. "Thermal Conductivity of Some Wearing Materials." *Physical Rev.* 18: 356.
Roos, Willem. 1929, January 1 (issue date). U.S. Patent 1,697,779. "Rudder for Ships." (U.S. application filed June 15, 1925. Roos also received British Patent 249,730, issued April 1, 1926.)
Rose, Kelby James. 2014, August. "The Naval Architecture of Vasa, a 17th-Century Swedish Warship. Ph.D. Dissertation, Anthropology, Texas A&M University. https://oaktrust.library.tamu.edu/handle/1969.1/153363.
Rose, Susan. 2004. "Mathematics and the Art of Navigation: The Advance of Scientific Seamanship in Elizabethan England." *Transactions Royal Hist. Soc'y.* 14: 175–84.
Rothman, Dan. 2013. "Spanish Treasure and the Canada Townships." http://www.newbostonhistoricalsociety.com/phips.html.
Roy, Rolins Thomas. 2019, March. "On Speech Intelligibility and P.A. Systems." https://www.acousticbulletin.com/wp-content/uploads/2019/03/speech-intelligibility-pa-systems_-rolins-t-roy.pdf.
Royal Museums Greenwich. 1685 (estimated date made). "Two Views of an East Indiaman of the Time of King William III." Paintings by Isaac Sailmaker. https://www.rmg.co.uk/collections/objects/rmgc-object-13164.
———. 1772 (estimated date made). "Defiance (1772)." Drawing by unknown artist. https://www.rmg.co.uk/collections/objects/rmgc-object-384408.
Royal Signals. 1905. "Royal Signals." Chapter IV, Heliograph (page 48 of the 1905 Signalling Handbook). https://royal-signals.org.uk/Datasheets/the_heliograph.php.
Royal Society of Arts. 1783."A Bounty of Twenty Guineas was given to Mr. Spalding, of Edinburgh, for his Improvement on the Diving Bell, 1776." *Transactions of the Society Instituted at London for the Encouragement of Arts, Manufactures, and Commerce* 1: 220–38. https://catalog.hathitrust.org/Record/000675024

Royal Society Picture Library. 2024, January 26 (access date). "How to Walk Underwater (Mar. 6, 1689)." https://pictures.royalsociety.org/image-rs-15313

_____. 2024a, January 26 (access date). "Halley's Diving Bell (1839)." https://pictures.royalsociety.org/image-rs-14064.

Royce, Patrick M. 1998 [1982]. *Royce's Sailing Illustrated: The Sailor's Bible Since 1956*. Royce Publications.

Roymech. 2024, January 26 (access date). "Coefficients of Friction." https://www.roymech.co.uk/Useful_Tables/Tribology/co_of_frict.htm.

Roys, Thomas. 1861, January 22 (issue date). U.S. Patent 31,190, "Improvement in Harpoon-Guns."

_____. 1866, April 24 (issue date). U.S. Patent 54,222, Improvement in Rocket-Harpoons."

Rubanova, Alina. 2012. "History of Compressors." https://archive.control.lth.se/media/Education/DoctorateProgram/2012/HistoryOfControl/Alina-Surge.pdf.

Rubio, H. 2024. "Analysis of the First Treatise on Machine Elements: Codex Madrid I." *Foundations of Science* 29:19–40.

Ruschenberger, William S.W. 1838. *Narrative Of A Voyage Round The World, During The Years 1835, 36, And 37*. Bentley.

Russell, Matthew A. 2004. "Comet: Submerged Cultural Resources Site Report, Channel Islands National Park." National Park Service. https://books.google.com/books?id=Fpo1E6LganQC&printsec=frontcover&dq=comet+submerged&hl=en&newbks=1&newbks_redir=1&sa=X&ved=2ahUKEwjPpZ_P4PuDAxXaEVkFHS3wCZsQ6AF6BAgJEAI

Russell, Paul Anthony. 2021. *Applied Mechanics for Marine Engineers* (7th ed.). Bloomsbury Publishing.

Russell, Peter Edward. 2000. *Prince Henry "the Navigator": A Life*. Yale University Press.

Ryan, James. 1831. *Treatise on the Art of Measuring*. Collins, etc.

Sager, Eric W. 1996. *Seafaring Labour: The Merchant Marine of Atlantic Canada, 1820–1914*. McGill-Queen's University Press.

Sahrhage, Dietrich, and Johannes Lundbeck. 2012. *A History of Fishing*. Springer Berlin Heidelberg.

Sailing Duyfken. "How to Steer a Tall Ship with No Wheel, but a Whipstaff." https://www.youtube.com/watch?v=5VdtDse07hw.

Salvors. 2021, January 1 (Revision 2). *U.S. Navy Salvor's Handbook*. Published by direction of Commander, Naval Sea Systems Command.

Sandman, Alison. 1999. "Educating Pilots: Licensing Exams, Cosmography Classes, and the Universidad de Mareantes in 16th Century Spain." In *Fernando Oliveira and his Era: Humanism and the Art of Navigation in Renaissance Europe (1450–1650)* (Proceedings of the lX International Reunion for the History of Nautical Science and Hydrography).

Saunders, Robert. 1871, August 29. British Patent 2270. "Improvements in Steering Ships and Vessels" (sealed February 23, 1872), in Patent Office. *English Patents of Inventions, Specifications1871, 2239–300*. HM Stationery Office. https://books.google.com/books/download/English_Patents_of_Inventions_Specificat.pdf.

Sawatsky, David. 2018, November 2. "Oxygen Toxicity and CCR/Rebreather Diving." https://www.diverite.com/uncategorized/oxygen-toxicity-and-ccr-rebreather-diving.

Sawatsky, K. David. 2008. "Oxygen Toxicity—How Does It Occur?" (Originally published in *Diver Magazine*, December 2008/January 2009). https://www.diverite.com/articles/oxygen-toxicity-how-does-it-occur.

Sawatsky, K. David. 2009, Feburary–March. "Oxygen Toxicity—Signs and Symptoms." (Originally published in *Diver Magazine*, February/March 2009). https://www.diverite.com/articles/oxygen-toxicity-signs-and-symptoms.

Schaefers, Edward A., and Dayton L. Alverson. "Second World Fishing Gear Congress." *Commercial Fisheries Review* 26(5): 1–11.

Schafer, Rollie. 2001. "Finding the Way and Fixing the Boundary: The Science and Art of Western Map Making, as Exemplified by William H. Emory and His Colleagues of the U.S. Corps of Topographical Engineeers." *Military History of the West* 31(1): 1–26.

Schaffer, John. 1857a. US Patent 16,935, "Capstan" (patented March 31, 1857).

Schaffer, John. 1857. US Patent 15,954, "Capstan for Steamboats" (patented August 25, 1857).

Schäuffelen, Otmar. 2005. *Chapman Great Sailing Ships of the World*. Hearst Books.

Schlereth, Hewitt. 1975. *Commonsense Celestial Navigation*. Regnery.

Schmidt, Peter G., Jr. 1959. "The Puretic Power Block and its Effect on Modern Purse Seining." In Hilmar Kristjonsson (ed.), *Modern Fishing Gear of the World*, 400–13. Fishing News (Books) Ltd.

Schnitzer, Laura Kate. 2012, January. "Aprons of Lead: Examination of an Artifact Assemblage from the *Queen Anne's Revenge* Shipwreck Site." M.A. Thesis, Maritime Studies, East Carolina University. https://thescholarship.ecu.edu/bitstream/handle/10342/3842/Schnitzer_ecu_0600M_10612.pdf?sequence=1&isAllowed=y.

Schoff, Wilfred H. 1914. "Parthian Stations by Isidore of Charax: An Account of the Overland Trade Route Between the Levant and India in the First Century B.C." https://www.parthia.com/doc/parthian_stations.htm.

Scientific American. 1849, August 4. "Combined Capstan and Windlass." *Sci. Amer.* 4(46): 1.

_____. 1914, August 1. "Kite Signaling of the French Navy." *Sci. Amer.* 111(5): 81.

Scott, Frank. 2009. "Robert Forbes and Frederic Howes and the Evolution of the Double Topsail." *The Mariner's Mirror* 95(1): 52–61. doi: 10.1080/00253359.2009.10657083.

Scott, Genio C. 1869. *Fishing in American Waters*. Harper & Brothers.

Scuba ("Nevin"). 2009, December 19. "Scuba Air Compressors Explained." https://www.scuba.com/blog/scuba-air-compressors-explained.

scubakim. 2011, November 22. "Japanese Female Divers (Ama)." http://www.scubakim.com/?p=1188.

Seaboard. 1920. "Mechanical Stokers Not Much Used at Sea." *Seaboard: A Maritime Reporter and Nautical Gazette* 98: 745.

Seaton, John. 1850. *Great-Circle Sailing Made Easy*. Houlston & Stoneman.

Seeker, S.V. 2010. "Argonaut Jr. 2010: Recreating Simon Lake's 1894 Wooden Submarine." https://www.submarineboat.com/argonaut_jr_2010.htm.

Sefton, John. 2010, March 18. Post, "Warrior HMS (Ironclad)." https://shipstamps.co.uk/forum/viewtopic.php?t=10075.

Seherr-Thoss, Hans-Christoph, et al. 2006. *Universal Joints and Driveshafts: Analysis, Design, Applications*. Physica-Verlag.

Seiko Museum Ginza. 2024, February 26 (access date). "The Stopwatch and the Chronograph Part 1: The Birth of the Stopwatch." https://museum.seiko.co.jp/en/knowledge/story_06.

Self, Douglas. 2020, August 20. "Pneumatic Networks." Museum of Retro Technology. http://douglas-self.com/museum/comms/pneumess/pneumess.htm.

_____. 2021, June 24. "Voicepipes and Speaking-Tubes." Museum of Retro Technology. http://www.douglas-self.com/museum/comms/voicepipe/voicepipe.htm.

_____. 2024, March 3. "Acoustic Location and Sound Mirrors." http://www.douglas-self.com/museum/comms/ear/ear.htm.

Seller, John. 1694. *Practical Navigation*. London.

Senate, United States. 1935. *Hearings, Special Committee Investigating the Munitions Industry* (74th Congress, 1st Session).

Sephton, James 2011. *Sovereign of the Seas: The Seventeenth-Century Warship*. Amberley Publishing.

Shannon, Claude. 1948. "A Mathematical Theory of Communication." *Bell System Tech. J.* 27: 379–423 (July) and 623–56 (October).

Shaw, Jim. 2019, Spring. "American Innovation in the Shipping Industry." *PowerShips* 309: 28–37.

Shawyer, Michael, and Avilio F. Medina Pizzali. "The Use of Ice on Small Fishing Vessels." FAO Fisheries Technical Paper 436. https://www.fao.org/3/y5013e/y5013e00.htm#Contents.

Sheppard, Warren W., and Charles Carroll Soule. 1922. *Practical Navigation*. World Technical Institute.

Sherman, Robert. 1999, October 26. "NR-1." (Originally created by John Pike.) https://man.fas.org/dod-101/sys/ship/nr-1.htm.

Shipwrecked Mariner, The. 1868, January. "Reminiscences of Sea Life." *The Shipwrecked Mariner* 15: 34–40.

Short Blue Fleet, The. 2023. "2—Development of the Fleeting System." https://shortbluefleet.org.uk/written-history/chapter-2-development-of-the-fleeting-system.

_____. 2023a. "3—The First Use of Ice at Sea for the Preservation of Fish." https://shortbluefleet.org.uk/written-history/2a-the-first-use-of-ice-at-sea-for-the-preservation-of-ish/

Siebe, Henry. 1874. *The Conquest of the Sea: A Book About Divers and Diving*. Chatto and Windus.

Siegwert, Roland, et al. 2011 [2004]. *Introduction to Autonomous Mobile Robots*. MIT Press.

Signal Corps Association. 2024, January 26 (access date). "Key Dates in Signal History." http://www.civilwarsignals.org/pages/signal/signalpages/comtime.html.

Sigsbee, Lt. Cmdr. Charles Dwight. 1880. *Deep-Sea Sounding and Dredging: A Description and Discussion of the Method and Appliances Used on Board the Coast and Geodetic Survey Steamer, "Blake."* GPO.

Simmonds, Peter Lund. 1879. *The Commercial Products of the Sea: Or, Marine Contributions to Food, Industry, and Art*. Griffith and Farran.

Singh, Prabjeet. 2021. "Spoilage of Fish-Process and Its Prevention." https://aquafind.com/articles/spolage.php.

Singh, Sadu. 2012. *Theory of Machines: Kinematics and Dynamics* (3rd ed.). Pearson Education India.

Singh, Shailendra, et al. 2014, August. "Hyperbaric Oxygen Therapy: A Brief History and Review of its Benefits and Indications for the Older Adult Patient." https://www.hmpgloballearningnetwork.com/site/altc/articles/hyperbaric-oxygen-therapy-brief-history-and-review-its-benefits-and-indications-older.

Sipe, F. Henry. 1974. *Compass Land Surveying*. McLain.

(Sir John) Soane's Museum London. 1829, March 2. "Letter from John Knowles F.R.S. Commenting on the Grounds of His Opposition to the Candidacy of Captain Charles Phillips." http://collections.soane.org/b5403.

_____. 1829, March 5. "A Printed Letter Signed by Francis Chantrey, John Franklin, John Richardson, William Edward Parry, and A. Copland Hutchison of the Royal Society, Rebutting Mr Knowles's Objections to the Candidacy of Captain Charles Phillips." http://collections.soane.org/b5290

Siranah (pseud.). 2008, October 31. 'The Sumner Line of Position." https://www.siranah.de/html/sail040t.htm.

_____. 2018, December 8. "Correction Tables for Sextant Altitudes." https://www.siranah.de/html/sail040h.htm.

_____. 2021, February 5. "History of Celestial Navigation " http://www.siranah.de/html/sail040d.htm.

Skene, Norman L. 2001. *Elements of Yacht Design*. Sheridan House.

Smiles, Samuel. 1874. *Lives of the Engineers: Harbours. Lighthouses. Bridges. Smeaton and Rennie*. J. Murray.

Smith, Gordon. 2011, July 31. "The Victoria Cross Action." https://www.naval-history.net/WW1Memoir-Hales.htm.

Smith, Laura Alexandrine. 1888. *The Music of the Waters. A Collection of the Sailors' Chanties, or Working Songs of the Sea, of All Maritime Nations. Boatmen's Fishermen's, and Rowing Songs, and Water Legends.* Kegan Paul, Trench and Co.

Smithsonian Institution. 2022, June 29. "Katanda Bone Harpoon Point." https://humanorigins.si.edu/evidence/behavior/getting-food/katanda-bone-harpoon-point.

Smolle, Christian, et al. 2021, January 8. "The History and Development of Hyperbaric Oxygenation (HBO) in Thermal Burn Injury." *Medicina (Kaunas),* 57(1): 49. https://www.ncbi.nlm.nih.gov/pmc/articles/PMC7827759.

Snow, Elliot. 1909. "Voice Pipes." *Proc. U.S. Naval Institute* 35: 769–888.

Sobel, Dava. 1995. *Longitude: The True Story of a Lone Genius Who Solved the Greatest Scientific Problem of His Time.* Penguin.

Sobuj, Md Sohanur Rahman. 2019, April 26. "Thermal Properties of Polymers." https://textilestudycenter.com/thermal-properties-polymers.

Spearman, James Morton. 1844. *The British Gunner.* Parker, Furnivall, & Parker.

Spencer, J.E. 1953, July–September. "The Abaca Plant and Its Fiber, Manila Hemp." *Economic Botany* 7(3): 195–213.

Sperry, Elmer A. 1914, November 13 (filing date). British Patent Application 135,871, "Gyroscopic Apparatus." https://patents.google.com/patent/GB135871A/en?q=G05D1%2f0206&sort=old.

———. 1920, November 30 (issue date). U.S. Patent 1,360,694, "Navigational Apparatus." (Application filed November 13, 1914.)

———. 1929, October 8. U.S. Patent 1,730,951. "Automatic Steering for Dirigible Craft." (Application filed July 7, 1922.)

Sperry Gyroscope Co. 1944, June. *Gyro-Compass Mark XIV, Mod. 1.* Instructions 17–1400 D. https://maritime.org/doc/gyromk14/index.php.

Sperry Gyroscope Company. 1912. *The Sperry Gyro-Compass and Navigation Equipment.* Sperry Gyroscope Company.

Stafford, Brian. 2022. *The Great Windships: How Sailing Ships Made the Modern World.* XLibris AU.

Stalkartt, Marmaduke. 1781. *Naval Architecture or the Rudiments and Rules of Ship Building.* J. Boydell, J. Dodsley, J. Sewell.

Stang, Ambrose H., and Lory R. Strickenberg. 1921, September 15. "Results of Some Tests of Manila Rope." *Technologic Papers of the Bureau of Standards* No. 198. GPO. https://nvlpubs.nist.gov/nistpubs/nbstechnologic/nbstechnologicpaperT198.pdf.

Staniforth, Mark. 1987. "The Casks from the Wreck of the 'William Salthouse.'" *Australian J. Hist. Archaeology* 5: 21–28.

Steel, David. 1794. *The Elements and Practice of Rigging and Seamanship.* 2013 Online version posted by San Francisco Maritime National Park Association. https://maritime.org/doc/steel.

———. 1805. *The Shipwright's Vade-Mecum.* P. Steel.

———. 1806. *The Art of Rigging* (2nd ed.). P. Mason.

———. 1812. *The Elements and Practice of Naval Architecture* (2nd ed.). Steel & Co.

Steele, Richard. 1790. *The Town Talk, The Fish Pool, The Plebeian etc.* John Nichols.

"Steerage Passenger." 1832. *The Quid; or Tales of My Messmates.* W. Strange.

Sterling, Christopher H. 2008. *Military Communications: From Ancient Times to the 21st Century.* ABC-CLIO.

Stevenson, Charles H. 1899. "The Preservation of Fishery Products for Food." GPO. (Extracted from U.S. Fish Commission Bulletin for 1898, 335–563.) https://spo.nmfs.noaa.gov/content/preservation-fishery-products-food.

Stewart, Joseph. 2011. *Exploring the History of Hyperbaric Chambers, Atmospheric Diving Suits and Manned Submersibles: the Scientists and Machinery.* Xlibris US.

stexboat. 2013, May 1. "The Rhodian Law." http://www.stexboat.com/books/maritime_law/Rhodian_Law_Sea.htm.

Stimson, Alan. 1985. *The Mariner's Astrolabe: A Survey of 48 Surviving Examples.* Instituto de Investigação Científica Tropical.

Stinton, Bob. 2024. "The Truth about Wool as Drysuit Insulation." https://www.divedui.com/pages/the-truth-about-wool-as-drysuit-insulation.

Stolle, Georg. 1917, October 30 (issue date). U.S. Patent 1,245,058, "Submersion Control for Divers' Armor with Flexed Members."

Sutherland, William. 1717. *The Prices of the Labour in Ship-building Adjusted,* etc. "D.L."

———. 1729. *Britain's Glory. Or, Ship-building Unvail'd,* etc. Vol. 1. A Bettesworth, etc.

Swanick, Lois Ann. 2005, December. "An Analysis of Navigational Instruments in the Age of Exploration: 15th to Mid-17th Century." M.A. Thesis, Anthropology, Texas A&M. (Online publication: 2006, April 12.) https://oaktrust.library.tamu.edu/handle/1969.1/3235.

Sweeney, Frank J. 2010. "Barrels." In Kit L. Lam (ed.), *The Wiley Encyclopedia of Packaging Technology.* Wiley.

Szondy, David. 2017, July 18. "US Navy Tests Signal Lamp-Based Ship-to-ship Texting." https://newatlas.com/us-navy-signal-lamps-fltc-texting/50523.

Tait, James. 1907. *Tait's New Seamanship.* J. Brown & Son.

Tanner, Zera Luther, and John Bell Blish. 1899, February 11. British Patent 564 of 1899, "Sounding Apparatus for Purposes of Navigation. https://patents.google.com/patent/GB189900564A/en.

Taquet, Marc. 2013. "Fish Aggregating Devices (FADs): Good or Bad Fishing Tools?: A Question of Scale and Knowledge." *Aquatic Living Resources.* 26: 25–35.

Taunt, Lt. Emory H. 1883. *Young Sailor's Assistant in Practical Seamanship.* Navy Department.

Taunton, William George Henry. 1841, December 11 (patent sealing date), British Patent 9176 of 1841, "Improvements in Machinery for Raising Weights." (enrolled June 11, 1842), in MacIntosh, *The Repertory of Patent Invention, Enlarged Series* I: 266 (Jan.–June 1843)

_____. 1845, January 5 (patent sealing date), British Patent 10,495 of 1845, "Improvements in Machinery for Revolving Windlasses, Barrels, Spindles, Shafts and for Pumping." Enrolled July 25, 1845 in *English Patents of Inventions, Specifications.* HM Stationery Office, 1855.

Tawresey, Alfred P.H. 1926, November. "Reducing Errors in Navy Signaling." *Sci. Am.* 135(5): 338–39.

Taylor, E.G.R. 1957. *The Haven-Finding Art: A History of Navigation from Odysseus to Captain Cook.* Abelard-Schuman.

Taylor, Elizabeth (executrix). 1762, December 6. British Patent 782, "Set of Engines, Tools, Instruments and Other Apparatus, for Making Blocks, Shivers, and Pins."

Taylor, John Ellor. 1876. *The Aquarium: Its Inhabitants, Structure & Management.* Hardwicke.

Taylor, Walter. 1775, November 28. British Patent 1110, "Bushing Cast-Iron or Other Metal Shivers for Ships' Blocks."

_____. 1781, June 5. British Patent 1295, "Construction of Shivers of Pulleys for Ship Blocks."

_____. 1786, November 17. British Patent 1567, "Grinding Grain, Making and Greasing Ships' Blocks."

Tetley, Laurie, and David Calcutt. 2007. *Electronic Navigation Systems.* Taylor & Francis.

Tetzlaff, Kay, et al. 2021, July 9. "Going to Extremes of Lung Physiology—Deep Breath-Hold Diving." *Frontiers Physiology* July 9, 12: 710429. https://www.frontiersin.org/articles/10.3389/fphys.2021.710429/full.

Theberge, Albert E. 2014, April, 7. "George Belknap and the Thomson Sounding Machine." https://www.hydro-international.com/content/article/george-belknap-and-the-thomson-sounding-machine.

Thompson, Emily E. 2022. *Storytelling in Sixteenth-Century France: Negotiating Shifting Forms.* University of Delaware Press.

Thompson, Mark. 1991. *Steamboats & Sailors of the Great Lakes.* Wayne State University Press.

Thompson, Silvanus Phillips. 2005. *The Life of Lord Kelvin,* Volume 2. AMS Chelsea Publishing.

Thomson, David Whittet. 1942, October. "Robert Fulton and the Nautilus." *Proc. U.S. Naval Institute* 68(10, Whole 476). https://www.usni.org/magazines/proceedings/1942/october/robert-fulton-and-nautilus.

Thomson, Sir Charles Wyville. 1873. *The Depths of the Sea.* Macmillan and Company.

Thomson, Sir William. 1876, September 1. British Patent 3452 of 1876, "Deep Sea Sounding Apparatus," in *English Patents of Inventions, Specifications1876, 3441–600* (1877).

_____. 1885, October 14. British Patent 12,240 of 1885, "Navigational Sounding Apparatus," in *Patents for Inventions: Abridgments of Specifications,* 1897, 15: 1885. See also U.S. Patent 377,696.

Thorp, Robert. *Mersey Built: The Role of Merseyside in the American Civil War.* Vernon Press.

Thursfield, J.B. 1904. "British Naval Maneuvers." In T.A. Brassey (ed.), *Naval Annual*, Chapter III. J. Griffin & Co.

Tierie, Gerrit. 1932. *Cornelis Drebbel (1572–1633).* H.J. Paris. https://www.drebbel.net/Tierie.pdf.

Tinniswood, J.T. 1945. "Anchors and Accessories, 1340–1640." *The Mariner's Mirror,* 31(2): 84–105. doi: 10.1080/00253359.1945.10658910.

Tipping, Colin. 1994, February. "Cargo Handling and the Medieval Cog." *The Mariner's Mirror* 80(1): 3–15.

Tisdale, Lt. Cmdr. Marlon S. 1922. "Communications Afloat." *Proc. U.S. Naval Inst.* 48: 35–47.

Tizard, Captain T.H. 1920. "Use of Sumner Lines in Navigation." *Nature* 105 (2644): 552–53.

Toghill, Jeff. 2003. *The Navigator's Handbook.* Lyons Press.

Tønnessen, Johan Nicolay, and Arne Odd Johnsen. 1982. *The History of Modern Whaling.* University of California Press.

Towne, Henry Robinson. 1883. *A Treatise on Cranes.* E.S. Dodge.

Towson, John Thomas. 1852. *Tables to Facilitate the Practice of Great Circle Sailing, and the Determination of Azimuths* (4th ed.). Hydrographic Office, Admiralty.

Treffner, Ivar. 2022, May. "A 17th Century Fluit Wreck in Gulf of Finland." M.A. dissertation, Dept. of Archaeology, University of Tartu. https://www.researchgate.net/publication/362311602_A_17th_Century_Fluit_Wreck_in_Gulf_of_Finland.

Triewald, Martin. 1736, December 31."A Letter to the Reverend John Theoph. Desaguliers … Concening an Improvement of the Diving Bell." *Philosophical Transac. Royal Soc'y* (London) 39 (444): 377–83. (Letter written 1732, November 1.) https://royalsocietypublishing.org/doi/pdf/10.1098/rstl.1735.0077.

Trowbridge, P. 1888. *Report on Power and Machinery Employed in Manufactures,* Census Office, Dept. of the Interior. GPO.

Tucker, Emily Corinne. 2015, November 20. "Diving Into The Past—A Modern Freediver Visits the Ama." https://www.deeperblue.com/diving-past-modern-freediver-visits-ama.

Tully, A.P. ("Tony"). 2003, July. "NAGATO's Last Year: July 1945–July 1946." http://www.combinedfleet.com/picposts/Nagatostory.html.

Turner, Gerard. L'Estrange. 1980. *Antique Scientific Instruments.* Blandford Press.

Twede, Diana, 2005. "The Cask Age: The Technology and History of Wooden Barrels." *Packaging Technology & Science* 18: 253–264.

Twiss, Sir Travers, ed. 1874. *The Black Book of the Admiralty,* Appendix—Part III. Longman & Co.

Tyler, Tom. 2022, October 26. "Shoulder Guns." https://whalesite.org/whaling/whalecraft/Shoulder_Guns/Shoulder_Guns.html.

_____. 2022a, October 26. "Swivel Guns and Swivel Gun Harpoons." https://whalesite.org/whaling/whalecraft/Swivel%20Guns/Swivel%20Guns%20and%20Irons.html

_____. 2022b, October 26. "Harpoons." https://whalesite.org/whaling/whalecraft/Harpoons/Harpoons.html.

_____. 2022c, October 26. "Two Flue Irons." https://whalesite.org/whaling/whalecraft/Two%20Flue/Two%20Flue%20Irons.html.

_____. 2022d, October 26. "Single Flue Irons." https://whalesite.org/whaling/whalecraft/Single_Flue/Single_Flue.html.

_____. 2022e, October 26. "Toggle Irons." https://whalesite.org/whaling/whalecraft/Toggle_Irons/Toggle_Irons.html#anchor472595.

Tyzack, Benjamin Cowle, et al. 1832, August 3. British Patent 6292 of 1832, "Certain Improvements in Windlasses etc.," in *The Repertory of Patent Inventions* 14: 264–67 (1833). (The patent number given by Harland is incorrect.)

Unger, Richard W. 1973. "Dutch Ship Design in the Fifteenth and Sixteenth Centuries." *Viator: Medieval and Renaissance Studies* 4: 387–411.

United Services Journal. 1840. "Colonel Pasley's Operations at Spithead." *United Services Journal and Naval and Military Magazine* January, 72–83; February, 149–64; March 319–33.

United States Coast Guard. 1948, May–June. "Testing and Development: Notes on Boat Falls and Blocks." *The Engineer's Digest* 56: 15–16, 22.

United States Hydrographic Office. 1874. *Instructions for Using Sir William Thomson's Deep-Sea Sounding-Machine for Sounding with Steel Wire, etc.* GPO.

United States Navy Diving Manual. 1916, July. *Diving Manual.*

_____. 1979, January. *United States Navy Diving Manual, Vol. 1 Air Diving,* NAVSEA-0884-LP-001-9010.

Ure, Andred. 1843. *A Dictionary of Arts, Manufactures, and Mines, Etc.* Longman, Brown, Green, and Longmans.

Vallejo Gallery. 2022. "U.S. Navy Diving Air Pump Mark III by Morse." Description was at https://www.vallejogallery.com/item_mobile.php?page=item_page&id=3100; only photo is now available, at https://www.vallejogallery.com/2022/list_mobile.php?&query=&id=117&artist_id=.

Valleriani, Matteo. 2012. "Galileo's Abandoned Project on Acoustic Instruments at the Medici Court." *Hist. Sci.* 1: 1–31.

Vanin, Gabriele. 2022, September 5. "The Beginnings of Celestial Navigation: Early Techniques and Instruments" (preprint). https://www.researchgate.net/publication/363331638_The_beginning_of_celestial_navigation.

Vaughan, H.S. 1913. "The Whipstaff." *The Mariner's Mirror* 3: 230–37.

Veit, Chuck. 2012. *Raising Missouri: John Gowen and the Salvage of the U.S. Steam Frigate Missouri.* lulu.com.

Verbuni, Frank, et al. 2019, November 13. "Why Halley Did Not Discover Proper Motion and Why Cassini Did." *J. Hist. Astronomy,* 50(4): 383–97. https://doi.org/10.1177/0021828619877967.

Victoria and Albert Museum. 1899. Catalogue of the Naval and Marine Engineering Collection in the Science Division of the Victoria and Albert Museum.

Viele, John. 1996. *The Florida Keys: A History of the Pioneers,* Volume 3. Pineapple Press.

Villiers, Alan. 1953. *The Way of a Ship.* Charles Scribner's Sons.

Vitruvius. 1st Century BCE. "Ballistae." *De Architectura,* Book X, Part 11. https://lexundria.com/vitr/10.11/mg

Voith, Walther. 1933, August 15 (issue date). U.S. Patent 1,922,606, "Method and Means for Propelling and Steering Water or Air Ships." (Application filed September 22, 1931.)

Wagner, Jörg Friedrich. 2005, January. "From Bohnenberger's Machine to Integrated Navigation Systems, 200 Years of Inertial Navigation." In Dieter Fritsch (ed.), *Photogrammetric Week '05,* 123–34. Wichmann. https://www.researchgate.net/publication/239922979_From_Bohnenberger's_Machine_to_Integrated_Navigation_Systems_200_Years_of_Inertial_Navigation.

Wainwright, Richard. 1889. "Naval Coast Signals." *Proc. US Naval Inst.* 15(1): 61–74.

Wakefield, Julie. 2005. *Halley's Quest.* Joseph Henry Press.

Wakeley, Andrew, et al. 1755. *The Mariner's Compass Rectified.* W. and J. Mount, T. Page & Son.

Walker, Cmdr. William. 1863. *The Magnetism of Ships, and the Mariner's Compass* (2nd ed.). Virtue, Brothers and Co.

Walker, John. 1997, May 5. "Lunar Perigee and Apogee Calculator." http://jgiesen.de/planets/moon/pacalc/PeriApCalc.html.

Walton, Harry. 1968. *The How and Why of Mechanical Movements.* E.P. Dutton.

Walton, Thomas. 1907. *Present-Day Shipbuilding.* C. Griffin, Ltd.

War Department, United States. 1924, July 31. *Rigging.* Training Regulations 185–5. GPO.

_____. 1944, August 30. *War Department Technical Manual TM11-2043, Telephone TP-3, 16.* GPO.

Ward, Kevin R., et al. 2013, January. "Chemical Oxygen Generation." *Respiratory Care* 58(1): 184–95. http://rc.rcjournal.com/content/58/1/184.

Warshaw, Matt. 2005. *Encyclopedia of Surfing.* Harcourt.

Wärtsilä 2024. "Rotary Pump." *Wärtsilä Encyclopedia of Marine and Energy Technology.*

https://www.wartsila.com/encyclopedia/term/rotary-pump.

Wärtsilä. 2024a. "Voith-Schneider Propulsor (VSP), Cycloidal Propeller." *Wärtsilä Enccyclopedia of Marine and Energy Technology.* https://www.wartsila.com/encyclopedia/term/voith-schneider-propulsor-(vsp)-cycloidal-propeller

Waterman, J.J. 1987. *Freezing Fish at Sea: A History.* H.M. Stationery Office. https://search.worldcat.org/title/682088085.

Waters, David W. 1988. "Reflections Upon Some Navigational and Hydrographic Problems of the XVth Century Related to the Voyage of Bartholomeu Dias, 1487–88." *Revista de Universidade de Coimbra,* 34: 275–347.

Watson, J.W., and D.W. Kerstetter. 2006, Fall. "Pelagic Longline Fishing Gear: A Brief History and Review of Research Efforts to Improve Selectivity." *Marine Technology Soc'y J.* 40(3): 6–11.

Watson, Thomas A. 1882, October 24 (issue date). US Patent 266,567, "Magneto-Telephone."

Webb, Paul. 2023, August. "1.4 Mapping the Seafloor." Introduction to Oceanography. https://rwu.pressbooks.pub/webboceanography/chapter/1-4-mapping-the-seafloor.

Webb, Robert Lloyd. 2011. *On the Northwest: Commercial Whaling in the Pacific Northwest, 1790–1967.* UBC Press.

Werrett, Simon. 2010. *Fireworks: Pyrotechnic Arts and Sciences in European History.* University of Chicago Press.

Westcott, George Foss. 1932. *Handbook of the Collections Illustrating Pumping Machinery, Part I Historical Notes.* (Science Museum, South Kensington). H.M. Stationery Office.

Western Australian Museum. 2024, January 28 (access date). "Batavia's History." http://museum.wa.gov.au/research/research-areas/maritime-archaeology/batavia-cape-inscription/batavia.

Whall, Hugh. 1976, October. "Trailer Sailors: The Asphalt Cruisers." *Motor Boating & Sailing,* 52–53, 68–69.

Whitaker, Omar B. 1926, April 20. U.S. Patent 1,581,147, "Automatic Steering Device."

White, Sir William Henry. 1894. *A Manual of Naval Architecture for the Use of Officers of the Royal Navy, Officers of the Mercantile Marine, Yachtsmen, Shipowners, and Shipbuilders.* J. Murray.

Whitney, Edwin H. 1896. "The Development of the Ship Windlass." *Cassier's Magazine* 9: 113–36.

Whitworth, Joseph, and George Wallis. 1854. *The Industry of the United States in Machinery, Manufactures, and Useful and Ornamental Arts.* G. Routledge & Company.

Wilcocks, J.C. 1875. *The Sea-Fisherman* (3rd ed.). Longmans, Green, and Co.

Williams, J.E.D. 1992. *From Sails to Satellites: The Origin and Development of Navigational Science.* Oxford University Press.

Williams, Robyn. 2013, May 3. "Corrosion on Shipwrecks." https://www.abc.net.au/radionational/programs/scienceshow/corrosion-on-shipwrecks/4666098.

Wilson, C.H., and E.E. Rich. 1967 [1941]. *The Cambridge Economic History of Europe from the Decline of the Roman Empire: Volume 4, The Economy of Expanding Europe in the Sixteenth and Seventeenth Centuries.* Cambridge University Press.

Wilson, David Gordon. 2011. "Understanding Pedal Power." Technical Paper 51. https://ocw.mit.edu/courses/ec-711-d-lab-energy-spring-2011/15286019b5a06fde2d43af6c00f460d4_MITEC_711S11_lab1_pedal.pdf.

Wolters, Timothy S. 2013. *Information at Sea: Shipboard Command and Control in the U.S. Navy, from Mobile Bay to Okinawa.* Johns Hopkins University Press.

Woodcroft, Bennet, 1854. *Titles of Patents of Invention, Chronologically Arranged from March 2, 1617 … to October 1, 1852* (multiple volumes).

Woods, Michael, and Mary Boyle Woods. 2000. *Ancient Agriculture: From Foraging to Farming.* Runestone Press.

Worker, Joseph G., and Thomas A. Peebles. 1922. *Mechanical Stokers: Including the Theory of Combustion of Coal.* McGraw-Hill.

Wright, Edward. 1657 [1599]. *Certain Errors in Navigation Detected and Corrected.* Joseph Moxon. Online edition by Early English Books Online Text Creation Partnership. https://quod.lib.umich.edu/e/eebo/A67154.0001.001/1:6.15?rgn=div2;view=toc.

Wrixon, Fred B. 1998. *Codes, Ciphers & Other Cryptic and Clandestine Communication.* Black Dog & Leventhal.

Wulandari, Puspita Septim, et al. 2014. "Pre-Service Teachers' Conceptual Understanding of Rolling Friction Coefficient." *AIP Conference Proceedings 2014,* 020060. Published online 21 September 2018. https://doi.org/10.1063/1.5054464.

Yao Ren. 2017, May 16. "Capstan Equation." http://yaor.me/capstan-equation.

Yong. 2018, April 20. "How Asia's Super Divers Evolved for a Life At Sea." https://www.theatlantic.com/science/archive/2018/04/bajau-sea-nomads-diving-evolution-spleen/558359/

Young, Andrew T. 2020. "Dip of the Horizon." https://aty.sdsu.edu/explain/atmos_refr/dip.html.

———. 2021. "Distance to the Horizon." https://aty.sdsu.edu/explain/atmos_refr/horizon.html.

Young, Arthur, and James Brisbane. 1863. *Nautical Dictionary.* Longman.

Zamparo, Paola, et al. 2002, September. "How Fins Affect the Economy and Efficiency of Human Swimming." *J. Exper. Biol.* 205(17): 2665–76. https://pubmed.ncbi.nlm.nih.gov/16341874/.

Zehner, Joe, and David Scoggin. 2020, January 25. "Remembering LASH." https://www.maritime-executive.com/editorials/remembering-lash.

Zhang, Ce, and Jianming Yang. 2020. *A History*

of *Mechanical Engineering*. Springer Nature Singapore.

zu Mondfeld, Wolfram. 2005. *Historic Ship Models*. Sterling.

Zupko, Ronald Edward. 1985. *A Dictionary of Weights and Measures for the British Isles: The Middle Ages to the Twentieth Century.* American Philosophical Society.

Author's Preexisting Work

As noted in the front matter, portions of this book were previously published in the *Grantville Gazette*, which ceased publication in July 2022. That online magazine presented fiction set in and nonfiction relating to the fictional literary universe created by the late Eric Flint's alternate-history sci-fi novel *1632*. In that novel, a new timeline is created when a fictional West Virginia town (Grantville) is moved from the year 2000 to Thuringia, Germany, during the Thirty Years War. The nonfiction considered the limited knowledge and resources that would have been available to the townspeople, and proposed how they might cope. Naturally, the fictional aspects have been omitted from the present work.

Cooper, Iver P. 2011. "The Wind Is Free; Sailing Ship Design, Part 1: Propulsion." *Grantville Gazette* 21.

Cooper, Iver P. 2011. "The Wind Is Free; Sailing Ship Design, Part 2: Seaworthiness." *Grantville Gazette* 22.

Cooper, Iver P. 2012. "Cold Comforts: Natural Refrigeration in the 1632 Universe." *Grantville Gazette* 40.

Cooper, Iver P. 2016, September. "Life at Sea in the Old and New Time Lines, Part 3: Shipboard Lighting and Fire Prevention." *Grantville Gazette* 67.

Cooper, Iver P. 2017, May. "Life at Sea in the Old and New Time Lines, Part 4: Lights Across the Waters." *Grantville Gazette* 71.

Cooper, Iver P. 2021, May 1. "Tethered Balloons and Kites in the 1632 Universe, Part 1." *Grantville Gazette* 95.

Cooper, Iver P. 2021, July 1. "Tethered Balloons and Kites in the 1632 Universe, Part 2." *Grantville Gazette* 96.

Use of the material from the last two articles is with the kind permission of Lucille Robbins, Eric Flint's widow and the chief executive officer of the publisher of the *Grantville Gazette*, 1632, Inc.

of *Mechanical Engineering*. Springer Nature Singapore.

zu Mondfeld, Wolfram. 2005. *Historic Ship Models*. Sterling.

Zupko, Ronald Edward. 1985. *A Dictionary of Weights and Measures for the British Isles: The Middle Ages to the Twentieth Century*. American Philosophical Society.

Author's Preexisting Work

As noted in the front matter, portions of this book were previously published in the *Grantville Gazette*, which ceased publication in July 2022. That online magazine presented fiction set in and nonfiction relating to the fictional literary universe created by the late Eric Flint's alternate-history sci-fi novel *1632*. In that novel, a new timeline is created when a fictional West Virginia town (Grantville) is moved from the year 2000 to Thuringia, Germany, during the Thirty Years War. The nonfiction considered the limited knowledge and resources that would have been available to the townspeople, and proposed how they might cope. Naturally, the fictional aspects have been omitted from the present work.

Cooper, Iver P. 2011. "The Wind Is Free; Sailing Ship Design, Part 1: Propulsion." *Grantville Gazette* 21.

Cooper, Iver P. 2011. "The Wind Is Free; Sailing Ship Design, Part 2: Seaworthiness." *Grantville Gazette* 22.

Cooper, Iver P. 2012. "Cold Comforts: Natural Refrigeration in the 1632 Universe." *Grantville Gazette* 40.

Cooper, Iver P. 2016, September. "Life at Sea in the Old and New Time Lines, Part 3: Shipboard Lighting and Fire Prevention." *Grantville Gazette* 67.

Cooper, Iver P. 2017, May. "Life at Sea in the Old and New Time Lines, Part 4: Lights Across the Waters." *Grantville Gazette* 71.

Cooper, Iver P. 2021, May 1. "Tethered Balloons and Kites in the 1632 Universe, Part 1." *Grantville Gazette* 95.

Cooper, Iver P. 2021, July 1. "Tethered Balloons and Kites in the 1632 Universe, Part 2." *Grantville Gazette* 96.

Use of the material from the last two articles is with the kind permission of Lucille Robbins, Eric Flint's widow and the chief executive officer of the publisher of the *Grantville Gazette*, 1632, Inc.

Index

156–57, 163–64, 168, 170–71, 206, 213, 225, 239–40
Preston, Grant 15
Priestey, Joseph 217
Prince Henry 36
Ptolemy 45

Raleigh, Walter 176
Ramsden dividing engine 29
Rapson, John 132–33, 138–40
Regiomontanus 51
Reinhold, Erasmus 45
rhumb line 9
Ritchie, Edward 15, 16
Rodney, Capt. George 14
ropes 80–84; *see also* chain

sailhandling 59–60
Saint-Hilaire, Marcq 54
salvage: cannon 225–26; law and 227–29; lifting a wreck 223–25; locating a wreck 222–23; preservation and deterioration of wreck 226–27
Schaffer, John 108
Scilly disaster (1707) 13, 17
Scoresby, William 19
Seattle cargo hook 112
Shannon, Claude 173
ship propulsion: internal combustion engines 62; oars 57; sails 37–60; steam 60–62
shipping containers: amphora 63; casks 63–64
ships: *Agincourt* 105; *Alarm* 145; *America* 104; *Black Prince* 109, 206–7; *Blake* 6; *Centaur* 79; *Charlotte Dundas* 61; *City of Tokio* 109; *Clermont* 61; *Colorado* 214; *Comet* 97–99; *Constellation* 181; *Constitution* 82; *Coronation* 127; *Cutty Sark* 82; *Defiance* 106; *Deutschland* 23; *Devastation* 80; *Doncaster* 101; *Dorsetshire* 205; *Duntrune* 101; *Eastport* 109; *Fiumicino* 5 167; *Formby* 108; *Fort Hindman* 109; *Foudroyant* 134; *Frigido* 141; *Gateway City* 68; *Glory* 17; *Gotheborg I* 59; *Great Eastern* 141; *Hampton Court* 127; *Holder* 170; *Ideal-X* 68; *Junella* 171; *Kalmar Nyckel* 128; *Karmay* 170; *Kearsage* 109; *Kronan* 127; *Le Royal Louis* 127; *Lord Dundas* 19; *Lucerne* 108; *Madoera* 115; *Marco Polo* 9; *Marion* 109; *Mary Rose* 119; *Massachusetts* 214; *Melita* 182; *Minotaur* 140; *Montgomeryshire* 115; *Najaden* 134; *New Mexico* 143; *New York* 214; *Niagara* 140; *Nipsic* 109; *North Star* 133; *Northern Wave* 171; *Odenwald* 141, 142; *Omaha* 109; *Ossipee* 109; *Pennsylvania* 62; *Polyphemus* 189; *Porter* 214; *Preussen* 82; *Prince Consort* 225; *Pushkin* 171; *Queen City* 110; *Queen Elizabeth*-class battleships 62; *Ronson* 104; *Sea Hawk* 116; *Selandia* 117; *Shenandoah* 109; *Sovereign of the Seas* 104; *Str. S.T. Crapo* 62; *Swan* 65; *Syracusia* 82; *Thomas W. Lawson* 58; *Trenton* 109; *Truelove* 164; *Unicorn* 134; *Utah* 143; *Vasa* 119, 121–8, 224; *Verbania* 66; *Victoria* 183; *Victory* 88, 104, 180; *Ville de Paris* 140; *Warrior* 66, 109; *Zeehaen* 59
Shorter, Edward 101
Siebe, Augustus 237
Smeaton, John 249
Smith, John 121
Snefru, Pharaoh 57
Sperry, Elmer 24, 142–44
steering gear: autopilot 141–44; axiometer (helm indicator) 135–6; cycloidal propeller 148–50; Flettner rotor 147–48; Flettner rudder 140–41; oar (stern) 118; paddlewheel used as 147; propeller 144–45; quadrant 130–32; Rapson slide 132–33; rudder, quarter 119; rudder, sternpost 118–20; steering wheel 128–34; tiller 120–21; waterjet 146–7; Wesselink barrel 134; whipstaff 121–28; *see also* transmission
stoking (for steam engines) 60–61
submarines: *Argonaut* 240–41; *Argonaut Junior* 240; *Deutschland* 24`; *Drebbel* 238–39; *Gymnote* 240; *Hunley* 239; *Ictineo I* and *II* 239; *Narval* 240; *Nautilus* 239; *NR-1* 241; *Plongeur* 239, 255; *Resurgam* 239
submersible: early autonomous 238–241; tethered manned 238; wheeled 240–41; *see also* submarines
Sumner lines 53–54

Tait, James 110
Tanner and Blish 4–5
Tartaglia, Niccolo 223–24
Taunton, William George Henry 95, 97, 99
Taylor, Elizabeth 79
Taylor, Walter 79
telescope 176
Thomson William (Lord Kelvin) 4, 14
timekeeping 33–35; chronometer 34, 52–53; nocturnal 34; planisphere 34; sandglass 11, 33; stop-watch 11; water clock (clypsedra) 11
Towson, John Thomas 9
transmission: belt drive 135; chain drive 135; gear train 134–5, 135–8; screw-type 138–40
Trithemius, Johannes 174
Truxton, Thomas 181
Tyzack, Benjamin Cowle 93–96

Unicode 172
union purchase 112

Vegetius 246
Vernier, Pierre 27
Very code 193, 199
Voith, Friedrich 150
VSOP87 47

Wagenaer, Lucas 3
Watson, Samuel 11
Weymouth, George 27
winches 87, 88–89; brace 100–101; reel 99–101
windlasses 87, 89–92; Clark, Chapman 111; pump-handle (pump-break) 92–101; reel winches 99–101; steam 109
Wright, Edward 8, 45, 48